No Such Thing As
Doomsday

Underground Shelters

And How To Prepare For

Earth Changes, War and Other Threats

By

Philip L. Hoag

Yellowstone River Publishing

Emigrant, Montana

United States of America

Yelllowstone River Publishing

P.O. Box 206

Emigrant, Montana 59027

U.S.A.

(406) 333-4707

(800) 327-7656

Manufactured in the United States of America

Library of Congress Catalog Card Number 96-090037

ISBN 1-888865-00-8

Dedication

I dedicate this book to the spirit of hope. Hope opens the mind to possibility and to the vision of a brighter future. Hope inspires people to decisive action and preparedness. Preparedness facilitates a passage through troubled waters and into this brighter future. I also dedicate this book to all of those who have not counted the cost of speaking the truth and taking a stand for principle, honor and freedom.

"Let the chips fall where they may."

Foreword

Do you want to build an underground shelter or put together a disaster preparedness program? Do you feel overwhelmed? This book can help you overcome inertia and get the project rolling. Without experienced advice, doing something for the first time can be like navigating in the dark. **No Such Thing As Doomsday** will give you the direction you need. It is the only book of its kind in an area where few civilians have traveled. When I first ventured into underground shelter building, the only viable sources of information were Arthur Robinson's book on steel tank shelters and Cresent Kerney's book on expedient shelters. The scope of this book goes far beyond the excellent work of Robinson and Kerney. **No Such Thing As Doomsday** breaks a trail into previously unexplored areas of advanced shelter building.

No Such Thing As Doomsday can save you time by saving you mistakes. Time is rather short! If you didn't agree you probably wouldn't be reading this book. In writing this book, I condensed the good and bad lessons learned from the construction of numerous underground shelters. This book will save you time by helping you achieve a direct route to your preparedness goal.

No Such Thing As Doomsday makes complex information understandable. This book brings engineering concepts down to earth where they can be comprehended and applied. **No Such Thing As Doomsday** provides the prospective shelter builder with the information he needs for making intelligent decisions.

Good management is an essential part of any successful project. **No Such Thing As Doomsday** will give you an understanding of the interpersonal dynamics involved in a preparedness project. This book includes information on how to organize a group and avoid costly disputes. It shows you how to maximize project effectiveness by harnessing group effort. **No Such Thing As Doomsday** provides essential information on the predictable psychological reactions people have to disasters and how to deal with these reactions.

If you really use the information found in this book, you should excel beyond the author's level of achievement. So have fun and be innovative. Great challenges produce memorable experiences.

Introduction

History teaches us again and again that disasters, wars and political misfortune befall people. However, man can defend himself against these threats. There is no absolute protection from modern weapons of mass annihilation, the effects of natural disaster or a totalitarian intrusion upon our lives and liberty. Preparation mitigates disaster. Preventive and protective measures enable us to mitigate the impact of disaster and thereby secure the best possibilities for the survival and well-being of family members, friends and neighbors. Without protective measures, we as individuals and our nation as a whole are vulnerable to blackmail, intimidation, control, injury and death.

Preparedness is a form of alternative life insurance which can provide an invaluable return. This type of policy doesn't pay a financial dividend, but in time of natural or man-made disaster, it provides a priceless return. It can insure the well-being of your family and loved ones.

The act of making survival preparations questions the belief system perpetuated by the ruling powers that "all is well" and "everything is under control." Conflicting like this with the prevailing status quo threatens the herd's sense of security. The masses cling to that which is familiar and are generally terrorized by the unknown and the possibility of change. Those who advocate disaster preparedness are, in essence, conveying a message that all is not well.

Those who make survival preparations, build underground shelters or defend traditional values are labeled "doomsdayers" or "extremists". This pattern is not unique to our times. Noah was mocked for years as he and his family built the Ark on dry land. If you pursue a path of preparedness, expect similar treatment.

A more enlightened point of view sees preparedness and sheltering as a sort of spare tire. No one would think of going on a long automobile trip without a spare in the trunk. Why should anyone living in a nuclear age consider it foolish to build a fallout shelter? They were popular and accepted in the 60s! International conflict, terrorism, and the proliferation of nuclear weapons has escalated since then. By all rational logic, preparedness should be considered an intelligent move.

The real "doomsdayers" are those who can't face reality so they resort to denial. These individuals see nuclear or natural disaster as a literal doomsday, an unsurvivable and terminal end. For those of us who have courage and a will to be, such a catastrophe is viewed as an unfortunate but survivable event, possibly a resolution and potentially a new beginning.

From a religious standpoint there is considerable support for the concept of sheltering and preparedness. For thousands of years various prophets have spoken of a coming planetary event. This foretold event has been suggested to involve some sort of worldwide disaster such as Armageddon, nuclear war, plague, famine, pestilence, polar shift or earth changes.

From a scientific standpoint, there have been recent geological indications suggesting the potential for massive earthquakes on the west coast in the near future. There has also been scientific evidence showing that the ice caps are starting to melt and, of course, there is the hole in the ozone layer and other environmental problems.

The evidence supporting the threat of a coming planetary event is not limited to what we have discussed here. There are many seemingly unresolvable problems which face this nation and the world today including the national deficit, abortion, crime, the inner cities and international politics. If you lump all these situations together, the growing magnitude of the problem points to an impending resolution in the near future. Like the body's immune function which combats uncontrolled infection, nature has always exhibited an ability to reestablish balance and order.

Preparedness, by all rights, should be the acceptable calling of the hour!

Acknowledgments

I would like to acknowledge the support and assistance my wife Arlene gave me during the many months it took to complete this book. Without her editing help this book could never have been completed. I would also like to thank the many other people who in some way contributed to this work including: my parents, Alan Engelbart, Christopher and Constance MacDonald, Michael Middleton, Mark Grenier, Michael Githens, Steve Hicks, Sharon Packer, Ed York, Edward Dratz, Patrick Wolberd, Susan Hamilton, Karl Belanger, Mike Macavoy, Tom with Digital Color and anyone else I have forgotten to list.

The cover and numerous illustrations in this book were done by Roxanne Duke.

Disclaimer

This book was written with the intent of providing information regarding preparedness and shelter building. It is sold with the understanding that the publisher and author are not a professional engineering service, nor are they legal consultants. If engineering or other expert services are required, the services of such competent professionals should be obtained. This book may not cover all the information available on the subject of preparedness and underground shelter building. The reader should avail himself or herself of all other information sources, become knowledgeable in these areas and tailor the concepts found in this book to meet their own unique situation. Additional sources of information are listed at the end of this book. The author and publisher have made every effort to make this book complete and accurate. Be advised that there is always the possibility that this book may contain mistakes in content. Thus, this book should be used only as a general guide and not as the ultimate source on shelter building information. The author and Yellowstone River Publishing shall have neither liability nor responsibility to any person or entity with respect to any loss or damage caused, or alleged to be caused, directly or indirectly by the information contained in this book. If you do not wish to be bound by the above, please return this book to the publisher for a full refund.

Table of Contents

About the Author

Philip Hoag lives in the state of Montana with his wife Arlene and their five children. Philip has a long involvement in preparedness. He organized, designed and helped team-manage the building of a large underground shelter project. This project is one of the largest civilian-built underground shelters in the United States. Philip also organized a local volunteer fire departmant and later, with one of his associates, started a volunteer ambulance service which they currently operate.

Philip lectures and gives classes on the subject of shelter building and preparedness. He has written articles which have appeared in American Survival Guide. He has been interviewed on numerous radio talk shows. Philip does consulting in the area of shelter design, systems and organization.

Threats

"We are apt to shut our eyes against a painful truth .... For my part, whatever anguish of the spirit it may cost, I am willing to know the whole truth; to know the worst; and to provide for it."

Patrick Henry

A threat is the existence of some kind of menace which has the potential of inflicting harm on an individual. In order to make adequate survival preparations, it is essential that all legitimate threats are clearly identified and carefully examined. If a person doesn't understand a threat, he can't effectively maneuver around it. Understanding a threat enables the individual to formulate plans and make preparations which can mitigate the future impact of that threat.

Chapter 1
The Communist Threat

In March 1994, just before his death, Richard Nixon said, *"Those who suggest that because of its vast problems Russia should no longer be treated as a world power ignore an unpleasant but undeniable truth. Russia is the only nation in the world that can destroy the United States."*

When examining the external threat of war, the first consideration is the Russian Soviets and the second consideration is the Red Chinese.

Since World War II, the Soviets have been the only real global contenders for a military bout with the United States. With the so-called end of the cold war, most Americans have been led by the media to falsely believe that the threat of confrontation with the Soviet Union and the threat of global nuclear war are now over. The suggestion from the media is that the world has entered a new political era in which the U.S. and Russia would establish a global partnership to maintain peace.

What's really going on in Soviet Russia? Is the old empire really being dismantled and is communism really dead? Due to the Soviet's deceptive character and the historically closed nature of Soviet society, it is very difficult to decipher what is really going on in Russia today. The Soviets have spent years developing a very sophisticated propaganda and disinformation apparatus. What makes things worse is that many of the old Communists have changed their uniforms and are now leaders at various levels in the new "democratic" government.

The other complicating factor is that the real power, be it in the East or the West, keeps itself hidden. Discerning the real situation in the Soviet Union is a difficult and complex task.

The Russian-Soviet Strategy of Deception

During the past seventy years the Soviet leadership has skillfully used the strategy of deception against the West. On January 14, 1960, Nikita Khrushchev made a speech to the Supreme Soviet in which he said, *"The Soviets intend to conceal vast reserves of missiles and warheads, hiding them in places throughout the expansive Soviet Union where the Imperialists could not stop them. Later they could be launched in a nuclear war."* A more recent 1978 quote from an official in the Soviet Council of Ministers, stated *"Perestroika is expressly designed to enhance Soviet military capability and combat readiness."* The highest ranking defector from the Soviet bloc intelligence community, Chech General Major Jan Sejna, revealed that the Soviet master plan involved a surprise military attack on the U.S. and the Western allies which would occur after a period of un-

Would you buy a used car from anyone of these three?

precedented Soviet peace concessions to the West.

The so-called collapse of the Soviet empire is yet another application of this strategy of deception. It is fairly well established that this Soviet leadership has long studied and practiced the concepts of Sun Tsu. Sun Tsu lived in China some time around 500 B.C. and he wrote a definitive text entitled, *The Art of War.* The gist of Sun Tsu's philosophy is that all effective warfare is based on deception. This deception tactic is clearly reflected in a 1921 quote from Vladimir Lenin.

"Telling the truth is a bourgeois prejudice. Deception, on the other hand, is often justified by the goal."

The Soviet leaders have always believed that war with America is inevitable. In many of their writings, they refer to the "inevitable upcoming war." The following quote comes from the Soviet General Druzhinin: *"As long as imperialism exists with its anti-humane, reactionary, and militaristic essence, the danger of military attack on our Motherland remains a fact. In recent years the danger of such ag-gression has especially grown. We cannot wait for thunder to strike. We must initiate, leaning on youth of all ages. We cannot waste time. Such loss is irreplace-able and during war costs a great amount of blood. Take heed, my friends, and make your own con-clusions."* What conclusion do you think he was referring to?

August 1991 Coup

The basis for the media-perpetuated belief that communism was overthrown in Russia and the old U.S.S.R. is the supposed coup of August 1991 in which so-called hard-liners were reputed to have attempted to oust the liberal demo-cratic reformer, comrade Mikhail Gor-bachev. In 1991, the media definitely portrayed Mikhail Gorbachev as a demo-cratic free-market reformer battling with Communist Party hard-liners. This por-trayal contains certain deficiencies.

In all likelihood, Gorbachev is only a player in this drama. Granted, he held a big part in the cast, but he didn't neces-

Nomenklatura Productions

sarily write or have power over the script. This coup was most certainly orchestrated by the Russian Soviet controlling elite who operate behind the scenes to accomplish several of their objectives. First of all, the coup attempt gave the majority of the people in the West a false impression that a good guy-bad guy conflict was going on in Russia. Furthermore, an unspoken message was conveyed here that if the U.S. did not provide massive financial aid to the Gorbachev good guys, then the hard-liner Communist bad guys would take over and the dreaded cold war would resume again. The other objective accomplished by this phony coup was the purging of Kremlin bureaucrats who were opposing the entrance of Western oil companies and banking interests into the U.S.S.R..

Prime Minister Valentin Pavlov didn't want the international bankers to get their foot in Russia's door, so he strongly opposed the Western plan to transform the Russian economy with the help of Western capital. Pavlov also wanted to have the central planners in Moscow maintain control of the key Soviet assets and prevent the Western multinational oil corporations from buying Russian coal, oil and gas reserves. This produced a sharp reaction from the Western elite. Zbigniew Brzezinski, founding director of the Trilateral Commission, wrote an article which appeared in the July 14th, 1991 edition of the New York Times which said, *"The central Soviet bureaucracy has been the major source of Soviet stagnation and it would be absurd for the West to attempt to re-stimulate the Soviet economy by dealing with the Moscow bureaucrats who have so directly contributed to the present Soviet crisis."*

This message was echoed at the July 1991 G-7 economic conference held in London which Gorbachev attended. At this economic conference, a little more than a month before the coup, Gorbachev was given his marching orders. He was told no more loans and no more aid until he purged those who obstructed Western economic ventures into Russia. As a result, a phony coup was engineered as a means of killing two birds with one stone.

When you dig deeply, the evidence clearly indicates that the leaders who participated in the coup were dupes in a grander plot. On August 19, 1991, Soviet Prime Minister Valintin Pavlov and other senior officials were summoned to a meeting by Vladimir Kryuchkov, head of the KGB. The KGB head informed them that Gorbachev was sick and incoherent and that a group of armed extremists were preparing to overthrow the government. Based on this report, Prime Minister Pavlov and the other officials declared a state of emergency and set up a governing committee to run the country through the duration of the crisis. The so-called plotters thought their actions were patriotically supporting Russia and Gorbachev, but they had unknowingly been tricked into endorsing an emergency regime.

Meanwhile Gorbachev, in good health, was taking a leisurely vacation at Pitsunda in the republic of Georgia where he had a Crimean dacha on the Black Sea. By the way, this luxurious summer home is purported to have a huge swimming pool with a glass wall and roof enclosure which move out of the way at the touch of a button. Three days subsequent to the initial coup, after making a triumphal return to Moscow, Gorbachev announced that during the coup, he had been held incommunicado and the plotters had cut his phone lines. But according to Major Vladimir Degtyaryov, who commanded the regiment which was responsible for protecting Gorbachev, Gorbachev was never placed under house arrest and no KGB troops ever invaded or seized control of his dacha. Also, according to the *Moscow News*, Valentin Zanin, the head of the manufacturing facility which produced the communication equipment for Gorbachev's dacha, indicated that the phone lines could not just be cut and that isolating the president of the U.S.S.R. from communications was impossible. Zanin said that if anyone tampered with the primary communication system, a network of backup systems would automatically kick in.

In August 1991 President Gamsakhurdia, of the Soviet Republic of Georgia, openly stated that Gorbachev himself had arranged the faked coup as part of a long-range Marxist strategy.

The other aspect of this coup which is so puzzling is how a group of conspirators with the backing of the KGB and with one of the world's largest armies at their disposal could fail so miserably. The KGB has a long record of successful coups, including Budapest, Prague, Ethiopia, Cambodia, Kabul and four internal coups since the death of Stalin, to name a few. The typical ingredients of a real coup include a massive arrest of political opponents, seizing control of all communications, including telephone, radio, television and other media, and seizing control of transportation, especially airports and roads.

CIA Fails to Anticipate Coup

This particular coup took the C.I.A. completely by surprise and as a result the agency's competency was questioned. The reason the C.I.A. analysts were unable to anticipate that a coup was going to happen was that none of the typical indications of a coup, such as large scale troop movements, elaborate preparations or urgent communications, ever occurred. The obvious reason these coup indicators never occurred was that this coup was not a real coup, it was a media event. Consequently, the only real casualties were a few expendable political figures.

One of the most credible evaluations of this landmark "death of communism coup" came from the Russian defector Viktor Sheymov. Before his defection in 1981, Sheymov was a major in charge of communications and security at the KGB headquarters in Moscow. On August 19, 1991, during the coup, he appeared on the MacNeil News Hour and gave the following commentary:

"I've seen quite a few KGB orchestrated coups while being in the center and I must say that this is a

very unusual coup, in the sense—first of all, the timing. Most coups happen on a Friday or Saturday. This is a coup which happened on a Monday, which is very unfortunate for the participants because everyone is in town. So, it looks like the coup wasn't really intended to succeed in the first place. Secondly, there are certain signs of how a coup performs. The KGB knows that in order for any coup to succeed, one has to act extremely decisively, and that's what they do. You can recall a lot of examples like Ethiopia, Afghanistan and so on. In this coup, the head of the government was just isolated and not eliminated. They could have done much better. For instance, they could have arranged, let's say, a heart attack for Mr. Gorbachev. That would have been much more plausible and at least they would have a chance to succeed. I wouldn't exclude that Mr. Gorbachev could be behind the coup, because he could benefit more than anybody else from this coup. For instance, he went recently to London [G-7 Conference] where he was literally begging for help. He didn't get too much for it. His popularity within the country is plummeting and he needs some kind of support from his people. In this case he could force the people to ask themselves fundamental questions: what is the alternative? Mr. Gorbachev until yesterday, for most of the Soviet population, was a bad guy .... He is a good guy now that he is under house arrest."

The Post-Coup Cover-Up

In March of 1992, a parliamentary investigation into the role of the military, secret police and Communist Party during the abortive August 1991 coup was halted by order of Russian Parliament President Ruslan Khazbulatov. The chairman of the commission, Lev Ponamarev, said the investigation

"found that most of the current corps of former Soviet generals supported the coup and are still in positions of authority with the armed forces of what is now the Commonwealth of Independent States." Also according to a March 31, 1992 article in the *Washington Times*, Yeltsin's director of the Russian SVR, Yevgeniy Primakov opposed this investigation. Both Gorbachev and Yeltsin wanted to keep the spotlight of public attention away from exposing what really transpired during the 1991 coup.

The Coup Accomplishes its Financial Objectives

As a result of this coup the G-7 group gave Russia an infusion of economic aid. Following the coup, U.S. Secretary of State James Baker visited Moscow and softened Washington's conditions for economic aid and U.S. banks flooded the Soviets with loan offers. The Soviets got the money and the international bankers temporarily got their foot in the door. Business as usual — no real change.

After the coup Gorbachev made a spirited defense of the Communist Party and said that he was absolutely against anti-Communist hysteria. Behind the scenes, a continuity of power and control remained intact despite the change in leadership from Gorbachev to Yeltsin. An article in the April 26, 1993 edition of *Time* magazine had an interview from Russian Parliament President Ruslan Khazbulatov. Khazbulatov said the following: *"Entire sections of the former Communist Party Central Committee simply drifted into the President's [Yeltsin] administration, together with their archives, safes, and even their telephones. Under the guise of working for a democratic President, they are simply restoring the old Communist Party ways."*

Considering the evidence, the media-generated concept that the 1991 coup resulted in the death of communism does not seem credible. To the contrary, the coup seems to have given the Soviet leadership a lifesaving transfusion and consolidated power. The only thing the Communists suffered was a change of name and an increase of income.

How Trustworthy Have the Soviets Been?

A significant factor to be examined when considering the Russian-Soviet threat is the Soviet Union's past conduct. This is important because the American public has a short attention span and poor memory. In the past years, the Soviets have made repeated peace overtures to the U.S. resulting in numerous disarmament treaties which only the U.S. honored.

The M.A.D. Treaty

The M.A.D. treaty (mutual assured destruction) specified that neither the U.S. nor

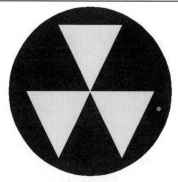

the Soviets would develop a national civil defense system. The idealistic logic involved here was that each nation would hold the other nation's civilian population as a nuclear hostage. Thus, if the Soviets launched a nuclear attack against the United States, the U.S. would retaliate in kind and annihilate the unprotected Soviet civilian population or vice versa. Under such a scenario, the consequences of initiating nuclear war would be too costly for either side. In order for the M.A.D. doctrine to work, both sides need to be trustworthy.

Once the U.S. signed the M.A.D. treaty, the Soviets started building a national civil defense system which now has the capacity to shelter the government leadership, the military and 70 percent of the industrial work force. Russia spends about $6 billion a year on civil defense. Their program involves about 100,000 full-time civil defense personnel and as many as 20 million part-time people. Russian children receive over 100 hours of civil defense instruction during the course of their pre-college schooling. Also, the Soviets have blast hardened much of their essential industrial base. Thus, in spite of any nuclear attack, their underground sheltered industry will continue to function.

The A.B.M. Treaty

In 1972 the U.S. and Soviets signed the A.B.M. (anti-ballistic missile) treaty, which limited the numbers and effectiveness of anti-ballistic missiles. The treaty allowed each nation to develop 100 anti-ballistic missiles. Immediately upon signing the treaty, the Soviets started building and deploying an A.B.M. system, but the U.S. leadership decided not to. As part of their A.B.M. system, the Soviets built the Krasnoyarsk radar system, which clearly violates the treaty. Also, the Soviets built and deployed far more than 100 A.B.M. missiles. Incidentally, one of the structures involved in this system is reported to be larger than the Great Pyramid.

Soviet Pushkino ABM Radar

Strategic Defense

The reason the Soviets fought so hard to get the Reagan administration to give up the Star Wars Program, a totally defensive system, was that they were in the process of

Soviet
Silo-based High Acceleration Interceptor ABM

deploying their own spaced-based strategic defense system. If the U.S. developed and deployed a strategic defense system, it would significantly complicate any attempt by the Soviet Union to launch a first strike against the United States. Civil defense and strategic defense are the two greatest deterrents to

nuclear war. The Soviets now have both and the United States has neither.

Glasnost, Perestroika and International Banking

During the 1991 Nobel Prize ceremonies Gorbachev said the following in his speech: *"To me it is self-evident that if Soviet perestroika succeeds, there will be a real chance for building a New World Order. If perestroika fails, the prospect of entering a new peaceful period in history will vanish."*

Glasnost and perestroika are effective tools that gained the Soviets advantages at the bargaining table which otherwise could hardly be gained on the battlefield. The Russian retreat from Eastern Europe was a deception intended to get the United States to lower its guard, remove its troops from Europe and reduce its global military strength. The bait which the Russians offered to the West was access to an open market in the Soviet Union. It appears that the one world multinational power-elite group jumped at this opportunity. The multinationals used their political control of the U.S. leadership to end the cold war with the Soviet Union and stop the development of a strategic defense system intended to protect the United States against a nuclear attack.

The Russian use of glasnost and perestroika is consistent with the application of the principals of Sun Tsu. Sun Tsu taught that weakening one's enemy is the purpose of war. This is not the first time the Soviets pulled this glasnost and perestroika game with the West. They did it from 1921-29 under Lenin, 1936-37 under Stalin, 1941-45 again under Stalin, 1956-59 under Khrushchev, and 1970-75 under Brezhnev. Periods of superficial Soviet reformation have always been followed by crackdowns, brutal purges and the resumption of Soviet espionage and military operations against the West. This concept is clearly reflected in Lenin's statements, *"We advance through retreat"* and *"They disarm, we build."* The American public seems to find a false security in the ceremonious signing of treaties. When facing an untrustworthy opponent, only a fool lets down his defense.

Western Aid and Loans — Economic Warfare Against the West

Western aid has propped up the crumbling Russian domestic situation and enabled the Russian-Soviet elite to chrome plate the barrel of their military machine. Loans to Soviet Russia are made without stipulations on how the money can be used. There is no way of preventing the Soviets from spending loan money on military equipment and development. According to an April 4, 1993, article in the *New York Times*, *"Some Russians are convinced that most aid has either been stolen or diverted by the Russian Mafia for private gain. Russian Vice President Aleksandr V. Rutskoi charged that 60 percent of the aid has been siphoned off by commercial structures."* U.S. banks give the Soviets unsecured loans at interest rates which are more favorable than the rates offered to the U.S. government and American business. It seems that the U.S. banks have some sort of long-standing, and inordinate, love affair with the Soviets.

The big U.S. lenders to the Soviets are First Chicago, Citibank, Chemical Chase Manhattan, Marine Midland, Bank of America and Manufacturer's Hanover. These banks don't usually mention Soviet-Bloc loans in their annual stockholder reports. If the Soviets were to default on their loans, this would significantly weaken or destabilize the Western economies. Right now the total outstanding loans from Western banks, the U.S. government , the World Bank and the International Monetary Fund, to the old Soviet Union, Russia, Eastern bloc and Third World countries is well over $1.2 trillion. If the loans fail, the privately owned corporation called the Federal Reserve will have to turn on the printing press to make up for the loss. This is the international debt bomb. The dropping of this bomb would most certainly result in massive inflation and the destruction of the Western economies. The magnitude of this international bad debt makes the savings and loan bailout seen like a small town bank scandal. This is a problem which the media does not focus public attention on.

Yet this lurking specter of a Russian-Soviet default is showing its appearance. According to a May 16, 1993, article in the *New York Times*, "Since last December, Russia has failed to meet interest payments on $4.2 billion of U.S. agricul-

9

agricultural credit guarantees. It is now $600 million in default. Russia has also failed to pay more than $200 million owed to 57 American companies." This Russian-Soviet insolvency shouldn't come as too much of a surprise to the Western banking elite. In December of 1988, Gorbachev made a speech at the United Nations in which he discussed the Soviet debt to the West and he said, **"Looking at things realistically, one has to admit that the accumulated debt cannot be repaid or recovered on the original terms."** Gorbachev went on to suggest, "write off the debt altogether," or allow "a lengthy moratorium of up to 100 years on debt servicing," or possibly **"limit their debt servicing payments"** and postpone the requirement of paying on the principal for **"a long period."** Besides postponing Russia's inevitable day of financial reckoning, Western aid has allowed the Russian elite to significantly bolster their war-making capability.

What About The KGB?

If, in fact, the cold war is over and Communism is dead, then the KGB and the GRU (military intelligence), would no longer be needed and should have succumbed to death by natural causes. An article in the November 15, 1992, edition of the *Washington Times* reported that U.S. counterintelligence con-

tends that Russian intelligence *"operations against the United States have shown little decline"* as a result of the collapse of the Soviet Union.

Russian analysts conclude that the Russian security service continues to operate without a clear-cut regulation of its activities and it defines its own goals and priority tasks. The reason it appears that the Russian intelligence service is unregulated and defining its own goals is that it is the real controlling power. In reality, the leader on the surface is a puppet on a string. The tail is not wagging the dog. The figurehead, be it Yeltsin or whoever replaces him, is just the tail that is being wagged by the real power. The real power is found within the old KGB, now operating under a new name.

According to a April 17, 1992, article in the *Washington Times* *"There is increasing evidence, say competent Russian-Western analysts and Russian democrats, that the security police remain powerful and are becoming even more so today despite dismantlement."* Yeltsin has done nothing to supervise the operations of the Russian intelligence service and their operations are about the same as operations were with the KGB before the breakup of the old Soviet Union. Yeltsin did subdivide the KGB into two branches, the FSR, which is involved in internal intelligence operations and the SVR, which is involved in international intelligence operations.

An article in the February 8, 1993, edition of *U.S. News & World Report* stated: *"Russian President Boris Yeltsin has cultivated the former KGB and even strengthened its authority. As a result, former KGB officers are in the vanguard of Russia's budding capitalist class—to the chagrin of democratic reformers and the consternation of Western counterintelligence agencies."*

The secret police have always been part of the history of Russian society. It is a massive and deeply rooted malignant tumor which is intertwined with the very tissue of Soviet society. For years the KGB has been both feeding off of and nurturing the Soviet Union's social, political and economic structures. This intimate relationship is like a vine unto its branches. The real controlling power of the Russian-Soviet empire does not reside in an executive office but behind the scenes in the Nomenklatura which populates all the key positions of authority in political and military structures and the secret police.

Consumer Shortage Crisis

In 1990 a crisis occurred in which Russian consumers started facing shortages of food stuffs and basic goods. Who ended up benefitting from the crisis? KGB Chief Vladimir A. Kryuchkov put the blame for the crisis on unnamed economic saboteurs. But under the banner of protecting the public's interest, the KGB expanded its powers and moved into control of key positions in the new free-market economy.

This action follows the archetypal pattern of creating a crisis and then taking power and freedom from the people under the guise of providing a solution. This action always results in consolidating power in the hands of the few.

The Ultimate Insider Position

What we see on the surface as the collapse of Communism is no more than a KGB engineered restructuring. This restructuring was implemented ahead of what the KGB saw coming as an inevitable collapse. Preempting this collapse with their own restructuring put them in control of the change and insured the KGB's continued position of power in the new structure. Thus, the KGB has the ultimate insider position. For quite a while the KGB had been moving vast amounts of capital into foreign banks and holding companies. This enabled them to buy up the most valuable real estate, petroleum products, uranium, gold and controlling interests in factories and plants at discount prices when the economic crisis hit. After the October 1994 collapse of the ruble, Yeltsin appointed an investigative commission, which under the circumstances should have been made up of economists but also included the director of the new Federal Counter Intelligence Service (F.S.K.), Sergei Stepasihn. This sounds like another government appointed Warren commission whose intended purpose was not to find truth as much as it was to bury it.

Aldrich Aims and the Russian Secret Police

The Aldrich Aims spy case certainly does not support the concept that the

cold war is over and the Russian foreign intelligence activities are benign. The damage done to U.S. intelligence by C.I.A. agent Aldrich Aims in his spy work for the old Soviet Union and also more recently Russia is said to be comparable in damage to Japan's successful attack on Pearl Harbor. Ames betrayed 10 legitimate agents that the Soviets arrested and executed.

KGB Concentration Camps

According to reports which appeared in the February 11 and June 30, 1993 issues of the Swiss newspaper, *Neue Zurcher Zeitung*, the old Soviet gulag-style concentration camp system is still in operation. The system is comprised of hundreds of camps containing a total of one million to two million prisoners.

Russian Domestic Hardships

The conversion of Russia into a democracy with a "free market economy" has ended in failure which may have been intended from the onset. The Russian economy is in shambles. Change produced economic crisis and domestic hardship. The people of Russia are frustrated. Hyperinflation in Russia is now running at about 20 percent a month. Nearly 40 million people in the former Soviet Union who are subsisting on pensions are being driven to near starvation by hyperinflation. People are selling clothes on the street in order to eat, workers in factories haven't been paid for months, and the streets are thick with young women who have resorted to prostitution for survival.

Russia's Gross National Product is expected to drop by 50 percent this year. It is estimated that 30 percent of the population

lives in poverty. Those not impoverished are loosing their savings and future security to runaway inflation. Hundreds of thousands of recently discharged Russian military personnel are now unemployed and are hardly able to feed their families. The Russian public is being led to believe that the cause of their misery is the free market economy and Western capitalist intervention in their economy.

History Repeating Itself in Today's Russia

An atmosphere of national frustration is fertile ground for manipulation. Any group or person with prudence and resource could exploit this Russian public sentiment and increase their position of power. History contains a similar example of the exploitation of national frustration in Germany during the 1930s. The social and economic conditions within Russia are ripe for change. History has shown that economically downtrodden nations are easily led to war. Russia today resembles the corrupt Weimar republic in hyperinflationary pre-Hitler Germany.

Recent polls taken in Moscow indicate that two out of three Russians would prefer to live under the old Communist system or under an even stronger version of the present Russian dictatorship. General Alexander Lebed, a famous commander from the 14th Russian Army in Moldavia. said, ***"Democracy in Russia is completely impossible. Most Russians don't care whether they are ruled by the Fascists or Communists or even Martians as*** *long as they are able to buy six kinds of sausage in the stores and lots of cheap vodka. What's wrong with a military dictator?"*

Zhirinovsky

Vladimir Zhirinovsky appears to have seized upon Russia's national identity crisis.

During a short period of time, Zhirinovsky has moved from being a relatively unknown to a position of national and international media prominence. It would appear that some powerful group perceived the mounting wave of frustration and positioned Zhirinovsky on stage. Zhirinovsky's platform advocates getting even with the West and restoring the Soviet Union to a position of international preeminence. This platform has made Zhirinovsky very popular with the Russian military. According to intelligence estimates, 50 percent of the armed forces and 80 percent of the elite military units support Zhirinovsky. Zhirinovsky's cold-hearted attitude toward America is clearly revealed in his statements and writing. ***"We will not gloat when California joins Mexico, when a negro republic is created in Mi-***

ami and when Russia takes back Alaska, or even when America dissolves into a commonwealth of new states. The factories will close down. There will be no medicine, no food, and you Americans will emigrate to Europe, to Japan and to Russia. Your Americans are lazy and burned out....Our army is still number one in the world and Bill Clinton will never hit the nuclear button to keep us out of Western Europe. He would lose too many votes when we destroyed New York and California. Clinton should forget about foreign policy and concentrate on playing the saxophone."

The media and the U.S. leadership tries to write off Vladimir as some sort of clown, but his platform clearly reflects the antagonism many Russians and the military hold toward the U.S.

The military is an institution of enormous power in Russia and it has not reconciled itself to the collapse of the Soviet Union. Zhirinovsky's appearance may be a calculated move by the military to harness and direct the nation's mounting wave of emotional unrest. The frustrations of the Russian people are real and could constitute a strong unifying force.

According to Dr. Lana Coss, professor of military strategy at the National War College, many Russians feel angry and betrayed by the United States.

There is tremendous disappointment among the general population with the United States due to the perception of unfulfilled economic promises. There is a perception among the Russian people that the United States benefitted from the collapse of the Soviet Union. This is another way of saying the U.S. caused it. For the first time there is popular animosity in Russia against the United States. In the past, the Russian people did not actually hate the United States, but they do now because they feel that the U.S. is benefitting from their misfortune. The general feeling is that the United States misled Russia into changing its economic system. The Russian people also feel that they fulfilled their end of the bargain and their reward was economic disaster. Many Russians actually suggest that Gorbachev is on the C.I.A. payroll.

Gorbachev and Convergence, an Unholy Alliance

"Further global progress is now possible only through a quest for universal consensus in the movement towards a new world order." Mikhail Gorbachev in his address at the United Nations, December 7, 1988.

With the end of the cold war and the advent of globalism, we are seeing a convergence of the agendas of the Western and the Soviet elite. The front man is the Soviet leader whose Communist reign of power included the bombing of civilians in Afghanistan. The new, media-reformed Gorbachev now has an office in the Presidio in San Francisco. News reports published in the U.S. regarding Gorbachev's presence and activities here are notably vague. On Wednesday April 21, 1991, a rare article appeared in the *San Jose Mercury News* regarding Gorbachev's activities in the United States, *"Friday he [Gorbachev] inaugurated the*

Gorbachev Foundation USA, which is moving into new headquarters in the Presidio in San Francisco. Gorbachev announced that the foundation is creating a national task force on U.S. base closing. It will be co-chaired by former San Jose Mayor Tom McEnery and former Democratic Rep. Mel Levine of Southern California." In the same article Gorbachev was asked the question, would he ever return to politics. Gorbachev answered, *"I never left politics."*

It seems rather peculiar that we have the ex-President of the Soviet Union running around the United States arranging for the closing of our military bases. If this situation had taken place twenty years ago, it would have immediately drawn media attention and a Congressional investigation.

Now with the supposed death of communism, we are seeing military cooperation between the U.S. and Russian Federation states in various treaties including Bridge to America and Partnership for Peace. This is being done under the guise of training for peacekeeping missions in third world countries. Unfortunately, it appears that the real intent is for the joint suppression of civilian uprisings. Evidence of this real intent appeared in a *"Working Paper"* published by the National Guard Bureau on June 24, 1994, which said: *"...for the past year and a half, the National Guard Bureau has worked with the Joint Staff and the U.S. European Command to establish National Guard partnerships linking the National Guards of selected U.S. States with Ministries of Defense throughout Central and Eastern Europe (CEE) and Newly Independent States (NIS) of the former Soviet Union. The Partnerships assist the participating nation's transition to democratic military institutions with peace-time utility in* **providing military support of civilian authorities***".*

One might like to think that this agreement only applies to our troops supporting the civilian authorities of the Newly Independent States, but this is a reciprocal agreement. If the U.S. government were to try and disarm the American public they would need the assistance of foreign troops (UN Peacekeeping Troops) because most American soldiers couldn't be depended on to use force against American citizens.

The Department of Defense allocated $20 million for the construction of a Military Operations Urban Terrain training facility (MOUT) at Fort Polk, Louisiana, to facilitate training for the Partnership for Peace program. U.S. military units are now training at MOUT facilities located at Fort Drum in New York State and Camp Lejeune in North Carolina. These facilities are also being used to train law enforcement personnel.

From August 6 to 28, 1994, 4,000 military personnel from the United States, Britain, Canada, and the former Soviet bloc nations, including Albania, Bulgaria, the Czech Republic, Estonia, Hungary, Kyrghystan, Latvia, Lithuania, Poland, Romania, the Slovak Republic, Slovenia, Ukraine and Uzbekistan took part in joint military maneuvers at Fort Polk. Part of their training included disarming civilian "militias." There were two joint Partnership for Peace military operations held in the United States in September of 1995 — one at Norfolk, Virginia, and another at Camp Lejeune, North Carolina, which included 200 Ukrainian soldiers. In October of 1995, 150 Russian troops from the 27th Guards Motorized Rifle Division of Orenburg, Russia, trained with the U.S. 1st Infantry Division at Fort Riley, Kansas. This was part of an exercise called Peacekeeper-95 in which the Russian and American troops operated jointly in a simulated peacekeeping operation.

In 1993 , U.S. Special Forces were sent to Tblisi, Georgia, to a covert mission to protect Shevardnadze. This included providing anti-terrorism training to Shevardnadze's security forces and providing them with weapons and equipment. In May of 1995, Joint Chiefs of Staff chairman John Shailikashvili visited the Soviet Republic of Georgia and pledged to upgrade their military's training, communications, medical services and supplies. This seems rather ironic since Shevardnadze has shown himself to be one of the most ruthless Communist bosses in all of the republics.

On July 4, 1994, FBI Director Louis Freeh signed a cooperation protocol with Russian law enforcement agencies which includes Yeltsin's reformed KGB. During the Branch Davidian incident, the FBI Counter Terrorism Center consulted with Igor Smirnov, an expert on psychological warfare from the Moscow Institute of Psycho-Correction in regards to using subliminal warfare measures against Koresh and his people.

The most profound and comprehensive perspective on convergence has come from a KGB officer named Anatoliy Golitsyn who defected to the United States in 1961. He has written two books, the first is *New Lies for Old* and the second is *The Perestroika Deception*. In *New Lies for Old,* Golitsyn predicted the creation of the Solidarity movement in Poland and the liberalization of the Soviet Union and Eastern Europe, including the reunification of Germany.

In *The Perestroika Deception,* Golitsyn predicted that the Soviets would introduce a false democratization as a means of disarming the West. Golitsyn suggests that at the right moment this false mask will be dropped and the Russians, in cooperation with the Chinese, will force the West into accepting their system as a second October Revolution, a one world government under their terms. He also predicted that the Russians will provoke an incident involving the detonation of a nuclear weapon somewhere in the West, possibly the U.S., that could not be traced back to them and that would create pressure for a world government. (See chapter on terrorism).

Golitsyn suggests that the final stages of convergence *"will be accompanied by blood baths and political re-education camps in Western Europe and the United States. The Soviet strategists are counting on an economic depression in the United States and intend to introduce their reformed model of socialism with a human face as an alternative to the American system during the depression."*

The ultimate goal of communism has been, and always will be, world domination. If the communist elite can get away with achieving world domination through the path of convergence they will. But nonetheless, the Soviets are willing and have thoroughly prepared, if necessary, to wage and win a global war against the West.

The Soviet Military

President Boris Yeltsin presents a confident image and reassuring message to the

American people that Russia will never go to war against the United States, but he does not control the military. The fact is that to a major extent Yeltsin must dance to the military's tune. It was only the military's support that pulled him through a coup attempt. Yeltsin can sign agreements to reduce the Soviet Union's nuclear, biological and chemical weapons stockpiles, but Yeltsin, according to a October 1994 C.I.A. assessment, has no control over the government agencies in charge of these arms. Some Russian analysts suggest that not only does Yeltsin not control the military, but right now the military does not even control itself.

At the present, Soviet military power appears on the surface to be unconsolidated, with no one political power structure within the military holding complete control. The Russian media seems to support this image. According to T.A.S.S., very little of what the U.S. and Russia agree on in arms control treaties gets implemented by the Russians because "there is very little central control over the armed forces of Russia right now." The frightening suggestion conveyed to the West is that the top man in Russia is not in control of the 30,000 nuclear warheads targeting the United States. Like the new Russian Intelligence Services, the military seems to be unsupervised by the nation's executive figurehead, Yeltsin, and directing its own agenda.

Russia is definitely violating agreements on the proliferation of missile technology and hardware. Russia is also violating treaties regarding chemical and biological weapons, including ongoing development and the concealment of weapons' production equipment. The question is, are treaty violations due to a lack of executive control over the military or is this just another deception intended to cover for an intentional and ongoing program to develop and produce weapons of mass destruction? Is the Russian military really operating without a central leadership or is this just another subtle weave in a tactic of extorting financial aid out of the West? The unsaid message conveyed by the Russian media is that the 30,000 nuclear missiles pointing at the United States can only be safely controlled if Yeltsin triumphs over the evil hardliners, and furthermore, Yeltsin's success hinges on Western financial support. In all reality, the image that the military lacks central command is probably just another good guy - bad guy deception tactic.

The imperialist Russian bear seems to be awakening from his post-cold war hibernation. In April of 1994, a Yeltsin spokesman explained that Russia's romantic embrace with the West was over and that Russia sees itself as a great power which has its own strategic, military and political interests different from those of the United States and Europe. At this time Russia is in the process of reconstituting the old empire. This is being done by a sophisticated combination of military force, divide-and-conquer tactics, and economic and political pressure to gain de-facto control of the former republics.

The Russian military announced that the Red Army will intervene as it pleases in former Soviet territory around Russia and that the army will put down internal uprisings. This intervention policy was certainly confirmed by the Russian attack on the city of Grosny. Russia and the other 21 former Soviet republics have joined together in a military alliance. Russia has more than 30 military bases in former Soviet territory, with Russian troops involved in peacekeeping in Moldovia, Georgia, and other former republics. In September of 1994, just before the Washington Summit, the Russian foreign intelligence service released a report calling for a unified defense and economic zone among all the states of the former Soviet Union except the Baltic States. It appears that they are in the process of reestablishing their nuclear umbrella over the former Soviet republics. The end result may well be the reformulation of the old East - West global power blocs.

Historically, the Soviets have coveted and have been vexed by the United States' military superiority. As a consequence, the Soviet leadership as far back as Stalin has pursued a goal of overtaking and surpassing the United States in military superiority. One incident in particular rubbed salt into the Soviet wound of military inferiority. The Soviets had to back down in the face of U.S. resolve and military superiority during the Cuban missile crisis in the 1960s. It is probable that the Soviet leadership made a resolution to galvanize their might so that they would never again be humiliated in any future confrontation.

Some Russian analysts suggest that Russia's conventional forces are in deplorable shape, due to the poor condition of the Russian economy. This conclusion that Russia now has a weak conventional military strengths smells of traditional Soviet deception. Lenin once said, ***"When we are weak, boast of strength...when we are strong, feign weakness."*** The similar text from Sun Tsu's book, *The Art of War,* reads, ***"When at your pinnacle of strength, feign weakness."***

The Soviet Space Program

If in fact Russia is bankrupt, how can they accelerate their space program? An article in the May 5, 1993, edition of *Aviation & Space Technology* magazine covered what they called *"a continuing surge in Russian space operations.... During the first quarter of 1993, Russia launched 12 new, unmanned military spacecraft."* During the same time period, the Russians also launched six purported "civilian" space missions. *"The Russian military space surge began in late 1992.... The 18 Russian space missions compare with eight for the U.S. during that period.... The new Russian activity shows the former U.S.S.R. retains a surprisingly potent military space capability."* Coincidentally, the time period that this expansion of Soviet space operations started was the same time that the Western media was proclaiming that the Soviet military had fallen apart.

Soviet Arms Production

Contrary to what the Western media may portray, Russian factories still produce 3,000 new T-80 tanks annually, replacing the old T-54 and T-70 versions. The Russians have about 50,000 tanks, which is 20 times larger than the NATO total. Thirty to forty percent of the Russian economy is still devoted to military production. Approximately 600 factories in Russia are involved in weapons production. Only about twelve of these have converted to the production of consumer goods. Modernization of Russia's nuclear force continues unabated with the development of a new advanced version of a mobile ICBM.

The media talks a lot about the START treaty and U.S. funding of the dismantling of Soviet nuclear weapons but an article with a conflicting message appeared in the June 3, 1993, edition of the *Washington Times*. This article recapped another article from the Russian magazine *Ogonek* which reported that in Russia and the other republics there are now 27 cities involved in strategic weapons production and development which have been closed to Westerners. Many of these facilities are purported to be located underground, encased in granite. The Russian government employs 755,000 people in these facilities. Between 1,500 and 2,000 of these people are scientists, the rest are technicians and workers.

Former Defense Secretary Dick Chaney refers to this fact as a "basic contradiction." Chaney also said: ***"When a nation is facing extraordinary economic hardship and bankruptcy, why does it continue to spend so much of its gross national product on the military?"*** The probable reason Russia continues to place such a high national priority on military weapons development and production is because they are planning on going to war in the near future.

Russian Subs Still On the Prowl

During the latter part of 1995, Russian Akula and Oscar-2 submarines have been shadowing U.S. Trident subs and aircraft carriers in the Pacific off the coast of Washington and in the Atlantic off the coast of Georgia. This is the first flurry of Russian sub incidents since 1987. These incidents involved Russian attack subs closely shadowing U.S. warships. These locations are near U.S.

Trident submarine bases. Two other sub incidents occurred, one near Hawaii and the other farther out in the Atlantic. The Hawaii incident involved a Russian Oscar-2 cruise missile sub and the U.S. aircraft carriers *Abraham Lincoln* and *Independence* and the other incident involved the carrier *America* in the Atlantic.

There is a certain boundary area around a warship which represents the ship's vulnerability. A warship's defensive weapon system has limits to its ability to effectively deter close-in attacks.

The Akula is probably the quietest sub in the world. The Akula can dive deeper and travel faster than the U.S. Los Angeles class attack submarines. Also, the Russians have recently developed a new torpedo with a speed of about 300 m.p.h. At this time, the fastest U.S. torpedo has a top speed of about 80 m.p.h.

The Appearance of Freedom

Many Westerners would suggest that the proliferation of independence movements in the Eastern bloc nations and old Soviet republics proves that the old Communist empire has disintegrated. The strategy which the KGB has historically used to destroy opposition and consolidate power in the Soviet Republics and the Eastern bloc satellite countries is to establish phony KGB controlled independence movements. Once established, these phony independence movements bring the underground opposition to the surface where it can be identified and eventually liquidated.

The exception is the Caucasus region, including Chechnya, Georgia and Azerbaijan.

The KGB cannot ignore the existence of a genuine and uncontrolled struggle for freedom. The real thing can be contagious and must be squashed with an iron fist.

Georgia

In December of 1991, when the new Commonwealth of States was formed, Georgia was the only republic that refused to join. Earlier that year, Zviad Gamsakhurdia had been elected President of Georgia taking about 87 percent of the vote. This was the first real election of a leader by the popular vote of the people in a Soviet republic. Gamsakhurdia's claim to fame continued when, in August of 1991, he was the only leader of a Soviet republic to openly voice his opinion that the coup to overthrow Gorbachev had in fact been faked by Gorbachev himself as part of a long range Marxist agenda.

Consequently, heavily armed leftist forces deposed Gamsakhurdia in December of 1991. And in October of 1992, Shevardnadze was elected to the new post of Parliament Chairman. Consistent with the Soviet style of democracy, in this election Shevardnadze ran unopposed and elections were not conducted in the six districts where there was strong support for Gamsakhurdia.

Since Shevardnadze has taken over control, the civilian population in the Republic of Georgia has been rapidly declining from 5 million to 3.8 million, but there are no indications that any sort of migration is taking place. Also, at Shevardnadze's request, the Russians have established military bases throughout Georgia. The question to be asked is whether the disappearance of the population is in any way related to the appearance of the Russian military.

In September of 1994, Madeleine Albright, the U.S. Ambassador to the UN., made an unpublicized speech in Moscow after a visit to Georgia and several other Republics. In her speech she made the comment "that peacekeeping is the new growth industry" [not related to the census]. She went on to say that the Russian peacekeeping force had come to Georgia, *on the invitation of Chairman Shevardnadze,"* and that these military forces *"have now become a neutral force"* [neutralizing citizens?].

Chechnya

Chechnya has had a history of rebellion for over 200 years. Because of their independent spirit, on February 23, 1944, Stalin exiled en masse 400,000 Chechens from their homeland to remote areas of central Asia. This exile order was not rescinded until 1957. Solzhenitsyn gave the following comment regarding the Chechens, [It is] ***"one nation which would not give in, would not acquire the mental habits of submission - - and not just individual rebels among them, but the whole nation to a man."***

Azerbajan

In 1991, Ayazan Mutalibov was elected president of Azerbajan. Mutalibov ran as the sole candidate in the election. He has been a Communist since 1963. His credentials include being a KGB general and the head of the Azeri Communist Party. He was a member of the Soviet Politburo during the Brezhnev era. In 1992, Mutalibov was run out of office by the angry citizens. Abulfez Elchiby, a staunch nationalists who had spent two years in a hard labor prison camp for his anti-Communist views, was elected to replace him. Elchiby had been Azerbajan's leading dissident since the 1970s. Unfortunately, in 1993 Elchiby's government was toppled and the parliament elected Ayazan Mutalibov chairman and acting president. On October 3, 1993, Mutalibov miraculously received 98 percent of the vote in a presidential election.

The Other Republics

Belarus

The present parliament in Belarus is dominated by ex-Communists. In June of 1994, Aleksandr Lukashenko, who has a long history in the Communist party, became the first elected president.

Kazakhstan

The parliament of Kazakhstan is firmly in Communist control. Nursultan

Nazarbayev was elected president in 1992. Nazarbayev is a Gorbachev ally and a past member of the Soviet Politburo. He headed the Communist party in Kazakhstan before its independence. Nazarbayev was the only candidate in the presidential election and his term was supposed to end in December 1996. But on March 11, 1995, Nazarbayev dissolved Parliament and proclaimed that he would rule by decree until new elections were held. On April 30, 1995, a referendum was held regarding extending Nazarbayev's rule until the year 2000. Nazarbayev received a miraculous 95 percent support.

Kyrgyzstan

Askar Akayev was elected president in 1991. As could be expected, Akayev was the only candidate in the election and he received 95 percent of the vote. Also 95 percent of the deputies elected to Parliament are former members of the Kyrgyz Communist party. Akayev's Communist credentials include a position on the Soviet Communist Party Central Committee, a position on the Soviet Congress of People's Deputies and a position in the Supreme Soviet.

Under pressure from public dissatisfaction, Akayev scheduled a referendum for January 1994 regarding whether or not he should continue in power until the end of his term. The results of this referendum gave Akayev a typically miraculous 96 percent support.

In July of 1994, Akayev proposed implementing curbs on freedom of the press because of criticism he was receiving from the "anti-democratic" press. U.S. Deputy Secretary of State Strobe Talbott describes Akayev as *a true Jeffersonian democrat*.

Latvia

Anatolijs Gorbunovs is the chairman of the Latvia Parliament. As could be expected, Gorbunovs was formerly a member of the Central Committee of the Communist Party of the Soviet Union and was also the Latvian Communist Party secretary of ideology.

Lithuania

In 1992, the Lithuanian Parliament elected Algirdas Brazauskas chairman and acting head of state. In February, 1992, Brazauskas won the presidential election by a 60 percent vote. Brazauskas's Communist credentials include state economic planner, secretary of the Lithuanian Communist Party in charge of economic affairs, and in 1988 he became the Communist party boss.

Moldovia

On December 8, 1991, Mircea Snegur was elected president of Moldovia. Being consistent with Eastern bloc democratic standards, he was the only candidate and he took a miraculous 98 percent of the vote. Snegur was previously the president of the Moldovian Supreme Soviet, deputy chairman of the USSR Supreme Soviet and secretary of the Moldovian Communist Party Central Committee. In !993, Petru Lucinschi was elected speaker of the Moldovian Parliament. Lucinschi was formerly the first secretary of the Moldovian Communist Party, a member of the Central Committee of the Soviet Communist Party and a member of the Soviet Politburo.

Tajikistan

In 1991, Rakhman Nabiyev became president of Tajikistan. Nabiyev is a former Communist party first secretary. In November, 1992, Imamali Rakhomonov was appointed acting president. He is a long time "former" communist and a Nabiyev supporter.

Ukraine

In December 1991, Leonid Kravchuk was elected president of the Ukraine. He was formerly the Ukraine's Communist Party chief of ideology. In July of 1994, Kravchuk was defeated by Leonid Kuchma, who was previously the director of the Soviet Union's largest missile production facility. The prime minister, Vitaly Masol was formerly the Soviet Union's top economic manager.

Uzbekistan

In 1991, Islam Karimov was elected president. Before the election, Karimov curtailed the activities of all opposing political factions. Karimov won the election by taking 86 percent of the vote. Karimov opposed the separation from the Soviet Union as well as the implementation of democracy and a free market economy. A referendum was held which endorsed the extension of Karimov's presidential term. In December, 1994, Uzbekistan held parliamentary elections in which the Democratic party (formerly called the Communist Party) captured 70 percent of the seats.

The Eastern Bloc Nations

An analysis of Albania, Bulgaria, the Czech Republic, Hungary, Poland, Romania and Slovakia produces the identical result of former Communists being in key positions of power.

Nomenklatura- the Real Power Behind the Kremlin Façade

Most Westerners look at the Soviet Union through the lenses of Western glasses. Everything they see is interpreted based on their own cultural experience and reference to Western style government. The Soviet propaganda machine encourages this misconception by renaming its power structures "democratic" in unfair and inequitable comparisons.

In many ways Soviet society is very different from the West but in other ways it is very similar.

Based on the ideals of Karl Marx, the Russian - Soviet empire is supposed to be a classless society, but it is not. There are three classes of society in the Russian - Soviet empire — the Nomenklatura, the rising middle class and the masses.

The Nomenklatura is a small, cohesive and elite group of about 25,000 Slavic white Russians who govern the 274 million people of the Russian - Soviet empire. Like their Western counterparts, the Nomenklatura operate from the shadows, out of public view, living an almost anonymous existence. But one aspect of the Soviet elite is completely opposite that of its Western counterpart. In the West, wealth secures power and privilege; but in the Soviet Eastern bloc, power and privilege secure wealth.

In the so-called classless society, the government supplies <u>everything</u> which the upper and lower classes receive. Only the rising middle class has need or use for money. However, everything is not much for the lower classes. Most Russians eat, dress and live well below Western standards. Health care and education are free for the lower class, but extremely substandard. Life is hard. Shortages and waiting daily in long lines for basic staples and inferior consumer goods are a part of the average Russian's daily life.

On the other hand, the state supplies anything and everything the Nomenklatura wants and desires. The Nomenklatura are able to indulge, at the state's expense, in every desire denied to the rest of Soviet society. This includes international travel, Western books and videos, private educational institutions for their children, deluxe automobiles, chauffeurs, maids and butlers, special well-staffed and equipped medical clinics, spacious Moscow apartments furnished with Western-

made appliances and furniture, country dachas, access to special stores stocked with Western consumer goods and more. Nomenklatura means privilege in all aspects of Soviet life and this privileged system is supported by an ocean of state funds.

The rising middle class is composed of those who are attempting to rise out the mediocrity of the lower class. These people include those who rise to achievement in the sciences, arts and entertainment. The middle class can buy, with rubles, a portion of the things which the Nomenklatura can demand from the state based on their position.

The Nomenklatura include members of the Politburo and their families, senior Communist party heads, K.G.B. leaders, marshals and generals, top scientists, diplomats, Central Committee officials, and heads of the military-industrial complex. The Nomenklatura do not defect to the West. There is no reason or motivation. The system gives them everything they could ever want. Going to the West would only mean lowering their position and standard of living. The best analogy of the Russian Nomenklatura is the elite warrior class of ancient Sparta who were supported by a slave population.

Russian-Soviet Conclusion

The transition from a state owned and controlled economy to a so-called free market economy has resulted in a deepening economic crisis. In all reality, things are probably proceeding according to plan under the direction of Russia's real controlling power. The hopes of the Russian people for finding relief from domestic austerities and government oppression through representative government and a free market economy have been dashed. This disillusionment has led many Russians to conclude that a more authoritarian government is the only real solution to Russia's problems. Leading the Russian public to such a conclusion may have been the elite's objective.

Deficiencies exist in the media generated concept that the Soviet-Communist empire has collapsed, the cold war is over and the old enemy is now benign. If you put together the evidence, including the Soviet character flaw, revealing statements by the Soviet leadership and the ongoing military expansion, it seems obvious that world conquest is their agenda. At best, the Soviet elite are not putting all of their cards on the table. Given historical precedents, it is not realistic to expect this elite to willingly surrender long held power and their dream of world supremacy. The circumstances of life and the future may well bring the Russian-Soviet power structure to a position of desperate survival, where their only option would be some form of major military expansion. The Soviets could make their military move against the West either out of national desperation or in a more calculated manner, which fulfills their long-term plan for world conquest.

The big question, of course, is if the Soviets intend on going to war with the West when and how will they strike. If and when the Soviets initiate war against the West, they obviously will want as much of the odds as possible in their favor. Holding the advantage is an important part of the science of winning war. Being able to pick the time and place of the battle is a great advantage. As long as the Western economy has the capacity to extend financial aid to the Soviets, they will probably be content to sit back and wait. Meanwhile, they will continue to refine their war machine for the ultimate day while they proceed to put all their strategic ducks in a row.

This waiting strategy is clearly reflected in this provoking statement by Dimitry Manuilsky in a lecture he delivered in 1981 at the Lenin School on Political Warfare in Moscow: ***"Today, of course, we are not strong enough to attack:...To win we shall need the element of surprise. The bourgeoisie will have to be put to sleep. So we shall begin by launching the most spectacular peace movement on record. There will be electrifying overtures and unheard of concessions. The capitalist countries, stupid and decadent, will rejoice to cooperate in their own destruction. They will jump at another chance to be friends. As soon as their guard is down, we shall smash them with our clenched fist."***

Red China

According to the KGB defector Anatoliy Golitsyn, the Sino-Soviet split in the early 1960s was a strategic deception intended to further Communist objectives. In March of 1989, Golitsyn sent a memo to the CIA which said that China was one of the principal architects of the Communists' long-range strategy. He also stated that the Sino-Soviet split was a joint Sino-Soviet disinformation operation intended to insure the success of their "restructuring" strategy. According to Golitsyn, *"China is destined to become a primary Soviet partner in the future world government towards which Moscow and Peking are jointly proceeding."*

In March of 1966, the Chinese government made the following statement in a communique to Moscow: *"The great peoples of China and the Soviet Union will eventually sweep away all obstacles and unite on the basis of Marxism-Leninism and proletarian internationalism....The Soviet people may rest assured that once the Soviet Union meets with imperialist aggression and puts up resolute resistance, China will definitely stand side by side with the Soviet Union and fight against the common enemy."*

The phony Sino-Soviet split has seduced the United States government into giving the Red Chinese a most favored nation trade status along with technical and military assistance. The idea is that supporting your enemy's enemy helps maintain a balance of power. The other reason behind developing Chinese trade relations is that the global corporate interests think that they can eventually control China through the process of economic development, Westernization and the infiltration of her financial institutions. But it seems like they have a surprise coming.

In 1994 the Vice Chairman of the Chinese Communist Party, Mo Xiusong, was asked if the long-term goal of the Chinese Communist Party was still World Communism. Mo Xiusong replied, *"Yes, of course, that is why we exist".*

Military Build-up

While the U.S. is engaging in a comprehensive reduction in its conventional and nuclear forces, the Chinese are engaged in a massive modernization. The Chinese have increased their annual "published" defense budget by 75 percent since 1988.

Most of the media attention regarding the nuclear threat to the United States focuses on the intercontinental ballistic missiles in the "former" Soviet Union. China has an arsenal of intercontinental ballistic missiles capable of striking targets anywhere in North America. Red China has extended the range of its Silkworm I.C.B.M. from 6,000 to 9,000 miles. The Chinese have recently developed a mobile launch I.C.B.M. and have increased the accuracy of their strategic rocket programs through the purchase of advanced American super computers.

The Chinese have also been working on improving their fighter aircraft. Beijing obtained a U.S. F-16 fighter aircraft from Pakistan in exchange for M-11 medium-range ballistic missiles and is attempting to reproduce a version of this aircraft. China has also been indirectly purchasing advanced American military technology through Israel. This includes the technology for 120 millimeter guns, armored vehicles, anti-tank missiles, jet fighters and air-to-air missiles.

Human Rights

The Chinese human rights issue goes far beyond Tiananmen Square. Their human rights abuses include the ongoing genocide in Tibet and human organ harvesting as revealed by human rights advocate Harry Wu. Condemned prisoners routinely have their executions arranged to accommodate the schedules of organ harvesting teams. Wu has also identified 1,100 forced labor camps throughout China.

Espionage

According to FBI deputy assistant director Patrick Watson, suspected cases of spying by China *"rivals that of the Soviet Union in its heyday."* Watson said that most of the spying is directed toward U.S. military and intelligence agencies.

Conclusion

With the assistance of American technology and hard currency profits made in sales to U.S. consumers, the Chinese are quietly pursuing their Marxist-Leninist objective of world domination by making high tech improvements to the world's largest standing army. Based on their human rights violations and their frank admissions toward a goal of world domination, it does not seem in the interest of U.S. security and moral values to vest Red China with a most favored nation trade status.

Note:

Some of the information for this chapter was drawn from *The New American* magazine. *The New American* magazine is one of the foremost sources of unfiltered information on Russian-Soviet activities, the Global agenda, and American politics. *The New American* magazine is located at 800 Rosevelt Rd., Building B, Suite 407, Glen Ellyn, IL 60137, telephone (708) 469-4611 or (800) 946-4611.

Chapter 2
A Soviet Nuclear Attack On The United States

Due to the long arm of the intercontinental ballistic missile, the United States is no longer protected from her enemies by the vast expanse of the Atlantic and Pacific oceans. There have been three occasions since World War II when the U.S. has been on the verge of war with the Soviets as the result of a conflict involving third world countries. The first was the Cuban missile crisis, the second was involving the Middle East during the Seven-Days War, and the third was in 1990 when India and Pakistan were on the verge of war.

The following is a fictional description of how a Soviet nuclear attack on the United States might realistically occur and what the result of such an attack would probably be. It is written in the present tense, as if it was actually happening, to give the reader a more powerful impression.

One feasible scenario for a nuclear surprise attack on the United States would be for Russia to plan an attack after the peaceful settlement of a gradually escalating, artificially created crises.

A bogus crises in Europe or elsewhere gives the Soviets a justifiable excuse for implementing the deployment and positioning of military forces and equipment necessary for attaining a pinnacle of war readiness. This crisis could be caused, for example, by heightened submarine activity shadowing NATO ships in the Baltic, Atlantic or elsewhere or the activation of military units in Eastern Europe. This is probably staged at a time when some sort of localized conflict is going on, such as Korea, Bosnia or the Middle East. Such a conflict ties up and concentrates a significant portion of the U.S. military force in a distant area.

The second element necessary is cold winter conditions so that Soviet tank units can move swiftly over the European countryside.

The NATO reaction to this artificially created crisis is to bolster its security. In reaction to the NATO move, the Soviets call for an emergency session of the U.N. Security Council. They vocally denounce NATO's security move and accuse the West of dangerously escalating international tensions. The Soviets then announce that they intend to deploy their own defensive forces.

Now the U.S. and NATO countries are labeled the bad guy aggressors who started a problem. Western diplomats anxiously try to convince the Soviets that this is all a big mistake. Every frantic attempt is made to appease the Soviets and deflate the crisis. The Soviets utilize their publicly announced military deployment to position strategic forces.

Uncertainty is a key strategy which the Soviets play upon the West. The uncertain and apologetic Western leaders are bending over backwards trying not to initiate hostilities.

Western Anxieties

Subside

All of a sudden, a surprise diplomatic breakthrough is achieved, (ideally sometime shortly before Thanksgiving, Christmas or New Years). A Summit meeting is scheduled for the near future. Western tensions relax. The president takes the U.S. military forces off alert and the Western Allies celebrate peace and the holiday.

Shortly after this de-escalation of the crisis, while its forces are still deployed, the Soviets prepare to launch a surprise attack once the U.S. has its back turned.

E.M.P.

Suddenly, without warning, nuclear detonations occur simultaneously, high in the atmosphere over different sections of the North American continent and Western Europe. These detonations originate from pre-positioned orbiting satellites armed with atomic weapons.

These high altitude bursts do not affect the surface of the earth with any form of destructive heat, overpressure or radiation, but a phenomena occurs which is called Electro Magnetic Pulse, or E.M.P. The Electro Magnetic Pulse radiates out across North America and Western Europe, collecting on power transmission lines and communication antennas. For a fraction of a second, the E.M.P. spikes the power grid and receiving antennas with a huge surge of voltage. In the process of seeking ground,

Soviet Orbital Anti-satellite Pellet Blast Weapon

this surge is so quick and great that it jumps across conventional surge protectors, lightning protection, and even open circuits where the power switches are turned off, burning out and rendering useless all sensitive electronics, computers and communication equipment.

Simultaneously, Soviet space-based weapon systems start decapitating U.S. reconnaissance satellites.

Confusion & Uncertainty

The confusion created by the disruption of the national communication network is considerable. The destruction of the U.S. spy satellites blind U.S. intelligence analysts. Military intelligence agencies are not able to give the president a definitive answer as to what the Soviets are really doing.

While the Soviets are launching I.C.B.M.s at U.S. targets, the Soviet head of state calls the President of the United States on the hot line to apologetically inform him that there has been an accident in space. The Soviet leader explains that a reactor on their orbiting space station has exploded. Not wanting to be responsible for the start of World War III, the uncertain President ignores advice from his military advisers to launch our I.C.B.M.s at Soviet targets.

The President doesn't have long to wait because Soviet submarines off the Atlantic coast launch ballistic missiles at Washington D.C. The time from launch

Military Targets

● **Primary Targets**   ⬤ **Secondary Targets**

to target impact is only three minutes. With the institution of a war alert, the president, his family and their Secret Service entourage are wisked away by helicopter. The helicopter is just crossing the Fairfax County line into Virginia when it is struck by the shock wave from one of the numerous air bursts over the nation's capital. The aircraft instantly disintegrates with the wreckage scattering over a wide area of countryside.

Strategic Targets in the U.S.

In the United States there are about 6,000 primary targets. This includes military related targets which would be involved in the launch a retaliatory counterattack. The initial Soviet I.C.B.M. attack concentrates on these high priority first-strike targets which are critically important for a Soviet victory. These targets include: I.C.B.M. silos and launch centers, air force, navy and army facilities, key military support industries, command and control centers, and political infrastructures.

Once all the designated priority first-strike military targets are hit, satellite reconnaissance is used by the Soviet military staff to determine the effectiveness of target destruction. Then further strikes are initiated where the initial attack failed or destruction of a target was incomplete. When all the military targets are properly destroyed, the Soviet military staff concentrates their attack priority on industrial targets which support the military effort including key military support industries, port facilities, fuel refineries, fuel storage facilities, power generating plants, chemical plants and communication facilities.

Destruction of U.S. Targets

Without warning, a significant portion of our retaliatory force is destroyed before it is deployed. Our bombers, which comprise the air wing of our TRIAD, are all off alert due to presidential directive. Virtually all of them are destroyed in their hangers. Each U.S. I.C.B.M. missile silo is struck simultaneously with two ground-penetrating warheads. Massive amounts of soil is drawn up into the resulting mushroom clouds, producing radioactive fallout particles which are spread for hundreds and thousands of miles before settling to earth.

Ninety percent of our own I.C.B.M.s are destroyed in their silos by the first wave of missiles. The ten percent which escaped the first strike are successfully launched from their silos but most of these are intercepted by the Soviet anti-ballistic missile system. Those that penetrate Soviet defenses only inflict minimal damage, quantitatively speaking, to military targets. The Soviet popu-

Soviet GALOSH Anti-ballistic Missile

lation, due to their extensive civilian civil defense program, suffer negligible losses.

Soviet I.C.B.M.s with ground-burst weapons impact and destroy U.S. military runways and civilian airports capable of accommodating the emergency dispersal of military aircraft. Military bases and Navel facilities are struck by I.C.B.M.s with air-burst weapons. The explosion of these weapons produces no radioactive fallout but there is optimum destructive effect of surface structures due to the resulting massive wave of overpressure and the thermal effect of the fireball.

One-third to one-half of the 39 U.S. strategic submarine fleet is destroyed in port. Of the remaining 19 or 20 U.S. submarines, 10 will have to travel for two to five days in order to be within striking range of Soviet targets. The other 10 U.S. subs are within striking range of Russian targets and successfully launch their missiles. Again, those sub-launched missiles that managed to penetrate the Soviet anti-ballistic missile defense system only inflict minimal damage, both to their targets and to the Soviet population due to civil-defense preparations. The other element which quickly comes into play are Russia's 250 attack submarines, which in short order decimate most surviving U.S. strategic submarines.

Meanwhile the Soviet strategic submarines are launching their missiles, which have greater range than those of their U.S. counterparts. The longer range missiles allow the Soviet subs to launch from protected waters. Thus it is very difficult for the surviving elements of the United States Navy to destroy much of the Soviet nuclear submarine fleet.

U.S. Casualties

Due to the lack of a U.S. anti-ballistic missile system and civil defense preparations, all missiles strike their appointed targets in the U.S., causing massive destruction and loss of life. The Soviet first strike has resulted in the destruction of most of the large cities in the United States. The Soviet nuclear attack on the U.S. produces about 67 million casualties due to the effects of the blast and high exposure to radiation. This represents about 25 percent of the population.

During the next two years another 121.5 million Americans starve to death, bringing the total U.S. casualties to about 188 million, or approximately 70 percent of the population. (Strategic and civil defense could have reduced U.S. fatalities to as little as 7,000). Russia lost no more than 5 percent of her population as a result of I.C.B.M.s which managed to get through the anti-ballistic missile system. Also, none of the Russian citizens starve following the war, thanks to the massive Soviet strategic grain reserves which had been purchased from the United States with credits from Western banks.

Little is left of U.S. nuclear forces to retaliate after the Soviet first strike. Damage to U.S. military command and control and communications links have further restricted an attempt to retaliate.

Soviet intelligence and Spetnaz teams, pre-positioned in the United States start implementing a pre-established plan to create confusion. Soviet agents have been monitoring the daily routines of key members of the military and civilian leadership for months. Simultaneously, Spetnaz teams begin assassinating these key leaders and anyone else who has the misfortune of being around or in the way.

Other teams have been assigned the task of disrupting communication, power transmission and transportation. Power transmission lines and generation facilities at key points are sabotaged. Key bridges on thoroughfares out of major cities and on the interstate highways are blown up.

Teams equipped with shoulder-fired missiles wait in ambush on a wooded hill adjacent to the government's super secret Iron Mountain shelter complex in West Virginia. A few helicopters which managed to successfully escape Washington, D.C., are loaded with their cargo of Pentagon officials, congressmen and cabinet members. As the helicopters approach the seeming safety of Iron mountain, they are met by a crossfire of surface-to-air missiles.

The ensuing war is composed of many nuclear and non-nuclear battles which continue for two to four years. The war involves conventional forces both in Europe and an invasion of American soil.

The Continuing War

During this protracted combination nuclear-conventional war, chemical warfare is used by the Soviets as extensively as nuclear weapons. Approximately one-third of the Soviet arsenal is composed of chemical weapons. The Soviets maintain chemical warfare units which have 80,000 to 100,000 well trained soldiers. Biological warfare is not used in conjunction with conventional Soviet troop operations because it is slow acting, difficult to control, and could thus endanger Soviet troops. Biological weapons are used very effectively to eliminate resistance in hostile areas not occupied or controlled by Soviet troops. The overall objective of the Soviet war effort is not to completely destroy the United States, but to take control of as many aspects of American society as possible.

Evidence supporting this mixture of nuclear and conventional warfare in a third world war is found in the Soviet book, *Tactics and Combined Arms Combat*, Moscow, Military Publishing House.

"It is believed that nuclear weapons as the main means of destruction will be deployed only for the destruction of the most important objectives; all other targets are neutralized and destroyed by artillery aviation and the fire of tanks and other weapons. Use of nuclear weapons against insignificant, secondary objectives contradicts the very nature of this weapon.

"Although the nuclear strike should be delivered at the beginning of the preparatory fire in order to achieve surprise, this should not be stereotyped. Nuclear weapons can also be employed in the middle, at the end, or at any other period of the preparatory fire.

"Nuclear strikes do not represent some kind of isolated act, but a component of combat. The operations of tank and motorized rifle are closely coordinated with them. Nuclear strikes and troop operations represent a uniform and inseparable process, joined by a common concept."

This scenario of an extended war involving both conventional and nuclear forces follows an overall Soviet strategy with the following objectives: destroy the most threatening enemy forces, destroy the main links and nodes in the national command and control authority, do not destroy large areas or create nuclear deserts, use minimum weapons yield to prevent target overkill, prepare to strike the most important targets twice, and do not attack and destroy all targets (because it is not possible or desirable).

This nuclear exchange is followed by an invasion of conventional forces involving sea and air battles, tanks, paratroopers and infantry. Airborne units strike across the Bering Sea and invade Alaska. With Nome, Prudhoe Bay, Fairbanks and Anchorage, the Soviets have secured the strategic oil reserves to support the ongoing war effort.

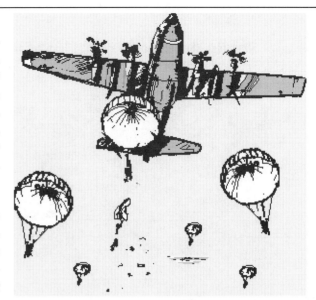

A combined sea, air and paratroop assault on Seattle, Tacoma, Victoria and Vancouver secures the port facilities and gains control of I-90 and I-5 North and South. The Soviets secure their beachhead in the Northwest.

Armored units begin moving east on I-90 to Ellensburg, Washington. At this point part of the advancing force turns south on Interstate 82. The objective of this unit is to secure the chemical weapons storage at Umitila, the bridge over the Columbia River and the intersection of I-82 and I-84. Securing the intersection of Interstate 82 and 84 opens the back door to Portland. Military units move east on I-84 to flank resistance to a Soviet military push south from Seattle toward Portland and Longview down I-5.

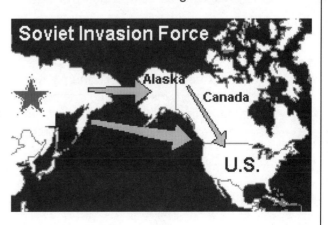

Soviet Air, Sea & Land Invasion

From Elsenburg, military forces continue moving east to Vantage, securing another bridge over the Columbia River and opening the door for a push to Spokane.

Meanwhile, in other parts of the country, an invasion force from Cuba establishes a beachhead on the southern Florida coast. Communist forces pre-positioned in Mexico, move across the border into Texas. Elements of the Soviet Red Banner Fleet secure New Orleans. This establishes key access to the central U.S. up the Mississippi River. The Soviets also mount a land, air and sea attack to secure the St. Lawrence Seaway.

More Information On Nuclear War

The Nuclear First Strike

Most Americans cannot fathom the possibility of an unprovoked strategic attack by the Soviets. Rejecting the possibility of nuclear war may be comforting and helpful for normal existence, but this societal self-delusion may prove a fatal liability.

In war, he who chooses the time and place of battle holds the best advantage for victory. Fighting on a proactive or offensive basis holds the greater advantage for victory. Fighting on a reactive or defensive basis is an encumbrance.

The purpose of launching a first strike on the United States would be to take advantage of the element of surprise and catch U.S. forces off guard. The Soviets have significant capability for a preemptive attack on the United States. The Soviets have concentrated their nuclear firepower into an enormous intercontinental ballistic missile force which is optimized for a first strike against the Unites States. The initial and primary objective would be to destroy the United States' defensive and retaliatory capability. In theory, once adequate industrial targets were dispatched, population centers could be targeted at will until the U.S. government capitulated with an unconditional surrender.

The United States would never launch a preemptive attack on Russia because the U.S. is unprepared to deal with the resulting dire consequences. The United States has neither civil nor strategic defense and thus our civilian populations would be completely exposed to Soviet nuclear retaliation. Even if the U.S. possessed a national strategic defense system, the act of preparing for a nuclear first strike would require the movement of U.S. forces into battle positions and the necessary evacuation of civilians from metropolitan areas. These actions would tip the Soviets off and the attack would no longer be a surprise, unless there was a devious strategy employed to cover or justify such preparations.

On the other hand, Russia has both strategic and civil defense. In the event of a surprise attack, the Russians could quickly shelter a significant portion of their civilian population and neutralize most I.C.B.M.'s launched against Russian targets with their operational missile defense system.

The Old Doomsday Scenario

Back in the 1950s, government planners developed a simplified scenario for an all-out nuclear war. This doomsday model for nuclear war suggested that both the U.S. and Soviet strategic missile forces would unleash their entire arsenals in the first day of a nuclear war. In spite of the development of more sophisticated war plans by both the U.S. and Soviets, this doomsday model seems to have struck an emotional cord deep in the psyche of the American people. Many Americans still retain and believe in this outdated model which they keep buried deep in the mind's filing cabinet in a file called "nuclear war."

This doomsday scenario is also the source of a common misunderstanding regarding how long people would have to stay in fallout shelters in the event of a nuclear war. It is true that if all the weapons in the U.S. and Soviet arsenals were detonated on the first day of a nuclear war, all that a person would have to do is stay in a shelter for 14 days, after which time the radiation would have been reduced to relatively safe levels. But the problem here is that the newer, more realistic model for nuclear war calls for a protracted nuclear and conventional war lasting for two or more years. The clear message here is in the event of a nuclear war, don't think that all you will need to do is cover your head for 14 days. You would probably have to use the fallout shelter on numerous occasions for months at a time.

Gradually Escalating War Scenario

Based on a 1987 report by the Federal Emergency Management Agency (F.E.M.A.), many government planners advocate that a nuclear war with the Soviets could not start with a surprise attack. This school of thought suggests that the possibility of a surprise attack is very small, because the Soviet first strike would require preparations including moving naval forces into position, deploying military forces, reinforcing troops in occupied countries, reinforcing homeland defenses, and preparing all missile and aircraft bases for maximum operation. Due to U.S. satellite surveillance, these operations could not be done in secret. In other words, the logic here is that neither of us could surprise the other.

Not everyone in the intelligence community agrees with this assessment. Another point of view suggests that the reason F.E.M.A. downplays the potential of a surprise nuclear attack in favor of a gradually escalating scenario is that the latter is the only scenario F.E.M.A. could possibly cope with.

India-Pakistan Conflict & the F.E.M.A. Relocation Plan

F.E.M.A. has no contingency plans for sheltering the civilian population in the event of a surprise nuclear attack. Recent history has already shown that it is questionable that F.E.M.A. can or would start relocating civilians out of populated areas in the event of a gradually escalating scenario. Back in the 1990s, the United States experienced a gradually escalating war scenario with the Soviet Union over an imminent war between India and Pakistan.

Bush eventually managed to bring about a diplomatic solution to the crisis, but the public was never told that the U.S. and the Soviets were both on high alert. The F.E.M.A. policy of evacuating civilians out of metropolitan target areas was never implemented. Meanwhile, the United States and Soviet Union were both positioning strategic forces in preparation for a possible global conflict.

According to an Associated Press article which appeared in the March 22, 1993, edition of the *Great Falls Tribune*, *"The world was on the edge of a nuclear confrontation between India and Pakistan in early 1990 because of the tug-of-war over Kashmir."* Senior intelligence officials referred to this showdown in the spring of 1990 as *"the most dangerous nuclear confrontation of the postwar era."* The article went on to explain that, *"the Bush*

administration kept the conflict secret, failing even to disclose it to key members of Congress.... The nuclear confrontation was defused by the intervention of President Bush's personal envoy, who was sent to India and Pakistan to negotiate a standdown between the two countries."

Government Warning in the Event of a Nuclear Attack?

The next logical question is would the government alert the public if the Soviets launched I.C.B.M.s at the U.S.? Probably not! The reason I say this is that a last minute alert probably would not save many lives. An alert would cause immediate civil disorder and gridlock on the highways out of the major cities. The other reason not to alert the public is that there is no public shelter system for them to run to.

Nuclear Weapon Delivery Systems

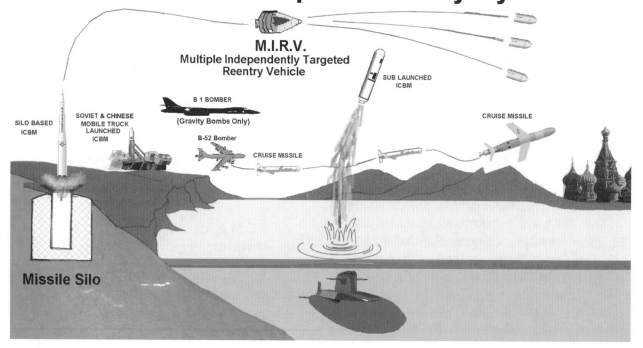

Chapter 3

Terrorism & The Third World Threat

"Somewhere, sometime — but in this decade — somebody...is going to set off a nuclear weapon in deadly earnest." Rear Admiral Edward D. Shaefer Jr., Director of U.S. Naval Intelligence.

It is inevitable that terrorist organizations will get, or now already have nuclear, biological and chemical weapons and the means to deliver them.

Nuclear Terrorism

Nuclear terrorism is enough of a real threat today to justify building a shelter. The possibility that a nuclear weapon will be used by terrorists increases annually. There are tens of thousands of nuclear weapons in national inventories, large amounts of weapons-grade uranium and plutonium potentially available to the black market, increasing dissemination of technical knowledge, and the existence of many countries that sponsor terrorism. The probability that terrorists will obtain possession of a functional nuclear weapon is increasing.

There is little or no defense against a terrorist attack. Small nuclear weapons could easily be brought into the country, strategically placed somewhere in a major U.S. city

and detonated. Nuclear terrorism by an Arab group would most likely be targeted against a large American city like New York, because of its large Jewish population, or Washington, D.C., because it is the nation's primary center of government. A tramp freighter with a nuclear weapon hidden onboard, unbeknownst to the crew, could easily be sailed into New York harbor and detonated. This would result in great loss of life and damage in close proximity to the detonation.

The threat of nuclear terrorism may be limited to some of the nation's larger cities, however, all Americans would be affected by the government's predictable reaction to nuclear terrorism. At the least, this reaction would involve a further reduction of civil liberties and an increase of the government's already broadbased police powers. At the worst, the president would use the incident to declare a national emergency. This would result in a declaration of martial law and the suspension of the Constitution under the guise of preventing further terrorists attacks. Such a governmental power grab would probably result in another civil war.

The problem with nuclear terrorism is that the U.S. government wouldn't

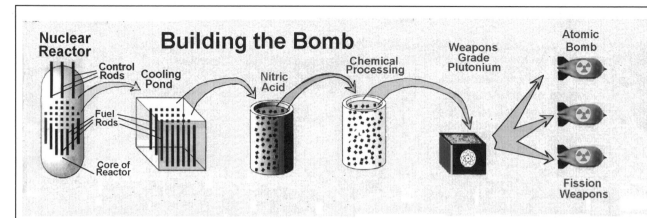

Building the Bomb

Nuclear Reactor — Control Rods — Fuel Rods — Core of Reactor — Cooling Pond — Nitric Acid — Chemical Processing — Weapons Grade Plutonium — Atomic Bomb — Fission Weapons

necessarily know who was responsible for it and who to retaliate against. It would also seem logical from the philosophy of Sun Tsu that if the Soviets wanted to inflict damage on their enemy America, they would initiate or support the action through a third world surrogate. Once the nuclear terrorism occurred, the Soviets would decry the act and deny any responsibility. They would probably even send a medical aid team to the devastated U.S. city.

Nuclear weapons and technology are rapidly being proliferated throughout the third world countries. Many of these new members to the nuclear club are not friendly to the U.S. North Korea, Iran, Libya and Iraq are several examples. In fact, U.S. intelligence agencies say that North Korea has developed a nuclear weapon and has also obtained I.C.B.M. delivery systems from Red China. These rockets include the 1,300-kilometer range Nodong-1 I.C.B.M., which is now operational, the 2,500-kilometer range Nodong II and the 5,000+ kilometer range Nodong III, which North Korea is still working on. The 5,000+ kilometer range Nodong III could reach the United States with a nuclear payload.

A 1995 C.I.A. report to Congress revealed that Iran obtained four ballistic missile launchers from North Korea. Despite U.S. protests, Russia is still going ahead with its sale of a nuclear reactor to Iran. U.S. intelligence experts suggest that Iran could have nuclear missile capability by or before the year 2000.

The potential ramifications of this situation in relation to Israel are dire, and history has shown that if Israel feels threatened, its inclination is to make a preemptive strike. Even more ominous is the fact that the last major Arab-Israeli conflict brought the United States and the Soviet Union to the very brink of world war.

The Black Market & Weapons of Mass Destruction

The dire economic conditions in the old Soviet empire have resulted in the smuggling of nuclear materials out to the international black market. In the fall of 1994, German police seized at least five shipments of weapons grade plutonium and uranium that were smuggled out of Russia during the previous year. Western intelligence speculates that former members of the East German intelligence agency, former K.G.B. officers, disgruntled members of the new Russian intelligence service and members of Russia's network of organized crime are involved in this smuggling. Germany's Federal Intelligence Agency has concluded that there have been more than 200 sales of nuclear material smuggled out of the former Soviet states and Warsaw Pact nations.

This is probably just the tip of the iceberg. Desperate economic conditions are placing enormous pressures on Russian military men who cannot support their families. The black market will pay up to $2.2 million for a kilogram of enriched uranium.

In some ways this trafficking in weapons of mass destruction is similar to the drug trade in the United States. The temptations of big money entice Americans, who are well off by Russian standards, into taking chances with the law and compromising principles. This pressure is much greater when the individual is destitute and desperate. The street value of weapons grade uranium 239 is $252,000,000 for nine pounds. Due to Russia's present degenerated quality of life, the sense of right and wrong is not much of a controlling factor.

The Export of Russian War Technology

There is a serious brain drain going on from Russia to North Korea and the hardline Arab states (Iran, Iraq, Libya and Syria). Russian scientists and technicians who are out of work can find big money abroad if they have skills in producing nuclear, biological or chemical weapons. In 1994, Russian law enforcement agencies stopped 60 Russian ballistic missile scientists and technicians at Moscow's Shermyetovo-2 airport as they prepared to fly off to North Korea.

The January 1993 issue of *Arms Control Today* reported that several Russian scientists have contracted to work in Algeria, four have left Russia to take jobs in India, 50 weapon and missile specialists remain in Iraq, 14 nuclear scientists are working in Iran and "many" Russian scientists are participating in weapons projects in Libya. In all reality, such intelligence figures only represent a fraction of the Russian scientists and technicians who have taken jobs abroad.

Reports indicate that the Russians can't account for 40 of the 30,000 nuclear weapons in their arsenal. Is it possible the these are already in the hands of radical Arab states?

The Export of American Technology

Just about any nation in this age can obtain sophisticated nuclear, biological or chemical tipped missiles. The basic technology has proliferated. Here in the United States more than half of the graduate students in technical fields are foreign citizens. Most American college students are not attracted to this type of difficult work and third world countries like India and Iran send their best students to the United States for their education. The U.S. proliferates and exports knowledge. Once given out, technical secrets become irretrievable.

The basic technical tools available today on the open market are substantially more sophisticated than those originally used to build the American stockpile of nuclear missiles and weapons.

The Clinton administration has scrapped virtually all national security controls on high tech exports. High end home computers can run programs for designing and manufacturing rockets. N.A.S.A. sells this software on the open market. In 1995 Clinton eased restrictions on the sale of supercomputers to foreign buyers as long as they were to be used for civilian purposes. The only flaw in this logic is that once nations like Red China or Russia take delivery on a supercomputer under the guise of civilian use, there is no way the U.S. can stop them from using it for a missile targeting and delivery system or a national air defense system.

Accurate guidance systems can also be purchased without restrictions. G.P.S. (global positioning) navigation equipment, which triangulates from orbiting satellites, has accuracy down to 28 feet anywhere on the planet. Put this together with a warhead and you have a ballistic missile.

The most difficult aspect involved in producing a nuclear missile is the warhead. Previously, producing the simultaneous detonation was one of the most challenging aspects involved with building an atomic bomb, but this has now been simplified by quartz timers on computer chips. The basic design for single stage uranium or plutonium bombs is public knowledge. One of the last and historically most difficult hurdles in building the bomb was getting the weapons grade nuclear material.

Black market access to the old Soviet stockpile seems to have lowered this hurdle for the aspiring nuclear terrorist. The last hurdle is separating the bomb grade uranium and producing plutonium. In the past this has involved very sophisticated technology, but new rather laborious methods have eliminated this hurdle also.

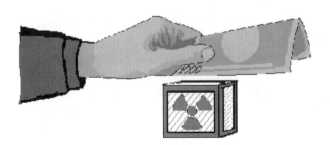

The Micro Neutron Bomb — the Ultimate Terrorist Weapon ?

Dr. Sam Cohen is the inventor of the neutron bomb. He also worked on the Manhattan project, which produced the first atomic bomb. He worked for the RAND Corporation as a nuclear analyst. Dr. Cohen served as a consultant for the Los Alamos and Lawrence Livermore nuclear weapons laboratories, the U.S. Air Force and the office of the Secretary of Defense.

Dr. Cohen is concerned that the Soviets have developed the technology for producing a cheap micro neutron bomb. Such a weapon would be the size of a baseball, with the destructive power of ten tons of TNT. It would emit a deadly pulse of penetrating neutron radiation for a radius of about one-third of a mile.

This technology has a dreadful potential for terrorism. A detonation in a metropolitan area could kill thousands of people. When detonated, neutron bombs generally produce a minimal blast and thermal effect, but they produce a substantial pulse of neutron radiation which penetrates hardened shelters and produces a destructive ionizing effect when it interacts with living tissue. In other words, such a weapon does not destroy buildings as much as it kills people.

Red Mercury

The key technological element necessary for making the micro neutron weapon is a substance called red mercury, which is a mercury antimony oxide. In 1968, the E.I. duPont de Nemours company produced a powdered red mercury compound. When this compound was irradiated and pressured into a gel, the density substantially increased. This provided the necessary precondition for producing the neutron explosion. Red mercury can be used to detonate a very small quantity of heavy hydrogen which would produce a massive pulse of deadly neutrons.

Recently, evidence has surfaced indicating the possibility of the development of such a micro neutron bomb. There is a lot of controversy as to whether or not this technology has really been developed.

Red Mercury and the International Black Market

Evidence suggests that red mercury is currently being manufactured in Russia and reportedly being sold at high prices on the international black market. Western governments are going out of their way to dismiss red mercury as a hoax. Western law enforcement officials claim they have confiscated substances passed off as red mercury which were nothing more than common mercury mixed with brick dust. In July of 1994, F.B.I. Director Louis Freeh made the following statement regarding illegal black market sales of illicit nuclear materials: *"Most were frauds where swindlers tried to sell harmless red mercury as more highly enriched radioactive material"*. Despite the denial of red mercury by Western governments, there are a growing number of weapon-hungry third world nations that are attempting to procure red mercury.

Russian Production of Red Mercury

Historically, the Russians have been leaders in high-pressure technology which is very relevant to the production of red mercury.

In 1991, a Russian company called Promekologia, headed by Boris Yeltsin's ecological adviser, was granted a license to develop and exploit the country's advances in weapons and energy. In February 1992, Yeltsin signed Secret Directive No. 75-RPS, which authorized the company to manufacture and sell red mercury in amounts up to ten tons per year.

Russian Admissions to the Micro Neutron Bomb

According to Dr. Cohen, the Russians have been working on low-yield neutron devices since the early 1950s.

In 1993, Russian General Y. Negin claimed his country had developed a low-yield nuclear weapon "in which a doubling of yield is achieved with a hundredfold reduction of weight compared to existing weapons." A Russian official named Evgeny Kerolev said that red mercury is so potent that a *"bomb the size of a grenade could blow a ship out of the sea."* Russian Minister of Atomic Energy, Victor Mikhailov, said: *"You can drop a couple hundred little bombs on foreign territory and the enemy is devastated."*

In an article which appeared in the Russian newspaper, *Pravda*, Mikhailov said that these *"little bombs"* were micro neutron bombs designed to wipe out people while leaving buildings standing. Mikhailov did not admit that Russia had actually developed such a weapon but he did say that *"such a weapon could appear by the year 2000."*

U.S. Involvement in Micro Neutron Bomb Research

Dr. Cohen says that in 1961 the U.S. tried to develop a low-yield pure fusion weapon at the Lawrence Livermore laboratories. The Los Alamos lab was involved in a similar project back in the early 1970s but neither project produced a workable weapon. Such a weapon would be very inexpensive because it would not require uranium or plutonium. It would only require small amounts of tritium. Indications from some of Dr. Cohen's old colleagues suggest that Los Alamos may have restarted pure-fusion research.

Dealing with this potential for nuclear terrorism should be one of this nation's highest national priorities, but U.S. government officials and agencies deny the potential existence of the micro neutron bomb. The U.S. position of denying the technical feasibility of such a weapon could prove to be disastrous to our national security. This is not the first time that the government and the scientific community said a bomb was impossible to build. In the late 1940s they said the hydrogen bomb couldn't be built, but shortly thereafter the Soviets deployed a hydrogen bomb.

Despite claims that a micro bomb is technically unfeasible, in 1993 Congress passed a bill banning research and development of micro bombs with yields less than five kiloton.

Biological Terrorism

Biological terrorism is a very real threat. Biological terrorism is the most cost-effective means for indiscriminate mass murder of human beings. There is nothing exotic about the technology. Several college-educated terrorists with access to technical libraries and financial backing could make large amounts of the biological weapon Anthrax. Anthrax can be easily isolated from the soil in some parts of the United States. One van loaded with tanks of Anthrax could drive around Manhattan and spray enough contaminates to kill several million people within one week. One cargo aircraft flying a pattern over the populated areas of central and eastern United States could result in 20 million deaths within one week. The problem with Anthrax is that once you feel the symptoms, it is too late for treatment. Anthrax's long incubation period would allow the terrorists to escape long before the victims even realized they had been attacked.

Nuclear materials may not be the only deadly items being smuggled out of the old Soviet Union. In all probability, the know-how and material for biological weapons are also available to the highest bidder. Transporting a warhead or a canister of nuclear material out of Russia can be difficult compared to slipping a small vile of a deadly disease out of the country.

Russian Biological Weapons Program

An article in the August 31, 1992 edition of the *Washington Post* stated that the United States was concerned that the Russian government was not fulfilling its promise to stop its extensive biological weapons program: *"Underlying U.S. concerns is the suspicion that the highly secret former Soviet program is not yet fully under President Yeltsin's control, and that elements of it have been hidden by Russian military officials who want to keep parts of the program intact....Besides the obviously military-run program that had provoked longstanding U.S. concern, the government was also conducting secret germ weapons research in the civilian Biopreparat facilities."* This information came from a Soviet defector who revealed that the Soviets had two biological warfare programs.

In February of 1992, Yeltsin admitted that *"the Soviet military had violated the 1972 treaty which barred the development, production or stockpiling of toxin and biological agents and any weaponry to deliver them."*

The February 1, 1993, edition of *Newsweek Magazine* contained an article which reported on the Soviet Union's biological warfare program: *"...a vast operation employing 25,000 people at 18 or more research and development facilities, six production plants and a major storage complex in Siberia. The goal was to take known pathogens and alter their genetic structures to make them resistant to Western drugs."* This information came from a defector named Vladimir Pasechnik, a microbiologist who was involved in the Soviet biological warfare program. *"Pasechnik maintains that a Soviet program to develop a genetically engineered, dry form of superplague, resistant to antibiotics, dates from 1984 and was a top priority."*

Conclusion

Ignorance and dependency tends to increase one's vulnerability to terror and terrorism. People can reduce their vulnerability through knowledge, practical protection, self sufficiency and by getting out of the way.

Chapter 4
Chemical and Biological Warfare

Chemical war is the use of poison gases and other toxic chemicals in time of war to kill or incapacitate an enemy. Modern nerve gases and chemical warfare agents are a by-product of insecticide research. They are composed of organic chemicals known as organophosphorus compounds that inhibit the production of cholinesterase. These chemicals are widely used for insect control and in higher concentrations they are lethal to mammals.

The Historical Use of Chemical Warfare

Chemical warfare was used extensively during World War I. In 1935 Italy used chemical warfare agents and caused nearly 15,000 casualties in its war against Ethiopia. The Japanese used chemical weapons in their war against the Chinese just prior to World War II. Since 1945 chemical weapons were used in Yemen in 1962, Red China in 1969, Cambodia in 1978 by the Vietnamese, North Vietnam in 1979 to repel the Chinese, Iraq in the Iran-Iraq war starting in 1980, the Soviets in Afghanistan from 1981 to 1990 and in the Ethiopian-Eritrean Civil War by the Soviet bloc forces starting in 1978.

The U.S. used CS and CN tear gas and a hallucinogenic called BZ during the Vietnam War and in Laos in 1975. In its war with Iran, Iraq used mustard gas and the nerve gas agents GA and Tabun. In 1986 Libya used chemical agents against Chadian troops.

Chemical aerial bombs, rockets and artillery shells are inexpensive to produce. Small and efficient processing plants can turn out chemical weapons by the ton.

Effects

Chemical agents usually cause burns, asphyxiation and neurological damage. Symptoms resulting from the exposure to chemical agents include sweating, tearing, excessive salivation, difficulty in breathing, nausea, vomiting, dizziness, weakness, convulsion and death.

The psychological effect of chemical agents on troops in the battlefield has proven to have a greater effect than the actual physical danger of the chemical weapon itself. Much of a chemical weapon's effectiveness lies in the panic and disorientation it produces in unprotected soldiers and civilians.

Desert Storm and Chemical Warfare

There were rumors circulating after Desert Storm that some U.S. troops had encountered a new type of chemical war gas which penetrated or compromised the current military-issue activated carbon N.B.C. air filters and masks. This chemical agent is purported to be a strain of hydrogen cyanide called, Prussian Blue. It is further alleged that it was developed in the U.S. by a company called Product Ingredient Technology (P.I.T.) of Boca Raton, Florida. This P.I.T. plant was financed and constructed by a Dr. Barbouti of Ishan Barbouti International (I.B.I.) who also built the Pharma-150 chemical-biological complex at Rabta, Libya. P.I.T. exported Prussian Blue to Iraq a year or so before the Persian Gulf War.

An informant, Peter Kawaja, who tried to alert the federal government about P.I.T. was raided in 1990 by eight heavily armed federal agents who removed evidence, tapes and documents not only implicating the involvement of foreign agents but also high ranking U.S. government officials. The evidence has been sealed by a federal magistrate under the guise of "national security".

The Ongoing Russian Chemical Weapons Program

The Soviet war machine places a high priority on the use of chemical warfare agents. As much as 30 percent of the Soviet Union's arsenal is related to chemical warfare.

The October 27, 1992 edition of the *Washington Post* had an article which gave some new insight into the Soviet chemical warfare program: "*A scientist who objected to what he calls Russia's ongoing development of chemical weapons has been jailed for allegedly revealing state secrets. The arrested scientist, Vil Mirzayanov, had earlier stated in an article appearing in the Moscow News that Russia has been pursuing research on a new, more toxic chemical weapon.*" This is a nerve gas called Novichok which is five times as deadly as conventional nerve gases. It is purported that 40,000 tons of Novichok is enough to kill all human life on earth. This alleged development runs counter to the public statements of President Yeltsin, who has urged a global ban on such chemical armaments. According to the *Washington Post* article, this new Russian chemical weapon is "*more toxic than anything in the U.S. arsenal. The new weapon was tested in early 1992 in Uzbekistan*".

What compounds the seriousness of this development is the fact that the Soviet military strategy relies heavily on the use of chemical warfare agents

Tactical Use and Limitations

The use of chemical warfare is limited by the excessive bulk of the chemical agents. This restricts the size of the area which chemical agents can be applied to. Weather, wind and the practical limitations of dispersal would generally limit chemical weapons to use against concentrated targets as opposed to large geographical areas. Chemical weapons can be very effective against troop concentrations, military facilities, fortifications and highly populated areas. Chemical agents do not pose much of a threat to a geographically dispersed civilian population.

Having chemical weapons in a nation's stockpile deters the enemy from using its weapons on that nation. The United States has unilaterally destroyed much of its chemical weapons stockpile, which has weakened the deterrent factor in recent years.

Known Chemical Warfare Agents

NERVE AGENTS

Tabun (GA) - cholinesterase inhibitor

Sarin (GB) - cholinesterase inhibitor

Soman (GD) - cholinesterase inhibitor

GP - cholinesterase inhibitor

Thickened Soman (GD or VR-55) - cholinesterase inhibitor (U.S.S.R.)

Thickened Soman (VX) - cholinesterase inhibitor (U.S.)

Yellow Rain - Unknown compound that causes bleeding and rapid death. May include mycotoxins produced by the genus Fusarium fungi.

Black Rain - Unknown compound that causes instant death; used by U.S.S.R. in Afghanistan.

Novichok - Recently developed cholinesterase inhibitor (U.S.S.R.). May affect human genes and thus damage could be genetically transmitted to offspring.

BLISTER AGENTS

Ethyldichlorarsine (ED) - blister agent

Lewisite (L) - irritates nasal passages, causes skin and membrane burns, poisonous.

Mustard (H, HD, HS) - causes skin and membrane inflammation, blindness

Phosgene Oxime (CX) - destroys skin and membrane tissue

BLOOD AGENTS

A blood agent is absorbed into the body through the lungs where it is then picked up by the blood and carried to the rest of the body.

Arsine Trihydride (SA) - causes gasping and choking, asphyxiation

Cyangen Chloride (CK) - causes convulsions, asphyxiation

Hydrogen Cyanide (AC) - causes convulsions, gasping, choking, asphyxiation

Hydrogen Cyanide (PB) Penetrates current issue U.S. military gas masks. Allegedly used against U.S. forces by Iraq during Persian Gulf War. Causes convulsions, gasping, choking, asphyxiation

CHOKING AGENTS

Chlorpicrin (PS) - causes severe coughing, lung edema, choking, asphyxiation

Chlorine (CL) - causes severe coughing, choking, skin and membrane burns, asphyxiation

Phosgene (CG) - causes severe coughing, choking, asphyxiation

TEAR GASES

Tear gases cause eyes to smart and tear and irritate nerves in mucous membranes including nose, mouth, throat and airway.

Brombenzylcyanide (CA) - long acting

Chloracetophenone (CN) - short acting

Chloracetophenone in Chlorpicrin (CS) -

Dibenz (CR) -

NAUSEA GASES

Adamsite (DM) - arsenic compound, causes sneezing, nausea and depression

Diphenylchlorarsine (DA) - causes sneezing, nausea and depression

OTHER

Buzz (BZ) - Hallucinogenic LSD derivative (U.S.)

Blue X Unknown composition. Incapacitating variously estimated for 1-2 and 8-12 hours (U.S.S.R.)

Protecting Yourself from Chemical Warfare Agents

The only way to protect yourself from chemical agents such as nerve gases, CN, mustard gas and others when you are outside in the open is the use of protective suit and a military grade activated carbon gas mask. The suit has to be airtight, and the mask must fit snugly and filter all air through canisters of chemicals that deactivate the chemical agent being used.

These suits and masks are used for hazardous materials handling and can be obtained from safety supply companies. The military has special chemically impregnated suits which are very expensive and difficult to obtain. When worn, these suits are hot, claustrophobic and clumsy. It is hard to manipulate equipment or to walk or run. Exertion produces more body heat, which increases the discomfort. Visibility is limited by the lens openings on the face of the gas masks which also tend to steam up. Hearing is reduced by the thick material of the hood.

The clumsy handicap which this protective equipment creates was verified by military experience in combat training exercises where soldiers had to wear chemical protection equipment. The obstructed vision resulted in poor visual target indemnification and 25 percent casualties to "friendly fire."

Airtight shelters with the right carbon air filtration filters are the best protection. Maintaining a positive air pressure inside the shelter, slightly above normal pressure, insures that chemical agents will not leak into the shelter.

Breath No Evil, by Stephen Quail and Duncan Long, is probably the most comprehensive book on the subject of chemical and biological agents available to civilians today. This book can be obtained from Safe-Trek Publishing, 90 Safe-Trek Place, Bozeman, MT 59715, (800) 424-7870.

History of Biological Warfare

The first recorded instance of biological warfare goes back to the 1300s. The Mongol army used catapults to hurl the bodies of plague victims over the walls into the city of Caffa in the Ukraine. The plague eventually spread from this city to Europe where it killed one third to one half of Europe's population. The early New England colonists intentionally distributed blankets from people infected with smallpox to the Indians. The white man's diseases, more than his technological edge, resulted in the decimation of the indigenous populations of the North and South American continents.

The Japanese had an extensive biological warfare research program during World War II. Evidence indicates that the Japanese used biological warfare in China against the nationalist forces there.

Plagues are not something strictly relegated to the ancient past. The influenza of 1918 killed over 20 million people throughout the world, including vast numbers in the United States.

Global Government & Biological Warfare

Dangerous new organisms have been produced in laboratories and at least one has been released upon an unsuspecting and defenseless world. There is substantial evidence indicating that AIDS is the direct and intentional product of the global government's population control program. It is well documented that the outbreak of AIDS was caused by the U.N. World Health Organization's immunization of central Africans with a contaminated smallpox vaccine. Every person who

was immunized developed AIDS. The origin and spread of AIDS has nothing to do with green monkeys. AIDS is a synthetically created virus.

It was discovered that between 1976 and 1985 some of the oral polio vaccines in the United States were made from a virus which was grown on the kidney tissues of African green monkeys. This resulted in the contamination of the vaccines with simian retroviruses. These retroviruses were found to cause leukemia and cancerous tumors in laboratory animals.

We are periodically seeing new reports of strange diseases and viruses appearing out of nowhere. The real possibility exists that some of these are perhaps being created and intentionally released into the world as part of the global population reduction scheme.

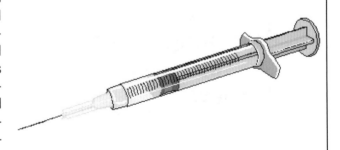

Gulf War Bio Warfare

Research done by Drs. Garth and Nancy Nicolson of the University of Texas M.D. Anderson Cancer center resulted in the discovery of micoplasma incognitus as the cause of Gulf War syndrome. Normal laboratory blood tests do not detect micoplasma incognitus. The only way to detect this micoplasma is to use a sensitive genetic marker analysis. Even with this method it is still difficult to detect because it is found mainly inside the cells and not in body fluids like a conventional bacteria.

Micoplasma incognitus causes chronic fatigue, recurring fever, night sweats, joint pain, stomach upsets, stomach cramps, headaches, skin rashes, heart pain, kidney pain, thyroid problems, and in extreme cases, autoimmune-like disorders. The effects of micoplasma incognitus are suppressed by the antibiotic doxycycline but it does not cure the underlying disease.

The big question is how did such a wide spectrum of U.S. service personnel, including some who never left the U.S., contract micoplasma?

Almost all military personnel who participated in the Gulf War were inoculated with one or more mysterious vaccines. The standard F.D.A. approval sequence for this vaccine, as determined by the Food, Drug and Cosmetics Act, was bypassed. The DOD could legitimately give these drugs as an "Investigational New Drug," but only after informing a person of the potential risks and benefits, and giving them the freedom to choose whether or not to participate.

Interviews of Persian Gulf War veterans indicates that immunizations were mandatory and were given without informing personnel of the risks involved. In some cases individuals were ordered under threat of court martial not to discuss the vaccinations they received with anyone including their physicians.

Persian Gulf War veterans were administered botulism toxoid, pyridostigmine and in some cases anthrax vaccine. One survey indicated that as much as 90 percent of the veterans have suffered illness since serving in the Gulf war, and as many as 10,000 have died as a result. Interestingly enough 700,000 service-related immunization records have inexplicably disappeared and now blood samples from some sick veterans are showing traces of a compound called squalene, a component of an experimental HIV immunization.

Starting in December of 1997, the U.S. Department of Defense started inoculating all members of the armed forces with anthrax vaccine. Tests have determined that this vaccine is only marginally safe and that it provides little or no protection from airborne anthrax. Airborne anthrax is the only form of anthrax which can be effectively used in bio-warfare.

Conclusion

The bioengineering of new viruses and the subsequent creation of incurable diseases poses a significant threat to life on this planet. Evidence would indicate their creation and application is part of a global population reduction program.

Required reading on this subject is the book *Emerging Viruses - AIDS & Ebola* by Dr. Leonard Horowitz, (800) 336-9266, Web<http://www.tetrahedron.org/horowitz.htm>. Information on Gulf War Sickness can be obtained from Captain **Joyce Riley** of the American Gulf War Veterans Association, (800) 231-7531, E-mail <gulfwar@flash.net> Web <http://www.gulfwarvets.com/links.htm>.

Chapter 5
Weather Modification

Weather modification technology is in use today by both the United States and the Soviet Union. Both the U.S. and Soviet projects involve the manipulation of the ionosphere and the alteration of the earth's magnetic fields. This technology seems to have both localized and global capabilities. Evidence indicates that this technology also has the capability of manipulating human behavior and mood patterns.

Nikolai Tesla

All of this seemingly Star Trek-like technology originated from a Serbian immigrant named Nikolai Tesla who came to the United States just before the turn of the century. This man, without question, proved to be the greatest scientific genius of this century. Tesla worked with both Westinghouse and Edison during the infancy of electricity and it was actually Tesla who developed alternating current.

Tesla also experimented with electromagnetic flux and studied the earth's gravitational field. During his research he discovered that the ionization of the atmosphere would alter when it was charged by radio wave transmissions in the low frequency range of 10 to 80 hertz. Tesla also discovered that he could cause both positive and negative ionization of the atmosphere by manipulating the radio frequency. Further studies indicated that with positive ionization, people and animals became tired and lethargic and with negative ionization the effect was one of feeling active and energetic.

Another interesting aspect of this technology is the effect which harmonious radio frequencies make when they impact air molecules. The molecules become excited and give off negatively charged electrons which readily combine with hydrogen and oxygen to produce water molecules. But even more profound is the fact that this type of radio wave also carries positively charged ions through the ionosphere into the magnetosphere. The positively charged ions then become trapped in the Van Allen belts, traveling between the Aurora Borealis and the Aurora Australis, right where we coincidentally have holes in the ozone layers.

Freon, the ozone killer, by its nature dissipates quickly. This inherent quality is a contradiction to the concept that a concentration of freon is creating holes in the ozone layers at the North and South Poles. However, when radio waves hit already unstable freon precipitates, they are so reactive that they cannot move on and dissipate into the magneto-sphere.

As far-fetched as this subject may seem, leaders in the elite global establishment have intimated the existence of this technology and its capabilities. In 1970 Zbigniew Brzezinski published a book titled *Between Two Ages*. Brzezinski is one of the founding directors of the Trilateral Commission. In this book he said, *"Technology will make available, to the leaders of major nations, techniques for conducting secret warfare, of which only a bare minimum of the security forces need be appraised.... Technology of weather modification could be employed to produce prolonged periods of drought or storm"*.

Further confirmation of the Western power elite's interest in weather modification was revealed in an article which appeared in the September 12, 1989 edition of the *Washington Post*. This article reported on the 1989 Tokyo conference on global environment. The president of the World Bank, Barber Conable, who is also a Trilateral Commission member, gave a speech in which he outlined the long-range goals of the one world elite. In this speech he said, *"while higher temperatures may cause 'a number of natural disasters,' they might also warm cold and unproductive lands in the north into productivity."*

There is very good evidence that the one world group and the Soviets have been jointly involved in weather modification over the Northern Hemisphere since the early 1970s. In 1971 it was alleged that the U.S. and the Soviets started cooperating in secret weather modification projects. One such project was named POLEX, Polar Experiment of the Global Atmospheric Research Program, and another in 1973 was called AIDJEX, the Arctic Ice Dynamics Joint Experiment. It is purported that these experiments involved attempts to melt the polar ice cap. Other U.S.

projects such as Nile Blue and Climate Dynamics involved the changing of the arctic ice pack. There was even an article in the December 16, 1980, edition of the *New York Times* discussing a joint U.S.-Soviet project involving the transmission of ELF waves from Antarctica designed *"to interfere with the earth's magnetic field."*

Soviet Woodpecker System

The Soviet weather modification project is referred to as the Woodpecker system. It involves the transmissions of extreme low frequency (ELF) waves at about 10 hertz using Tesla transmitters in Angarsk and Khabarovsk in Siberia, Gomel, Sakhalin Island, Nikolayev in the Ukraine, Riga in Latvia and also a site 60 miles south of Havana in Cuba. To give you an idea of the magnitude of this system, the facility near Havana, Cuba, is said to be maintained and operated by a staff of 2,500 Soviet personnel.

These transmitters generate electromagnetic transmissions that produce an ELF scalar grid over the United States. This is done by transmitting these low frequency scalar waves in pairs so that they converge at a predetermined point on the earth's surface and cause a disruption of the atmosphere. This technology can be used to alter the course of the jet stream and set up long-term weather blocks.

Long-term Weather Blocking

The long-lasting California drought in the 80s was caused by a massive ridge of high pressure 800 miles off the California coast which hovered for extended periods of time, blocking the usual flow of moist air coming in from the Pacific and pushing storms around to the north. Meteorologists who have analyzed this phenomena consider it to be one of the most unusual national patterns ever recorded, unique in the annals of weather recording. Such long-lasting centers of high pressure were unheard of until 1977. Evidence suggests that this was possibly caused by Woodpecker generated giant standing ELF waves which are transmitted by the Soviets intentionally to block the flow of normal weather patterns.

This phenomena of long-lasting centers of high pressure is not limited to producing drought. In 1993 the Midwest region experienced severe flooding which was a result of the wettest period in this particular area since rainfall record keeping began in 1876. This flood was a result again of what meteorologists called a blocking pattern.

The normal weather systems usually move from west to east across the U.S., but during the '93 flood the weather systems stalled for six weeks over the upper Midwest. A high pressure system over the eastern part of the United States was causing warm, moist air to move up from the Gulf of Mexico and dump moisture in the Midwest, where it met the jet stream. This weather pattern involved an unusual shift in the jet stream, which during the summer is usually weak and typically found much farther north in Canada.

This stationary high pressure front also blocked the path of cold Canadian air, resulting in record low temperatures in the Northwest. According to the September 1993 issue of *Storm, The World Weather Magazine*: *"It is extremely unusual for weather patterns to persist for so many weeks, bringing heavy rainfall to the same area almost on a daily basis. . . . The reasons for the weather patterns to become fixed, as they did in June and July, 1993, are unclear."*

GWEN

A system somewhat similar to the Soviet Woodpecker has been set up in the United States called the Ground Wave Emergency Network, or GWEN. This network was built under the guise, or possible dual use, of an emergency communication system that would not be interrupted by electromagnetic pulse during a nuclear war.

GWEN units are capable of altering the magnetic field within a 200- to 250-mile radius. The individual units themselves are made up of 300-foot tall towers which transmit radio waves through hundreds of bare copper wires which are each 300 feet in length. These wires are buried in the ground in a spoke pattern radiating out from the base of the tower. The wires interact with the earth like a thin-shelled conductor, radiating the radio wave energy for very long distances through the ground.

These units are located in Texas, Nevada, Maine, North Carolina, Maryland, Georgia, Oregon, Washington, California, Iowa, Wisconsin, Minnesota, Missouri, North Dakota, South Dakota, Pennsylvania, Mississippi, New York, Alabama, Virginia, Kansas, Nebraska, New Jersey, Arkansas, Massachusetts, Oklahoma, Colorado and Montana.

There is fairly significant evidence to suggest that these GWEN units in combination with the Soviet Woodpecker transmissions made a significant contribution to creating the extensive flooding in the Midwest in 1993.

It is interesting to note that the Russians are now openly marketing a small version of their weather engineering system called Elate, which can fine tune weather patterns over a 200-mile area. These units seem to have the same range as the individual GWEN unit. One of these Elate units operates at Moscow's Bykovo Airport.

The threats of weather warfare, totalitarian government and famine dovetail together. As we saw in the famine which the Soviets artificially created in the Ukraine prior to World War II, famine is an effective means of subjugating a people. By controlling food, you can control people. Weather modification can affect food production and eventually the available supply. Starving resisters out is much more effective than having to track them down and shoot it out with them. If you have not surrendered your weapons, you don't get a food ration coupon. Long-term food storage, well hidden, is the only insulation against famine and totalitarian oppression.

Tesla-Scalar Electromagnetic Weapons

An article in the spring 1993 edition of *Orbis Magazine* reported on another article which had appeared in the 1992 edition of a Russian magazine called *Military Thought.* This article, among other things, revealed So-viet involvement in the development of Tesla-scalar electromagnetic weapons: *"...The current civil-military consensus also includes an image of future war based on the development and deployment of advanced conventional munitions, direct-energy weapons, space-based strike weapons and anti-ballistic missiles, and third-generation nuclear weapons."*

This book can only briefly cover the subjects of weather modification and tesla-scalar electromagnetic weapons. An extensive well documented analysis of weather warfare and other excellent documentaries on the Soviets can be obtained from Mr. C.B. Baker, Youth Action News, P.O. Box 312, Alexandria, Virginia, 22313. Also, books concerning Nikolai Tesla and Tesla technology can be obtained from the Tesla Book Company, P.O. Box 121873, Chula Vista, CA. 91912, 1-(800) 398-2056.

Nick Begich has recently published a book regarding the U.S. government's Woodpecker type project in Alaska which is named High-Frequency Active Auroral Research Program (HAARP). This book is titled, *Angels Don't Play HAARP*, ISBN # 0-9648812-0-9. The book costs $14.95 and can be ordered from Earthpulse Press, P.O. Box 201393, Anchorage, Alaska, 99520, telephone (907) 249-9111 (voice mail).

Chapter 6
Earth Changes

Beyond the threat of man-made disaster is the concern about large scale natural disasters or cataclysm. The recent buzz word for this category of threat is called "earth changes." The concept of "earth changes" is a threat which is difficult to support in the parameters of modern logic, but when consideration is based on the full spectrum of earth's geological history, radical changes have been and will continue to be an inevitable part of life on this planet.

The climatic and geological stability which humanity has experienced during its history on earth has been the exception as opposed to the historical rule. Scientific evidence clearly shows that on numerous occasions the earth has actually changed the position of its magnetic poles, that northern latitudes have previously had tropical climates, that many dry areas were once under sea and that continental earth masses have made radical movements across the face of the globe. It is an accepted scientific fact that on at least three occasions in ancient history catastrophic events have occurred which have exterminated the majority of the life species on this planet.

People generally consider massive cataclysm as something isolated to ancient history. Mankind has inhabited this planet for a comparatively short span of time relative to the earth's existence. This period of time has been fairly stable in comparison to other periods of earth's history. This is the foundation of people's attitude that "it can't happen here."

Apocalypse

Prophecy is something which exists on thin ice when it comes to the Western scientific mind, but the religious basis of today's society rests upon scriptures which clearly foretell a coming planetary event. Numerous prophets, seers and religious scriptures throughout the ages, including the book of Revelation, suggest the coming of an apocalyptic planetary event. The interpretation of many of these prophecies suggest that some sort of "end time" will occur somewhere around the year 2000. Prophecies typify this event as a judgment, a cleansing of the earth, or a "karmic" reconciliation for want of a better word, which drastically reduces the earth's population.

The idea of earth changes should not be ignored as a possibility. The big question when contemplating this threat is what sort of cataclysm might occur. Many of the earth-change scenarios seem to suggest radical conditions on the surface of the earth for a period of time. These conditions include dangers in coastal areas, unusual temperatures, severe earthquakes, volcanic ash and extreme surface winds. The impact of such earth change activity could possibly be a 2/3 to 3/4 reduction of earth's population. If the future proves such speculation correct, the need for below ground sheltering is obvious.

Cosmic Events

As the reader may recall, during the third week of July, 1994, a comet composed of 21 chunks of ice rock stretching over 100,000 miles of space smashed into Jupiter while traveling at a speed of 130,000 miles per hour. Some of the projectiles which made up this comet were two to three miles in diameter. Scientists calculated that the comet impacted Jupiter with a force millions of times greater than the most powerful H-bomb.

An untold number of asteroids and a lesser number of comets transit through and around outer space in close proximity to earth. On numerous occasions during earth's history large objects have struck the earth.

Scientists attribute the extinction of the dinosaurs to a catastrophic event which occurred 65 million years ago. An object from outer space, either an asteroid or a comet, struck the earth at a spot on the present day Yucatan Peninsula. This object is estimated to have been 6 miles in diameter. Scientists speculate that the comet which collided with

Jupiter had 10 times more force than this space object that struck the earth 65 million years ago.

The impact of the object on the earth's crust had two distinct effects. First, the object's impact on the surface threw debris high up into the atmosphere and, second, the impact was so massive that it sent shock waves rippling through the core of the earth. These shock waves focused on a point on the other side of earth directly opposite the impact point, which is referred to as an antipode. This focusing of shock waves resulted in a rupture of the earth's crust at the antipode and a series of massive volcanic eruptions. The volcanic eruptions combined with the initial material thrown into the atmosphere on impact caused the planet to be shrouded in a dark cloud of dust and ash. Thus the surface of the earth was deprived of sunlight and the planet's temperature dropped radically, killing most life forms.

This theory is supported by the geological structures on the Moon and Mercury. Astronomers have discovered that at the antipode of the large craters on the Moon and Mercury there are volcanoes and unusual patches of broken crust. Scientists have also identified at least 130 significant impact craters on earth.

According to a meteor hazard report which was presented to Congress in 1992 by a NASA sponsored committee of international astronomers: There are about 300,000 asteroids in the 330 to 3,000 foot diameter range which cross the earth's orbit. Once every 5,000 years one of these objects collides with the earth producing a crater with a diameter of 1.2 miles or greater and an explosion equal to the detonation of hundreds or thousands of nuclear weapons. An asteroid entering the earth's atmosphere at a speed of eight miles per second will have a destructive force equal to the explosion of a quantity of TNT 100 times the object's volume. There are 2,000 or more objects greater than 3,300 feet in diameter which travel in close proximity to earth. Every 500,000 to 1,000,000 years one of these strikes the earth. This produces an explosion equal to the detonation of millions of atomic weapons and results in a crater 10 to 15 times larger than the object's diameter. This either kills or drastically affects all the life forms living on the planet at the time of impact.

More recently in earth's history, on June 30th, 1908, an asteroid 400 feet in diameter exploded while entering the atmosphere two miles above Boloto located in central east Siberia. The force generated by the explosion flattened and charred 800 to 1,200 square miles of forests and the noise of the explosion was heard 550 miles away in the city of Irkutsk. It has been estimated that the energy released during this explosion was equal to the force of the Mount St. Helens eruption.

Since January of 1995 there has been a rash of new comets discovered entering the solar system. One of these, a comet named Hale/Bopp, will cross the earth's orbit in the spring of 1997. This comet is 3,000 to 5,000 miles in diameter and apparently has a 3,600-year orbit pattern around our solar system. Hale/Bopp is moving toward the sun at a speed of 33,800 miles per hour and at this time it is not known how close its path of travel will be to the earth. This comet is infinitely bigger than Haleys' comet and the comet which struck Jupiter.

Earthquakes

During the course of recorded history, earthquakes have killed an estimated 74,000,000 people. Earthquakes have be-

come a more frequent part of life during the last two decades. For many Californians, the specter of the "big one" continually lurks in the shadowy recesses of the subconscious mind.

When we look at the threat of earthquakes, the question is not whether or not they will occur but rather when and where will they occur. I say this not from the position of being a fatalist but from a position of geological/scientific reality. Many people living in close proximity to the major faults are able to live their day-to-day lives without anxiety or concern because of either ignorance or denial. Unfortunately, this approach does not alter the reality of the inevitable threat.

In order to dispel ignorance, we need to understand the nature of earthquakes.

The first layer of crust which makes up the earth's surface is referred to as the lithosphere. This layer is about 40 to 60 miles thick and is composed of and segmented into 13 different major land masses and a number of minor ones. These land masses are referred to as plates. The plates are rigid and they float or drift on top of the next layer down in earth's crust, which is called the asthenosphere. The asthenosphere layer is about 60 miles thick and is somewhat fluid or plastic.

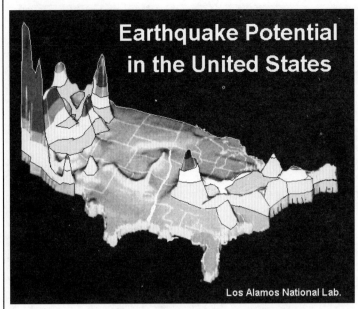

Earthquake Potential in the United States

Los Alamos National Lab.

The plates move and interact with each other. The boundary at which one plate interacts with another is called a fault. There are three different ways that plates interact with each other and thus there are three different types of faults. Plates either spread away from each other, converge on each other, or move parallel to each other.

When two plates spread away from each other this creates a rift or crevice. As the plates move apart and this rift spreads and opens up, molten rock eventually pushes up from the mantle of the planet's core and fills in the void. This process is not that apparent because most of it happens deep underneath the sea, in the 40,000 mile long mid-Atlantic ridge and a similar Pacific ridge. The Atlantic ridge is divided in half by a 30-mile wide rift or valley which runs the full length of the ridge. Scientific measurements indicate that this rift widens 1/2 to 2 inches annually.

When two plates converge on each other from opposite directions, this either causes one plate to go under another (which is called a subduction) or it results in both plates being thrust up into the air, which creates mountain chains such as the Alps and the Himalayas.

When two plates move parallel to each other in opposite directions, this is called a transform fault.

Earthquakes are a result of the movement of plates and the stress released during either subduction or transform fault interaction. The earthquakes which have the greatest effect on the earth's surface are a result of transform fault movement.

The earthquake threat in California is a result of transform faults, such as the San Andreas fault. The San Andreas fault exists between the Pacific plate and the North American plate. Both plates are moving parallel to each other and in opposite directions. This parallel movement has been going on for a long time. Geologists have found evidence which indicates that the fault, over time, has displaced as much as 450 miles.

One analogy which illustrates what actually occurs during an earthquake is the bending of a tree branch. When you bend a tree branch, the stress increases in the wood until the stress exceeds the strength of the wood fibers and then the wood fibers snap. The breaking snap is a violent release of the built-up stress. The two broken pieces then flex back into their original unbent and relaxed positions.

There are a number of factors which play into the effect an earthquake has upon a building. The first is the strength of the earthquake, which we will later discuss. The second is how close the building is to the epicenter of the earthquake. The epicenter is the point at which the released stress energy is greatest.

The third factor is the nature of the ground and soils which underlie the

SOME OF HISTORY'S MAJOR QUAKES

Richter Scale	Location	Date	Fatalities
6.7	Northridge, CA	1993	61
7.0	Loma Prieta, CA	1989	63
7.2	Kobel, Japan	1995	5,000
7.7	Tokyo, Japan	1983	107
7.7	Hokkaido, Japan	1993	200
7.9	Mexico City	1985	9,500
8.2	Tangshan, China	1976	250,000 +
8.2	Tokyo, Japan	1923	143,000
8.3	San Francisco, CA	1906	700
8.3?	San Francisco, CA	1838	?
8.3+	Carrizo Plain, CA	1857	? unpopulated area at time of quake
8.4 Mw (9.2)*	Alaska	1964	114
8.3 Mw (9.5)*	Chile	1960	5,700
?	Shaanxi, China	1556	800,000
8.7 to 8.8	New Madrid, MO	1811-12	few unpopulated area at time of quake

* Mw = Seismic Moment = energy released during a quake

building. Areas with underlying earth which is comprised of water-saturated, granular material are severely affected by earthquakes. Areas where the underlying earth is composed of bedrock will receive less structural damage even if they are in a closer proximity to the epicenter of the earthquake. This is due mainly to a process called liquefaction. The intense shaking associated with earthquakes, which is a by-product of the released stress, causes the water-saturated, granular material to change from a solid to a liquid state. As a result, the ground fails under structures and massive landslides occur.

The last factor is the nature of the construction of the building itself. The earth moves back and forth under buildings, pulling the foundation with it. The vertical portion of the structure wants to remain stationary due to inertia. As a result, many structures are separated from their foundations at the weak point where the wooden framing attaches to the concrete foundation. The same effect destroys brick and unreinforced masonry structures, like tall masonry chimneys. The effect of earthquakes on buildings is actually a very complex science, especially in relation to the dynamic forces applied to large commercial structures.

Tsunamis

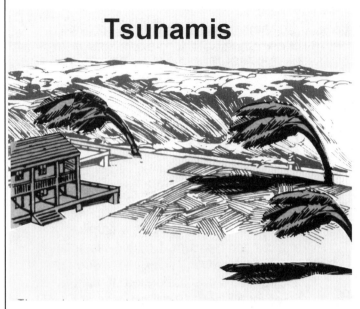

An important consideration here is an understanding of the measurement factors used in rating the strength of earthquakes. The common rating for earthquakes is called the Richter scale, which is determined by readings on a seismograph. It is a logarithmic relationship to the amount of ground movement an earthquake produces. A recorded magnitude of 7 on the Richter scale is 10 times greater in force than a magnitude of 6 on the Richter scale. A magnitude of 8 is 1,000 times greater force than a magnitude of 6. The Richter scale is not necessarily an accurate method of measurement for quakes over a magnitude of 7. The amount of area affected by larger quakes is not reflected in the Richter scale measurement system. Earthquakes over 7 with the same Richter scale rating can have different real magnitudes of impact depending on how much area is affected and how long the quake lasts. This has led to the development of the Seismic Moment system or the Mw, which gives a more realistic comparison of the real magnitude of earthquakes over a reading of 7.

So what does this mean for California? It is inevitable that the Los Angeles area and surrounding communities will be struck by an 8 magnitude earthquake. Some experts suggest that there is a 50 to 90 percent chance that Southern California will get hit with a quake equal to the 1857 and the 1906 quakes within the next 50 years. San Francisco has not been hit by a really big quake since 1906 and the previous big quake occurred 68 years before in 1838. Some experts suggest that San Francisco has a 30 percent chance of an 8 magnitude earthquake within the next 30 years.

The Pacific Northwest has not seen a major earthquake in 150 years. The situation in the Northwest is different than California. In the Northwest we have a subduction zone. The Juan De Fuca plate and two adjacent smaller plates subduct down under the edge of the North American plate. This geological plate formation is similar to subduction zones in southwestern Japan, southern Chile and Colombia, which have a history of producing earthquakes on the magnitude of 8 to 9.5 Mw. The last point to consider is wherever you have active or dormant volcanos, you have the potential for major earthquakes.

The Midwest could see another earthquake on the New Madrid fault, just east of the Mississippi River. Some experts suggest that enough stress has built up to produce a 7.6 earthquake. Due to the geology of the Midwest, the force of an earthquake travels farther out from its epicenter. This means that a Midwest 7.6 earthquake can probably do as much damage as a California 8.6.

Two good books on this subject are: *Quake Busters*, by Wayne W. Topping, published by the Topping International Institute, 2622 Birchwood Ave, #7, Bellingham, WA 98225, (206) 647-2703 and *The Coming Quake*, by T.A. Heppenheimer.

Gordon Michael Scallion has become the herald of the California "big one". Scallion puts out a newsletter called "The Earth Changes Report." Scallion's newsletter is a mixture of his predictions and scientific reporting. Scallion himself admits that a prediction is nothing that is set in concrete. Many of Scallion's predictions have not materialized but, on the other hand, he has made some which were right on the money. Scallion predicted the coming of hurricane Andrew several months ahead of time. This prediction included the wind speed, the latitude it would cross Florida and the date it would reach the Florida coast. He has also very accurately predicted some of the earthquakes in California, including general magnitude and the approximate epicenter location.

Scallion consolidates scientific information on the earthquake activity around the world. One of the more interesting facts which he has collected is documentation regarding the exponential increase in the occurrence of earthquakes over the last 20 years. Scallion has a long-term prediction for a massive earthquake in the near future, which he says will put most of California and part of the western United States underwater. He also suggests that this event will reshape the coastlines of America and that the United States will be divided by a new body of sea water which will rush in and fill the Mississippi River drainage, linking up with a swollen version of the Great Lakes.

Many people do not believe in psychic prediction. However, Scallion is a good source of scientific information and he has been correct on enough occasions that it would seem wise to keep an ear to what he is saying.

You can get information on Scallion's newsletter by writing to: *The Earth Changes Report*, RR#1 Box 391, Westmoreland, NH 03467, or calling: (603) 399-4616.

Another newsletter which provides earth changes information similar to Scallion's is the *Abrahamsen Report*, P.O. Box 919, Anacortes, WA. 98221-0919, (800) 891-5818.

Pole Shift

Pole shift is another earth change scenario which many people are not familiar with. The reason it's unfamiliar is that it hasn't occurred during historical times. However, scientific evidence shows that it has happened at least 80 times in the last 110 million years.

Here is a brief explanation of the scientific evidence regarding the shifting of the poles. When the plates spread at the mid-oceanic ridges, molten rock pushes up from the mantle. As this rock cools, the iron atoms in the basalt line up magnetically according to the pull of the earth's magnetic field. As the plates continue to move apart the basalt splits, leaving identical bands on each side of the mid-oceanic ridges. What scientists have found in examining the layers of basalt are alternating magnetic patterns in the iron atoms contained in the basalt. This indicates that when each layer of rock was cooling, the planet's north and south magnetic poles were in positions different from their current locations. In other words, the North Pole and South Pole have undergone numerous major shifts. One book that presents scientific evidence in support of the "earth changes" type scenario is *Pole Shift*, by John White, published by the A.R.E. Press, P.O. Box 595, Virginia Beach, VA 23451.

A quote from *Pole Shift* summarizes the dangers of this potential threat:

"What will happen during pole shift? The ultimate disaster! Enormous tidal waves will roll across the continents as oceans become displaced from their basins. Hurricane winds of hundreds of miles per hour will scour the planet. Earthquakes greater than any ever measured will change the shape of the continents. Volcanoes will pour out huge lava flows, along with poisonous gases and choking ash. Climates will change instantly, and the geography of the globe will be radically altered. If the pole shift is less than a full 180 degrees, the polar icecaps will melt rapidly, raising sea levels, while new icecaps will begin to build...."

Three Days' Darkness

Neither their silver nor their gold shall be able to deliver them in the day of the Lord's wrath. Bible, 'Zephaniah' 1:18.

Another very interesting book is called *Three Days' Darkness, Prophecies of Saints and Seers.* This book approaches earth changes from the prophecies of religious seers throughout the history of the Roman Catholic Church. Many different monks and nuns, separated by time, received strikingly similar prophecies of an apocalyptic event which would bring three days of darkness upon the earth and result in the death of 2/3 to 3/4 of the earth's population. This is another perspective of the same threat, and a valuable book to read. This book is available from the author Albert J. Hebert, S.M. at P.O. Box 309, Paulina, LA 70763.

Two magazines which cover the subject of earth changes are: *Nexus Magazine*, P.O. Box 177, Kempton, IL 60946-0177, (815) 253-6464 and *Atlantis Rising*, P.O. Box 441, Livingston, MT 59047, (800) 228-8381.

Chapter 7
American
Vulnerabilities

"More than 25 countries, many of them hostile to the United States and our friends and allies, may now have or be developing nuclear, biological, and chemical weapons, and the means to deliver them." C.I.A. *Director James Woolsey, July 28, 1993.*

Defense

Protecting citizens is the supreme responsibility of any government. The preamble to the United States Constitution calls for the government to provide for "the common defense." In recent years the United States has reduced its offensive military capability and dismantled purely defensive programs intended to protect the civilian population. Over the past five years the alarm systems which were intended to warn the public of an impending nuclear attack have been dismantled in major U.S. cities, including Washington D.C. National civil defense and strategic defense programs are the only real deterrents to a nuclear attack. In the United States today we have neither.

Anti-Ballistic Missile Defense

Most Americans don't realize that the United States military does not possess a means of stopping an I.C.B.M. launched at a U.S. target, be it launched intentionally or launched by accident. Over the past twenty years the number of nations with nuclear weapons capability has more than doubled— from six to thirteen nations.

The technology for good missile defenses has existed since the early 1960s. The United States signed the A.B.M. Treaty with the Soviet Union, which in essence stipulated that neither nation would develop a national anti-ballistic missile system. Each nation would be allowed to have 100 anti-ballistic missiles.

The currently available anti-missile technology far surpasses that which existed during the Kennedy and Johnson administrations. Our government has decided not to deploy an anti-missile defensive system in deference to the A.B.M. Treaty. Furthermore any defensive weapon systems with anti-missile capability have been intentionally crippled.

In 1991, the Democratic-controlled Congress passed the Missile Defense Act. This act called for a missile defense system to protect the United States and our forces overseas from ballistic missile attack. President Bush signed the act into law but we still have no defense against ballistic missile attack. This inaction by Congress is a violation of the law.

The Patriot Missile

The patriot missile, which became a household word as a result of Operation Desert Storm, is not able to intercept I.C.B.M.s. It does not have the velocity, accuracy or range necessary to destroy an incoming I.C.B.M. The Patriot is only capable of intercepting aircraft and tactical ballistic missiles.

The Patriot system is one example of intentional technological crippling. The Patriot radar and interceptor system was designed back in 1974. It was originally designed with the capability of seeing incoming warheads 150 miles away and could intercept and destroy them 30 miles away. In deference to the A.B.M. Treaty, the Patriot's radar power was cut so that it could only see warheads 50 miles away and its operating hardware was downgraded so its warhead was made intentionally inaccurate. Modifica-

tions were made to the Patriot missile so that it had to receive data on the incoming warhead indirectly.

During the 1980s some of the Patriot potency was restored, enabling it to intercept Iraqi Scud missiles in close proximity to their intended targets. Currently, the Patriot PAC-3 program is underway to upgrade the Patriot system capability but without undoing its major self-imposed handicaps. This is like a prizefighter going into a fight with one hand tied behind his back.

AEGIS Anti-Missile System

During the 1970s and 80s, the U.S. Navy built an air defense system for ships called AEGIS. Its radar is so powerful it can track debris in space and its computer can handle dozens of targets almost simultaneously. The reason the AEGIS system did not suffer the fate of the Patriot is that the Navy hid its real capacity from the Washington arms-control bureaucracy.

If the AEGIS missiles were equipped with fast boosters and an accurate fuse, each launch system could protect a 100-square-mile radius and intercept incoming intercontinental ballistic missiles 50 miles from their targets. One upgraded AEGIS ship positioned in New York harbor could simultaneously destroy 20 incoming nuclear warheads. Even the AEGIS system was intentionally crippled due to the A.B.M. Treaty. The A.B.M. Treaty does not allow the AEGIS missile system to receive target data directly from space-based sensors. Proponents of strategic defense have been urging Congress to fund an Aegis Cruiser Phase I modification of the M-2 missile software to provide endo-atmospheric ballistic missile defense.

THAAD

The Theater High Altitude Air Defense system, THAAD, will probably replace the Patriot missile. This new system involves a combination of ground-based radars and intercepters that could intercept incoming warheads beyond and on the edges of the earth's atmosphere. This technology has the capability of discriminating between warheads and decoys on the basis of how much heat they emit long before they even enter the atmosphere. THAAD is designed to accomplish interception at a high altitude and it directly rams the incoming missile at the moment of detonation.

The THAAD program has effectively replaced all of the other proposed missile defense systems. The THAAD system has greater capabilities than the Patriot missile, but it is still relatively limited in its range and it needs to be pre-deployed in order to be effective. This is a high-priced system, at approximately $16 billion. Some advocates of strategic defense think that this limited system is taking funds away from the deployment of an effective and comprehensive national missile defense system.

The deployment of existing advanced missile defense technology sits in limbo due to the U.S. government's narrow interpretation of the A.B.M. Treaty. In the spring of 1993, President Clinton tried to bypass the necessity of a full Senate vote (as required by the Constitution for all foreign treaties) and was intending to make administrative changes to strengthen the A.B.M. Treaty by executive order.

LEAP

LEAP, or Lightweight Exo-Atmospheric Projectile Phase I System, is an interceptor system which involves retrofitting nuclear attack missiles and thus transforming them into non-nuclear defense missiles. This LEAP system would have the capacity to intercept incoming inter-continental ballistic missiles both in the atmosphere and in space. This is an immediately deployable, low cost system. For about $2 billion dollars, three-quarters of the United States could be protected from missile attack by 1997. Furthermore, the LEAP program does not violate the terms of the A.B.M. Treaty.

B-1 / B-2 Bomber

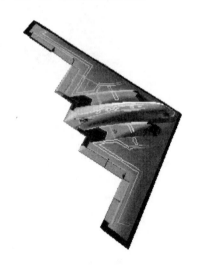

The American people tend to entertain a false sense of national security. This has its roots in a superficial understanding and an unwarranted faith in advanced U.S. military technology. After spending $65 billion on the B-1 and B-2 projects, the 1952 vintage B-52 bomber is the only thing in the U.S. Air Force arsenal capable of delivering a cruise missile or deploying large payloads of precision guided bombs to distant targets. In December of 1994, after spending $13 billion on the Tri-Service Standoff Attack Missile, the Pentagon canned the project. This missile was intended to enable the B-2 bomber to launch a precision guided missile and strike a target 100 miles away.

Bush Lets Down the Nation's Guard

President Bush was a major player in the unilateral disarmament of the United States. Bush grounded the airplane that served as the key war-time link between the president and the U.S. nuclear fleet and he took our strategic bombers off their alert posture. Another move by the Bush administration was to turn off the U.S. network of early warning, over the horizon, back-scatter radars that could detect the approach of enemy bombers 2,000 miles away. This radar network was the cornerstone of the North American air defense. This move leaves the entire U.S. bomber force vulnerable to surprise attack and increases the incentive for the Soviets to implement a first strike attack on the United States.

Bush also reduced U.S. strategic nuclear forces well beyond the requirements of arms control treaties. He had all nuclear weapons removed from army and air force units stationed outside the United States, including navy surface ships. The Bush administration unilaterally stopped the testing of all U.S. nuclear weapons, which will cause U.S. nuclear weapons to become unreliable within two years. The Bush administration also shut down the only factory in the United States that produced the essential ingredient of today's modern nuclear bombs, tritium. The problem is that tritium decays naturally and without a renewable source most of our nuclear weapons will become unusable in about ten years.

The most recent event in presidential disregard for the safety of U.S. citizens is Clinton's elimination of the national radiological monitoring program. This program provided city and county governments throughout the United States with calibrated radiation measuring devices. In time of a nuclear accident or war, this equipment allows local officials to detect and monitor radiation levels and advise the residents of this otherwise invisible danger.

Soviet Strategic Advantages

Reductions in the U.S. defense spending, combined with the continual expansion and modernization of the Soviet's strategic offensive forces, has resulted in U.S. military forces becoming vastly outnumbered and increasingly obsolete in comparison to their Soviet counterparts. One significant advantage the Soviets and now the Red Chinese have over the U.S. is their mobile I.C.B.M. With the exception of our sub-launched I.C.B.M., all of the American

I.C.B.M.s are land-based, fixed installations. These are easily identified by Soviet reconnaissance satellites and thus are precisely targeted. The U.S. scrapped the MX missile with rail mobility and the development of the mobile truck-mounted Midgetman missile. It is virtually impossible for U.S. intelligence to track and target the Soviet armada of mobile I.C.B.M. missile launchers.

During the period of time between 1988 and 1990, the Soviets have maintained the advantage over the United States in all areas of weapons production. The following is a list of weapons categories and the advantage which the Soviets have over the U.S.: I.C.B.M.s 12.6 to 1, bombers 9.3 to 1, submarines 2.1 to 1, S.L.B.M.s (sub-launched ballistic missiles) 2.1 to 1, S.R.B.M.s 1,950 to 0, warships 2.1 to 1, tanks 3.0 to 1 and armored vehicles 6.9 to 1. Also, the Soviet Union is purported to have about 5 million parachute infantry and the United States has about 30,000.

War and the American People

Many Americans have a hard time relating to war as a real threat. With the exception of the veteran community, this nation has been shielded from the direct effects of war. Most Americans consider war to be a threat that exists in another region of the world. Their exposure to war is a superficial melodrama played out on the television in the form of synthesized news programs and home entertainment videos.

Europeans, Africans and Asians are more realistic about war and the possibility that some day in the future it could touch their lives. Unlike the Americas, Europe, Africa, and Asia have experienced the destruction of World War II and other more recent localized wars.

The U.S. hasn't seen a war on its soil since the Civil War. The smoke of war tends to linger for successive generations in nations where the ravages of war have occurred. Wars' strangling grip on a nation leaves a lingering cultural imprint. Many Americans are oblivious to the threat of war and the possibility that its hand could touch them in this day and age.

Media Disinformation, Propaganda & Mind Control

One of America's greatest vulnerability is the ignorance of its people. This has its roots in the public's addiction to the controlled media. The conditioned nature and format of the mass media, including news and entertainment, significantly influences most Americans. The process is so subtle that most people can't even begin to grasp the reality of what a major role the media plays in the formation of their personal opinions and beliefs.

No potentate, dictator or religious leader in the history of man has ever held as much immediate and direct power to shape and mold the thoughts, opinions, beliefs and attitudes of the people as the tight conglomerate which controls the media. The news and entertainment media reaches not only into everyone's living room but also inside their heads.

The media tells us virtually everything we think we know about the larger part of the world which is outside the boundaries of our everyday travels. News in this nation today is a lot like processed food and we are all a product of what we take in.

The media is very subtle, thorough and effective in the way it can intentionally shift and direct public attention. The media repetitiously focuses attention in certain areas and at the same time blacks out or diverts attention away from other areas and issues. Other techniques, such as emphasis, choice of words, the announcer's tone of voice and facial expressions, illustrations and associations, to name a few, strongly affect the way people interpret that which they see, hear, and read.

A recent and classic example occurred on CNN during the reporting of the bombing of the federal building in Oklahoma City. One may have noticed that in the corner of the screen whenever commentators were reporting on the bombing, there was a logo. The logo was a silhouette picture of an AR-15 assault rifle. Now, wouldn't it have been more appropriate to use a logo such as the picture of a bundle of dynamite sticks with a burning fuse? In reality this was the effective use of the tragedy to paint a picture of guilt by association and gain mileage for gun control.

The entertainment portion of the media industry has as much influence on people's attitudes, opinions and beliefs as the news. If you examine the overall content of entertainment, political correctness is elevated as the acceptable norm. This is seen in the way script writers elevate or degrade the portrayal of certain characters or groups.

The most devious aspect of media mind control is not just their slanted and biased tactics, but rather their ability to control and establish the tacit boundaries for the acceptable spectrum of opinion. In order to insure reputability, the media must of necessity give the appearance of providing contrasting views.

One of the most classic examples of this control of the spectrum of opinion is the media talk show celebrities Larry King and Rush Limbaugh. Most people view these two polarized individuals as representations of the liberal and conservative camps. Limbaugh is an attempt to harness, direct and contain the aspirations of conservative Americans. Typically Limbaugh says a lot of good things, but nothing too deep. Limbaugh will never touch anything relating to the C.F.R., Federal Reserve, the Bilderbergers, the new world order, the drugs and murders behind Whitewater, the U.N., or the corroboration of the financial power elite to subjugate the U.S. Constitution and the American republic. Limbaugh goes out of his way to discredit anyone who entertains such ideas as a nut or a wacko. Limbaugh specializes in slinging meaningless surface mud at Democrats and endorsing NAFTA, which furthers global government at the cost of American national sovereignty.

It is a modern day bread circus. The gladiators of today are in the news arena. Not only do they manage to keep the masses entertained but also thoroughly confused and distracted from the real issues.

The Establishments' Mutt and Jeff

The message conveyed is that anything not represented in the spectrum which exists between these two is not credible, is crackpot and is not worth investigating. In reality, the seemingly contrasting views offered through the media are within the boundaries of the power elite's effective control. From the point of view of the power elite, it is heads I win, tails you lose.

Over the past fifty years the international financial elite have consolidated their control of the American media. The shadow of a hidden agenda can be perceived by observing the direction which the public is being led by media mesmerism. Some would argue that this is not orchestration but a manifestation of natural social evolution. To the contrary, the banking elite through corporate control, their foundations, think tanks and their political party, the Council on Foreign Relations, control all the leading positions in media and government. The reality is that the direction this nation is taking has been and continues to be carefully and intentionally orchestrated. This is by no means an accident.

The Soviets recognize this American media dependence and gullibility. The Soviet propaganda and disinformation machinery feeds the Western media and the media feeds the American public with effective disinformation that is designed to tell the people what they want to hear and believe. The essence of what most Americans want to hear is that everything is all right, the threat of Commu-

nism is gone and, finally, we have attained a lasting peace.

Political Motivation

Another significant American vulnerability is the ambition of its political leaders. The first priority of most political figures is their public image and reelection position. Consequently, American political figures tend to walk on eggshells in terms of not endorsing any controversial position which is defined as something threatening to the nation's financial power elite. Shaking a stick at the dragon is a sure way to enrage the beast and elicit media wrath. Disfavor with the national media is a proven formula for the destruction of a political career. On the contrary, not rocking the boat, endorsing the elite liberal party line and turning a blind eye to the criminal nature of what is really going on is a certain and proven recipe for the accelerated advancement of a political career.

Part and parcel with this pattern is the tendency of U.S. leaders to make too many concessions in negotiations with the Soviets. Every administration all the way back to Nixon has at one time or another bent over backwards in an attempt to secure treaties with the Soviets as a means of creating an artificial sense of peace and stability. The ceremonious signing of treaties reinforces the nation's sense of security and enhances the public's view of the current administration.

How Threats Can Be Used to Manipulate People

We have just reviewed some of the legitimate, major threats that we could be facing in the upcoming years. Nevertheless, threats can also be used as an essential part of a government's control of its people. Threats cause people to surrender differences, independence, rights and resources in deference to a leadership which offers a solution to the threat. A long time ago, the power hungry figured out that they could gain greater control over people simply by artificially creating a threat.

Pulling together and surrendering differences is not necessarily an evil function if the leadership has pure motives and the threat is real. But if the leadership has an ulterior motive to make gain, the end result is always tyranny and abuse. Even if a threat appears to be real, one should always analyze who stands to potentially benefit from the crisis.

Drugs

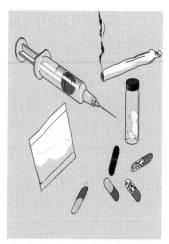

The war on drugs is an example of a threat which the United States has the ability to swiftly eliminate. This threat is not being dealt

with because it creates $150 billion in profits annually for Western banking interests and the resulting crime problem causes a calculated reaction in the general public. The media focuses the spotlight of public attention on the national epidemic of drug related crime and then leads the public to the conclusion that the only solution is to give the federal government broader police powers. What the media does not tell the American people is that whenever government is empowered, this means that personal liberties, freedoms and safeguards are eliminated.

The war on drugs is the justification for instituting a police state which of necessity includes gun control, no-knock gestapo searches of private residences, property seizure and a nation-wide federal police force, or a secret police.

It is fairly well documented that certain agencies and individuals associated with the federal government have been involved in drug trafficking all the way back to the Viet Nam War, but the major suppliers are not American. The truth is almost stranger than fiction and well suppressed by the justice system, drug enforcement, the State Department and the media. If you want to know the truth, I suggest you purchase a copy of a well-documented book called *Red Cocaine — The Drugging of America*" by Joseph D. Douglass, Jr., Clarion House Publishing, Box 88304, Atlanta, GA 30338, ISBN# 0-9626646-0-x.

Environment

The endangered environment is another example of a threat which is used to convince the public that government needs to take more power from the people. We probably all agree that the environment needs protecting. I fully support protecting the health of our environment. I recycle, I bend over backwards trying not to waste, and I love the outdoors. However, the hidden purpose of the people who started and steer the environmental movement is not to protect the environment.

The primary purpose behind the environmental movement is to erode property rights, condition the public into accepting government population control and to further the malignant spread of administrative law. Population control, the elimination of property rights, and administrative law are foundation stones of the new, one world United Nation's government.

War

The Essence of War — Government Power & National Identity

War has both visible justifications and invisible purposes. The visible justifications for war are defending a nation from military attack and defending national interests, but in reality these visible justifications are often not the real instigating causes of war.

The real purposes of war are not subordinate to the people of a nation. The real purpose of modern warfare is not to protect or defend a nation or its interests.

A nation is a group of people organized with a common goal and native identity. War is the glue which holds nations together. The threat of war is a catalyst which strengthens and develops national identity. War, or its possibility, provides an external necessity which pulls the people together and creates a united sense of purpose or national identity.

War stabilizes a nation's internal organization of political power. War causes the people to vest the leaders of the state with power and authority. No government can remain in power without the continuing presence of war or some substitute which provides an external threat. The absence of threat eventually leads to a loss of ruling power.

The capacity of a nation to make war and crisis is the most significant social power it can exercise. War-making and crisis, active or contemplated, are essential to the survival of the elite.

War — Economic Stimulation, Technological Development and Banking Profits

Another invisible purpose of war is economic. War creates an artificial demand outside the supply-and-demand system of a nation's normal economy. The demands of a nation's war-making institutions stimulate its economic metabolism through defense spending and research and development. This inevitably results in technological advances which spin off and advance other aspects of society. This can be clearly seen in the rapid technological advancement of aviation as a result of World War II research and development.

But most importantly war creates debt. Governments borrow money from banks to finance wars. Banking institutions profit from war. Creating war means creating profit. For economic purposes, war is a tremendous national growth factor and war facilitates the amassing of huge fortunes in the hands of the banking elite.

The Hatred of America

Why do the people of some foreign countries like Libya and Iran hate the United States? Are these people really as irrational and mindless as the media would have us believe?

Taking Libya as an example, in recent years the media has expended considerable time and effort convincing the public that Qadhafi's Libya is an international terrorist threat. What the media chooses not to focus on is the history of the Western abuse of Libya since the early 1900s.

Western domination of Libya started in 1911 when Italy colonized Libya. Subsequently Mussolini organized a mass migration with the intent of crushing nationalist resistance and ensuring Italian control over Libya's strategic oil reserves. By 1931, this war of resistance had taken more than 750,000 Libyan lives. After World War II, Britain, France and the United States maintained de facto control of Libya. Libya is an ideal strategic location for the control of North Africa and the Mediterranean. The U.S. spent $100 million building Wheelus airbase on the out-skirts of Tripoli. Western military and economic interest had unrestricted access to Libya's bases and oil. During the Korean War, Wheelus air base was used extensively by the US Air Force. When the Arab-Israeli war closed the Suez Canal, Libya ended up being the only uninterrupted source of middle eastern oil which brought further U.S. corporate investment into Libya.

King Idris proclaimed Libya's formal independence in 1951 and acted as a puppet for U.S. interests until 1969 when he was overthrown by Colonel Muammar Qadhafi. Qadhafi nationalized 51% of the foreign banks and the U.S. and Britian were kicked off their military bases in Libya. In 1973, Qadhafi nationalized 51% of all oil companies, including Exxon, Mobil, Texaco, Socal and Shell. Libya doubled the price of its crude oil and the majority of other OPEC nations followed suite. But Qadhafi's worst offense may have been giving support to another frustrated group, the P.L.O. This brought on the wrath of the Mossad and a powerful Israeli lobby.

Thus the reason Libya has been a target of U.S. media and the State Department is that it dared to stand up against Western abuse and the global economic interests, which control U.S. foreign policy. Consequently, the U.S. government has waged an undeclared war against Libya and the controlled Western media has continued to hammer the public with the concept that Libya is an instrument for international terror.

The western operations against Libya have included a number of assassination attempts on Muammar Qadhafi. One attempt included the 1986 bombing of Tripoli and Benghazi in which 100 people were killed.

Libya was accused of being involved in the 1988 bombing of the Pan Am jumbo jet over Lockerbie, Scotland, but other evidence indicates the possibility of anti-Libya provocateurs. Guess who? This accusation led predictably to U.N. economic sanctions against Libya in 1992.

I by no means justify the use of terrorism, but most of the time it is hard to see through all the media smoke and determine who is really terrorizing who. It is very important to understand both sides of a story and what types of frustration drive people and countries to desperate action. In my opinion, the world community should not limit its police actions to suppressing the symptoms of terrorism, but they should also deal with the root causes of terrorism which are much closer to home.

Conclusion

It is statistically accurate to say that American society today is in a state of rapid decay. A report filed by the U.S. Senate Judiciary Committee on March 1991 concluded that the United States was, *"the most violent and self-destructive nation on earth."* The report indicated that the population of the U.S. had grown by 41 percent since 1960 and during the same time period violent crime had grown by 516 percent.

America and her people are vulnerable for many reasons that we have discussed here, but there is a more significant vulnerability which underlies the many problems we face today. The underlying vulnerability has its roots in the erosion of the spiritual - moral values. This is the immune deficiency which opens the door to a host of opportunistic social ills. History clearly showed this pattern in the downfall of ancient Israel, Greece and the Roman Empire. The destruction of virtue

and moral value always leads to chaos and the destruction of civilization and order.

The prophets of the Old Testament warned the people of Israel not to practice the ritual of sacrificing their offspring to fire. The people ignored this warning and thus Israel lost its grace of protection. Subsequently Israel was overrun by Babylonian invaders and the people were carried off into slavery.

In this nation today, one out of three children conceived is murdered through the ritual of abortion. The magnitude of this crime far surpasses any previous holocaust. What makes the crime worse is that the victims are both innocent and totally helpless. This is all sanitized as legal and just by the so-called law of the US. government.

Americas' greatest vulnerability by far is moral - spiritual decay and the spilling of the blood of the innocents.

Not only can the protection of grace be lost , but it can also be gained. When Hiroshima was bombed by the first atomic weapon, everyone within an area of one mile of the epicenter of the explosion died with the exception of one household. This particular household was located only eight blocks from the epicenter and all 16 members of the household survived. Hundreds of "experts" and investigators have, throughout the years, studied this phenomena in an attempt to determine what difference contributed to this miraculous survival.

Only one thing was found to be different. This house was a Catholic parish in which all the residents lived a spiritual life and faithfully gave daily devotions.

Chapter 8
The Effects of Nuclear War

The reason for nuclear weapons is economics. Not only do they give the best punch per dollar, but they also produce the greatest destructive force per volume and weight. When considering the deployment and delivery of a weapon with a missile, this becomes an obvious and significant factor. The destructive force of a nuclear weapon is thousands of times greater than a conventional weapon of similar size and weight. It takes about 110 pounds of uranium for a fission reaction which releases the same amount of energy as the explosion of one million tons (one megaton) of TNT.

Nuclear energy is the energy contained in the center of the nucleus of an atom. When the nucleus is altered this energy is released.

Nuclear Fission

Nuclear fission is the release of energy through the splitting of the nucleus. When the nucleus is split, most of the protons and neutrons are released and form into lighter nuclei. Several neutrons are released along with a great deal of energy. These released neutrons strike other adjacent nuclei, splitting them and releasing more electrons and neutrons. These neutrons in turn produce still more fission, and a chain reaction quickly builds into a nuclear chain reaction. In order for the process of a nuclear chain reaction to occur, a substance is required that will easily undergo fission. Certain forms of uranium

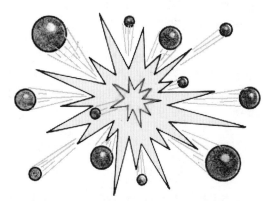

Splitting the Atom - Nuclear Fission

and plutonium, such as uranium 235 and plutonium 239 will do this and will support a chain reaction. Thus uranium 235 and plutonium 239 are referred to as fissionable materials.

Fission Weapons

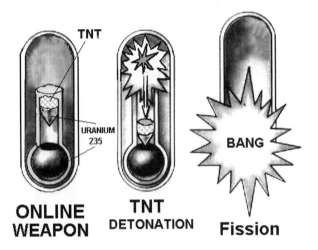

ONLINE WEAPON   **TNT DETONATION**   **Fission**

Fission weapons, which are also called atomic weapons, make use of this nuclear fission to produce an explosion. A chain reaction occurs whereby many nuclei simultaneously split, thus releasing an enormous amount of energy.

A fission bomb contains a quantity of conventional explosive. The detonation of the bomb is started off by an explosion of the conventional explosive that pushes two pieces of fissionable material together with great force. This starts the chain reaction which is completed in less than a millionth of a second. The result is the release of vast amounts of energy and an enormous explosion.

Fusion

Fusion is the exact opposite of fission. In nuclear fusion, the nucleus is not split. Nuclear fusion results from the forceful joining of two hydrogen nuclei. Given the same amount of nuclear material, fusion results in an energy release far greater than fission, but fusion is far more difficult to achieve.

Fusion Weapons

Fusion bombs are often called hydrogen bombs because they depend on the fusion of many hydrogen nuclei, specifically deuterium and tritium (which are both isotopes, or forms, of hydrogen). Deuterium and tritium are the essential materials used in a fusion weapon because they undergo fusion more readily than common hydrogen. In order to facilitate fusion, hydrogen nuclei must be confined and heated to temperatures of millions of degrees. For this reason, fusion weapons are also referred to as thermonuclear weapons.

The conventional explosives used to trigger atomic bombs are not powerful enough to set off a thermonuclear fusion explosion. The detonation of a fusion weapon begins with the detonation of a conventional explosive that sets off a fission reaction explosion. The explosion produces extremely high temperatures and pressures which causes the deuterium and tritium nuclei to fuse al-

Hydrogen Isotopes

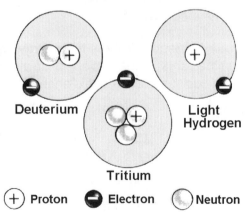

Deuterium   Light Hydrogen

Tritium

⊕ Proton   ⊖ Electron   ◯ Neutron

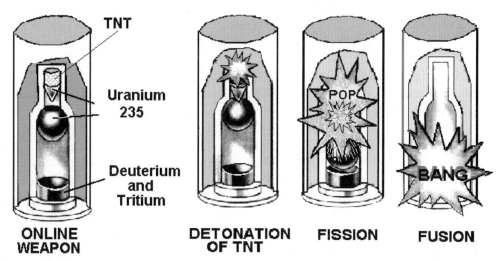

TNT

Uranium 235

Deuterium and Tritium

ONLINE WEAPON   **DETONATION OF TNT**   **FISSION**   **FUSION**

most instantaneously, resulting in a massive explosion.

Comparative Effects of Different Sized Nuclear Weapons

The amount of energy released by a nuclear weapon is referred to as its yield. The power of a nuclear weapon is expressed in terms of its relationship to equivalent power released in the detonation of tons of TNT. A one-megaton weapon equals the effect of one million tons of TNT. A one-kiloton weapon is equivalent, in energy released, to 1,000 tons of TNT. It should be noted that one large nuclear weapon is not as effective as numerous small nuclear weapons. For example, a ten-megaton weapon is ten times larger than a one-megaton weapon but the effective destruction capability of a ten megaton weapon is only twice that of a one-megaton weapon. In essence, ten one-megaton weapons would have five times more destructive effect than one ten-megaton weapon. The larger warheads are more fragile than smaller warheads. The other issue involved is the fact that the pinpoint accuracy of a warhead is dependent on a high speed when entering the atmosphere. Due to their size, the larger warheads are slower and thus less accurate. For the most part, the largest weapons in the U.S. and Soviet nuclear stockpiles are one megaton and the bulk are kiloton weapons. The Soviets do have some SS18 missiles with single, large warheads which they probably maintain for special targets.

Tactical Nuclear Weapons

Tactical nuclear weapons have much shorter ranges and much lower yields than conventional nuclear weapons delivered by I.C.B.M.s. They are mostly battlefield weapons, such as anti-aircraft missiles and artillery shells that are designed for use against enemy troops and tanks. The yield of a battlefield nuclear weapon may be as little as a tenth of a kiloton or as much as several kilotons.

Radiation

In order to understand the effects of nuclear war, we need to first understand the basic concept of nuclear radiation.

Elements are the basic building blocks of the physical world. Some of the more common elements which we are all familiar with are hydrogen, nitrogen, oxygen, iron, copper, etc. The smallest sub-component of an element which cannot be further divided by a chemical means is called an atom. Despite their extremely small size, atoms are made up of combinations of three yet smaller particles. These three smaller particles are called protons, neutrons and electrons. The protons have a positive charge, the electrons have a negative charge and the neutrons are neutral — they have no charge.

Most atoms hold together quite well and are therefore referred to as stable elements. A few atoms, and several which mankind have been able to create, don't hold together well. These unstable atoms tend to break down. When these unstable atoms break down, parts of the atom fly off into surrounding space in the form of energy. The spontaneous, uncontrollable breakdown of unstable atoms with a resulting energy release is referred to as radioactivity.

These particles of energy that fly off of unstable atoms are released from the nucleus of the atom and the released particles are referred to as nuclear radiation. Nuclear radiation is the spontaneous, uncontrollable release of energy from the nucleus of atoms of an unstable element.

Types of Radiation

There are four basic types of nuclear radiation:

Alpha Radiation

Alpha radiation consists of actual particles of matter that have a positive electrical charge, are relatively large and heavy, and travel at a very high velocity. Alpha radiation is not very penetrating. The damaging radioactive effect of alpha radiation cannot penetrate the outer layer of human skin. The separation of 2 or 3 inches of air or even a piece of paper will reduce or shield the negative radioactive effects of alpha radiation on living tissue.

Alpha radiation is only dangerous when it is ingested or inhaled into the body with water, food or air. Internal organs can be seriously damaged by internal contamination from alpha radiation. When alpha radiation particles are left on the skin for a long period of time, the emissions can cause burns similar to heat burns.

Beta Radiation

Beta radiation consists of actual particles of matter that have a negative electrical charge, are very light in weight, and travel at a velocity which is close to the speed of light. Beta radiation can radiate 20 to 30 feet through air. Beta radiation in fallout can cause skin burns if left in contact with the skin. The ingestion of beta radiation particles can result in fatal injuries to internal tissues and organs. Beta radiation is more penetrating than alpha radiation. Both alpha and beta are insignificant threats in comparison to gamma radiation.

Gamma Radiation

The third type of radiation is called gamma radiation. Gamma radiation is pure energy, similar to x-rays. Gamma rays are photons which have no measurable mass or electrical charge. Gamma radiation is often referred to as gamma rays. Gamma rays, in contrast to alpha and beta radiation, are highly penetrating and dangerous to living tissue. Gamma radiation is a serious threat and the major concern in relation to the extended fallout problem.

Neutron Radiation

The fourth type of radiation and the most deadly type to be exposed to is neutron radiation. Neutron radiation is an actual neutron particle with no electrical charge. During the initial detonation of a nuclear weapon, neutron radiation is released as a very penetrating pulse which occurs for a second or less. This type of radiation is only experienced within the immediate vicinity of point zero, approximately a 1.5-mile radius. Even with this momentary duration, neutron radiation is even more penetrating than gamma radiation. Neutron radiation is extremely penetrating, destructive and very difficult to shield against. Exposure to this pulse of neutron radiation can be very damaging or fatal.

The Direct Effects of Nuclear Weapons

There are seven potentially destructive effects which result from the detonation of a nuclear weapon and six of these are dangerous to the human body or body parts. These are the flash, initial nuclear radiation, the fire-ball, thermal radiation, blast, radioactive fallout and electromagnetic pulse. Electromagnetic pulse has no immediate negative effect on the human body but it can destroy electronic equipment necessary for life support in a shelter.

1. The Flash

The Flash lasts several millionths of a second. Sunglasses will not protect eyes from the flash, and the blink reflex is not fast enough to protect the eyes. Indirect exposure to eyes can produce temporary injury called flash blindness. This occurs no matter what direction a person is facing and is worse at night. Directly looking at the blast can cause permanent injury to the retina, referred to as choriretinal burns.

2. Initial Radiation

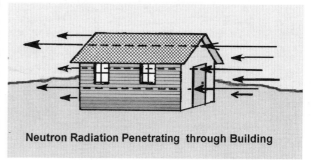

Neutron Radiation Penetrating through Building

Initial radiation is a term used to represent the total radiation (Alpha, Beta, Gamma and Neutron) released during the first 60 to 100 seconds of a nuclear detonation. The Neutron radiation has the greatest ability to penetrate heavy barriers. It can inflict biologically damaging effects ten times that of gamma radiation that is found in radioactive fallout.

3. Fireball

During a surface burst of a nuclear weapon, the fireball reaches temperature levels of millions of degrees Fahrenheit. All materials in the fireball are converted to a gaseous form. The pressures developed at ground zero are approximately one million times standard atmospheric pressure. In clear weather, the fireball from a 100-KT ground burst will be seen for more than 400 miles. An airburst fireball can be seen approximately 700 miles away. This fireball lasts less than 90 seconds. The fireball is about 30 times more brilliant than the sun at noon.

4. Thermal Radiation

With the initial detonation of a nuclear weapon, there is an emission of thermal radiation that is mainly in the form of x-rays. These are absorbed within a few feet of air and then re-emitted from the fireball as ultraviolet, visible, and infrared rays. This emission of thermal radiation is actually done in two separate pulses which last less than one minute and travel at the speed of light.

The first pulse has a very short duration and is mainly comprised of ultraviolet rays.

This first pulse can cause permanent or temporary damage to the eyesight of anyone looking in the direction of the explosion.

The second pulse lasts several seconds and is the main hazard. This second pulse of thermal radiation consists mostly of visible and infrared rays which ignite combustible material, cause skin burns, and seriously injure eyes.

Thermal radiation represents about 35 percent of the energy expended in a thermonuclear reaction. The effects of this phase can be reduced up to 65 percent if the atmosphere in the vicinity happens to be foggy or cloudy.

5. Blast

Structural Collapse due to Overpressure

The blast is a shock wave that lasts one minute. It is a wall of pressure that moves outward away from the fireball. The blast contains four destructive phases that include positive overpressure, reflected overpressure, blast wind or dynamic pressure, and negative overpressure.

a. Positive Overpressure

As a result of an air burst detonation of a nuclear weapon, a positive overpressure is generated. Positive overpressure is the amount of pressure generated by the explosion of the bomb which exceeds the natural atmospheric pressure.

Overpressure may be likened to the pressure a scuba diver experiences when he descends into deep water. Positive overpressure creates a uniform pressure on objects from all directions. It does not move or displace structures or objects it comes in contact with but crushes them.

At ground zero, a 100-kiloton air burst weapon detonation would create an overpressure of 300 pounds per square inch. At one half mile this would reduce to 20 p.s.i., and at one mile from ground zero this would reduce to 1 p.s.i. A ground burst weapon, on the other hand, can generate an overpressure of 140,000 pounds per square inch at ground zero reducing 60 feet away to 8,000 to 10,000 pounds per square inch.

Effects of Positive Overpressure In P.S.I.

P.S.I. = Pounds per Square Inch

1 p.s.i. = 144 pounds per square foot

1 p.s.i. - Broken glass and light damage to structures. Lacerations from flying glass.

2 p.s.i. - Severe damage to structures. Windows blown out, interior partitions in house blown out, furniture and contents swept out of the far side of the house, and masonry walls collapse. The blast winds associated with this range of overpressure are greater than most hurricanes. People in basement shelters would generally survive.

5 p.s.i. - Severe damage to structures. Those people who are exposed to the force will be thrown hard enough that if they impact a solid object, 99 percent will be killed. Houses collapse and crush most basement shelters or are blown off foundations. Vehicles are damaged enough that they can't be driven. Blast shelters would be required for surviving pressures of 5 p.s.i. or greater.

15 to 20 p.s.i. - Commercial buildings are severely damaged. 50 percent of the people suffer ruptured eardrums and injury to lungs.

At 20 plus p.s.i. damage to the physical body is significant. The pressure ruptures the walls of the abdominal and thoracic cavities, causing hemorrhaging, and air is actually forced into the circulatory system. The body's soft tissues and the skeletal structure suffer severe damage.

40 p.s.i. - 99 percent fatalities.

b. Reflected Overpressure

Reflected overpressure is a phenomena whereby the blast wave of overpressure strikes an object and is reflected off the object thereby amplifying the pressure of the original blast wave by several times. The equation which determines the amplification of this reflected force is based on the angle at which it strikes and reflects off an object. For example, a 60 p.s.i. pressure coming in contact with a surface at 90 degrees and reflecting off will produce 225 p.s.i. The same overpressure coming in contact with a surface at a 40-degree angle or incidence will produce a reflected overpressure of 150 p.s.i. And this same overpressure force striking a surface at an incidence of 10 degrees will produce a reflected overpressure of 75 p.s.i.

c. Blast Wind

Effects of Blast Wave

During the surface burst detonation of a nuclear weapon, a blast wind results which is a high speed wind that directly follows the blast wave and creates a tremendous amount of destruction. A 100-kiloton air burst weapon produces a 2,000 m.p.h. wind at ground zero, a 700 m.p.h. wind at one-half mile, a 240 m.p.h. wind at one mile, and an 85 m.p.h. wind at two miles.

d. Negative Overpressure

Negative overpressure is a suction phase which is only about 4 p.s.i. below atmospheric pressure. This follows directly after the passing of the positive overpressure.

6. Radioactive Fallout

The fireball created by a one megaton surface or groundburst vaporizes and fuses about 500,000 tons of dirt and debris with fission products and radioactive material.

This material is drawn up with an updraft of superheated gasses into the forming mushroom cloud. Then, as the heat diminishes, radioactive materials that

have been vaporized condense on the drawn-up particles of earth, which are also condensing. Eventually, all these particles, many of which have combined with radioactive materials, fall back to earth. Within 24 hours, half of this fallout, the heavier particles, will have settled down out of the air onto the earth's surface. These fallout particles are highly radioactive and give off strong rays. The lighter particles, which remain in the air for a longer time, settle down to earth much farther from the detonation site and end up losing much of their radiation strength while still suspended high in the atmosphere.

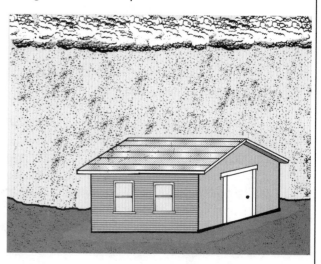

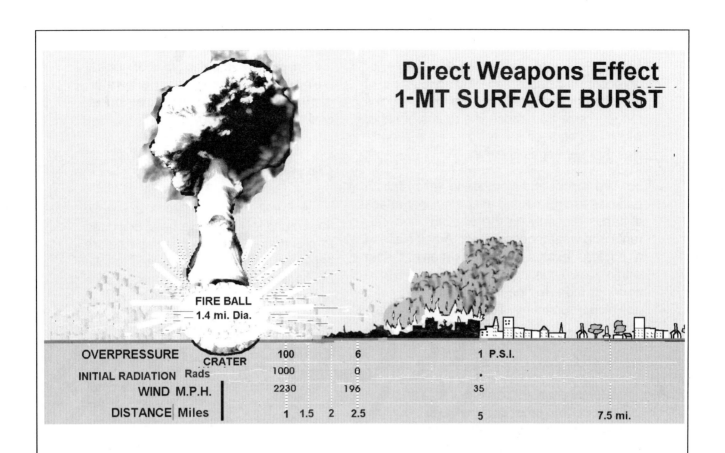

Direct Weapons Effect
1-MT SURFACE BURST

FIRE BALL
1.4 mi. Dia.

OVERPRESSURE		100	6		1 P.S.I.	
INITIAL RADIATION Rads		1000	0		•	
WIND M.P.H.		2230	196		35	
DISTANCE Miles		1 1.5 2 2.5			5	7.5 mi.

CRATER

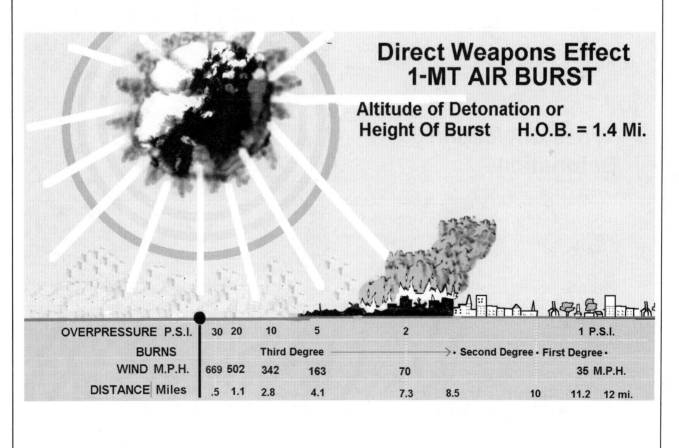

Direct Weapons Effect
1-MT AIR BURST

Altitude of Detonation or
Height Of Burst   H.O.B. = 1.4 Mi.

OVERPRESSURE P.S.I.	30	20	10	5	2				1 P.S.I.	
BURNS			Third Degree	→			Second Degree •	First Degree •		
WIND M.P.H.	669	502	342	163	70				35 M.P.H.	
DISTANCE Miles	.5	1.1	2.8	4.1	7.3	8.5	10	11.2	12 mi.	

Fallout is not a mysterious, invisible, or unrecognizable substance that appears out of the sky without warning. Fallout particles range in size from those like grains of sand, which can be seen easily, to very small particles that appear as fine dust.

The extent and distance to which fallout particles are distributed after a nuclear attack depends primarily on the existing wind currents and weather conditions. Areas close to a nuclear explosion might receive fallout within 1 to 30 minutes, but it might take 5 to 10 hours or more for the fallout particles to drift down on a community 100 or 200 miles away. Due to the prevailing winds which blow from west to east, higher accumulations of fallout would tend to cover the eastern United States.

After 28 days, 99 percent of this fallout on the ground will have decayed to biologically safe levels. The remainder of the very fine dust still in the atmosphere will end up in the stratosphere and will continue to particulate to the surface of the Earth at a rate of 10 percent per year. This fallout is non-hazardous by the time it returns to the earth's surface.

Types of Weapon Detonations

When a nuclear weapon is launched toward an enemy target, the weapon is programmed to detonate at a certain distance, proximity or altitude to effect a desired destructive effect. When the military command structure targets a particular type of enemy installation they predetermine how close to the target they want the weapon to detonate based on the target type. Most nuclear weapons are designed to explode in the air just over a target. The explosion point is sometimes called **point zero**, and the target area beneath it is called **ground zero**.

There are five basic types of nuclear weapon detonations: **penetrator, ground burst, surface burst, air burst, and high altitude burst.**

Penetrator

A Penetrator is a detonation below the ground level by a warhead designed to penetrate the earth's surface before exploding. This is typically used to take out hardened below-ground missile silos and command and control centers.

Surface Burst

A surface or ground burst is a detonation on or above the surface of the earth with the fireball contacting the earth's surface. This burst expends a

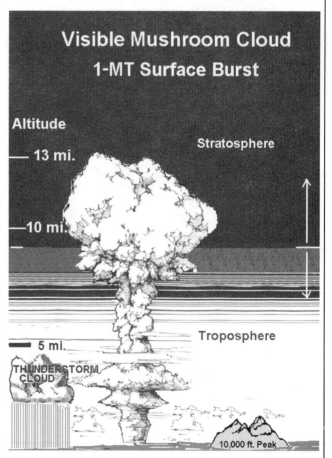

Visible Mushroom Cloud
1-MT Surface Burst

Altitude
— 13 mi.
—10 mi.
— 5 mi.

Stratosphere

Troposphere

THUNDERSTORM CLOUD

10,000 ft. Peak

large portion of its energy digging a crater, creating about 25 percent less thermal radiation than an air burst. Thermal radiation is deflected upward and outward, with the earth and man-made barriers absorbing much of the heat. Tons of earth are drawn up into the fireball creating, a maximum amount of fallout.

Surface bursts deliver a maximum coverage with high overpressures. They are typically used against hardened strategic military targets.

Air Burst

An air burst is the detonation of a nuclear weapon where the actual explosion happens above the ground, at an altitude of 100,000 feet or less, and the fireball does not reach the earth's surface. No crater is formed and there is very little ground shock. An air burst will deliver a maximum coverage of lower overpressures. This type of detonation is ideal for destroying soft targets, such as cities which cover a large area. The air burst produces about 25 percent more thermal radiation than a ground burst. This is because in a ground burst you have physical obstructions between the target and the fireball which absorb energy. No radioactive fallout occurs except for a small amount from the fragments of the bomb casing because no dirt has been drawn up.

High Altitude Burst

A high altitude burst is the detonation of a nuclear weapon where the actual explosion occurs at a very high altitude, not less than 100,000 feet and some times as much as 300 miles above the earth's surface. A high altitude burst does not produce any physically damaging effect on the earth's surface. High altitude bursts are intended to create electro-magnetic pulse (E.M.P.) which disrupts communications and destroys electronic equipment.

Electromagnetic Pulse (EMP)

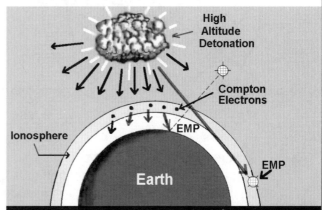

Electromagnetic Pulse (EMP)

Electromagnetic pulse (EMP) is a complicated phenomena associated with a nuclear explosion. EMP is a radiated electromagnetic broadband pulse which has a very high energy content. It is generated by the detonation of a nuclear weapon. This detonation produces an emission of high energy gamma rays. When the gamma rays interact with air molecules, a negative charge is caused by the electrons. This relative displacement of positively and negatively charged regions produces an intense electric field and radiates a radio frequency pulse which is intercepted by any and all conductors and transformed into voltage. This can be as much as 50 to 60 thousand volts per meter. Power lines, telephone lines, and antennas will collect and thus magnify this energy and cause connected equipment to fail.

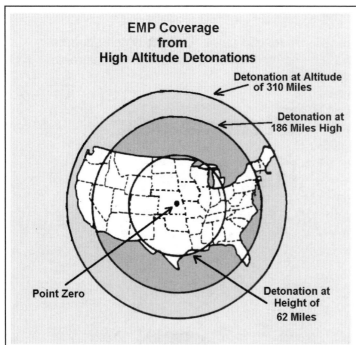

EMP Coverage from High Altitude Detonations

Detonation at Altitude of 310 Miles

Detonation at 186 Miles High

Point Zero

Detonation at Height of 62 Miles

Distinctly different types of source regions are created, depending upon whether the detonation is a surface burst, a high altitude, or an air burst. In the case of a surface burst, the air molecules of the immediate atmosphere severely reduce the distance to which gamma rays can penetrate. A typical low-yield surface burst may create an affected area one mile in diameter. Increasing the weapon yield by a factor of one thousand will increase the diameter of the source region only by a factor of about three.

In surface bursts, the damaging EMP effects are restricted to areas effected by high overpressure and thermal radiation. Consideration of EMP is extremely important for hardened blast structures in direct target areas which could experience the effects of a high intensity blast.

A high yield, high altitude nuclear detonation in the ionosphere (above the atmosphere) gives off gamma rays. These gamma rays, due to the lack of obstructing air molecules in this partial vacuum environment, travel many miles and cover a large area with radiation. In such instances, gamma radiation affects the atmosphere over a vast region.

As the gamma radiation penetrates into the atmosphere, it interacts with air molecules causing the expulsion of electrons from those molecules. These rapidly moving electrons are curved by the earth's magnetic field. This results in radiated synchrotron radiation which is the intense electromagnetic pulse.

Because of the great height, the EMP radiated from such a high altitude detonation will appear over a substantial portion of the earth's surface. The EMP can damage electronic equipment in regions far away which are unaffected by the direct effects of a nuclear weapon. In such cases, shelters far away from any direct target areas could sustain damage to their power systems, communications and sensitive electronic equipment.

In the electromagnetic sense, the spectrum and wave form of EMP differ significantly from any other natural or man-made sources, such as lightning or radio waves. The spectrum is broad, extending from extremely low frequencies into the UHF band. The wave form indicates a higher amplitude and much faster rise time than, for example, lightning. In terms of the rise time of the events, EMP is in the low nanosecond range and lightning, at its fastest, is at least ten times slower. EMP is also widely distributed, as opposed to the localized effects of lightning.

Although there are vast differences between the phenomena of EMP and lightning, both can cause the same type of damage and an analogy between the two is useful for assessing the threat of EMP in terms of a familiar phenomenon. Most damage from EMP occurs as energy in the form of strong electromagnetic fields is converted into very large currents and voltages when it impinges on cables or other conductors. Thus, like lightning, EMP can cause functional damage, such as the burn-out or permanent electronic damage to components, or operational upset, such as the opening of circuit breakers or the erasure of storage in the memory bank of a computer. It is this sort of potential damage that poses EMP as a serious threat that must be considered in the design of any shelter.

The Effects of Radiation on People

In order to understand the dangerous and deadly effects of radiation on people, one needs to understand the process of **ionization**. Ionization occurs when radiation impacts the molecules in living tissue. As a result of the impact, electrons, which have a negative charge, are separated or knocked out of their orbit around the nucleus of an atom or molecule, leaving it with a positive charge. Ionization is the separation of an atom or molecule that is normally neutral in charge into electrically charged components.

Radiation damage to living organisms results from changes induced in individual cells. Ionization has an effect similar to very small bullets penetrating living tissue and disrupting the cellular life process by destroying the chemical bonds between the cells. Some of the component parts essential to the normal functioning of the cell are altered or destroyed. This cellular disruption includes the formation of by-products which are poisons, the breaking up of chromosome, an increase in the permeability of cell membranes, the actual destruction of the cells themselves, and the disruption of the cell division process which is necessary for normal cell replacement. If the exposure to radiation is great enough, these changes in cellular activity will reduce or destroy the functioning of internal organs.

The organs found in the human body and the different types of cells all have varying degrees of sensitivity to radiation. One of the most notable effects of radiation on humans is the destruction of red blood cells. The following are other notable body components which are initially impaired or destroyed by ionization, in their order of sensitivity: lymphoid tissue and bone marrow; testes and ovaries; skin and hair; blood vessels; smooth muscle and nerve cells.

The destruction of lymphoid tissue causes the lymph glands to fail, which results in the impairment of lymphocytes and the body's ability to control and suppress infection.

Bone marrow experiences significant changes when exposed to radiation. The destruction of any bone marrow means the reduction or termination of blood cell production. With the loss of cell regeneration, victims waste away and eventually develop extreme atrophy.

Another notable symptom of substantial ionization is the ulceration of the lining of the intestines. This is due to the disruption of the cell replication process in the intestinal tissue. As a result, bacterial infection occurs as the ulcers are contaminated with fecal material. With the suppression of the production of lymphocytes (white blood cells), the body cannot contain the infection and in extreme cases the patient dies from unchecked bacterial infection.

Hemorrhaging is yet another common symptom of major ionization to body tissue. Ionization suppresses the replication of platelet cells in the blood, which are necessary for clotting. In cases of excessive ionization, this hemorrhaging results in anemia and death.

The combined loss of skin and covering tissues, white blood cells, and antibodies, results in the body becoming vulnerable to bacterial and viral infection. In this situation, a patient can die from the proliferation of bacteria which are normally harmless.

It is important for people to understand that individuals exposed to fallout or initial radiation do not become radioactive and are not dangerous to other people as long as they have been externally decontaminated. Assure people that radiation sickness is not contagious or infectious, and one person cannot "catch it" from another person.

Roentgens, Rads & Rems

The second area we need to familiarize ourselves with is the terms or units used to measure, quantitatively, exposure to radiation. There are three measurement units that are used when dealing with radiation, depending on the situation.

Roentgen is a term used in the measurement of gamma rays or x-rays but is not used with other types of radiation such as alpha, beta, or neutron. Roentgen is not a measure of a dosage that an individual may receive but rather is a measure of exposure in the air.

RAD stands for radiation absorbed dose and is a measure of the amount of all kinds of radiation — alpha, beta, gamma, and neutron. A certain dose in rads for each kind of radiation has a different effect on a living organism. One rad of gamma radiation will have a different effect on an individual than one rad of alpha or neutron radiation.

REM is short for Roentgen Equivalent Man. REM takes into account the different types of radiation and their effect on the human body. The measurement of a REM is derived by multiplying the "relative biological effectiveness" (RBE) by the number of rads. The relative biological effectiveness varies with the kind of radiation. The method of converting Rads into Rems is as follows:

During a massive nuclear attack on the United States, many millions of Americans would be exposed to doses of radiation. The human body can deal with a certain rate of radiation exposure. People can tolerate a large total quantity of radiation exposure if it is received in

Converting Rads to Rems

Radiation Type	RADS	X	RBE	=	REMS
Gamma	1	X	1	=	1
Beta	1	X	1	=	1
Alpha	1	X	20	=	20
Neutron	1	X	10	=	10

gradual amounts over an extended period of time because the human body has the ability to repair, on an ongoing basis, a certain amount of the damage caused by radiation exposure. On the other hand, a person exposed to the same large amount of total radiation during a short period of time would die. People don't all react the same way to similar doses of radiation exposure. The general symptoms which most people experience due to their short term exposure to various amounts of radiation are listed below.

Effects of Radiation Poisoning - Short Term Exposure

O-25 Rads - No apparent "short-term" effect. May be some blood cell changes.

25-100 Rads - Typically people with this level of radiation exposure experience a loss of appetite and a small amount of nausea and sickness for the higher end of this dose category. Blood changes are noticeable. Up to 25 percent of persons experiencing this level of exposure will be incapacitated, but none will die. The normal period of convalescence will be about 7 days.

100-200 Rads - Definite identifiable changes in blood cells. White blood cells and platelets are affected at low range doses. Victims at the higher end of this radiation exposure category will show the toxic symptoms of damage to the gastrointestinal tract, such as weakness, fatigue, nausea, vomiting diarrhea, fever and infection. Those in this higher level will also experience hair loss, livid skin spots, fevers, hemorrhaging and some will experience heart failure. About 25 to 100 percent of persons exposed in this range will be incapacitated and approximately 25 percent will die within 30 to 60 days. The period of convalescence will be up to 40 days. Treatment should include reassurance, rest, light diet, water and antibiotics. For extreme cases, transfusions of platelets and white blood cells will be helpful.

200-450 Rads - Symptoms the same as for the previous exposure category but much more severe. 100 percent of persons exposed will eventually be incapacitated. More than half of this group experience nausea and vomiting soon after exposure and are ill for several days. This period of sickness is followed by one to three weeks where the victims seem free from any apparent symptoms. At the end of this period of no symptoms, more than half of the victims begin to experience hair loss and most develop moderately severe illness caused by infection.

This dose is fatal to at least 25 percent of those exposed to the low end of this dosage and up to 50 percent fatalities at the top end of the dose range within 30-60 days. Recovery time for survivors will be several weeks to several months.

The treatment for patients with this level of radiation exposure is the same as the previous category. Diarrhea and vomiting are common symptoms of radiation poisoning and cause dehydration and an imbalance of electrolytes in the system. Maintaining the body's electrolytes and hydration is important for both survival and recovery. If this cannot be accomplished through oral hydration mixtures, because of vomiting, then I.V. fluids should be administered, if available. Blood transfusions will help recovery and improve survival rates. If they are available, sedatives can be administered for extreme cases.

450-600 Rads - Symptoms in this category of exposure are the same as the previous exposure rate but now the onset of symptoms is sooner and very severe. The primary signs of illness in this group are: mouth, throat and skin hemorrhages. It is also very common to see infections, such as sore throats, pneumonia and intestinal inflammation.

About 50-75 percent of the people exposed to this dosage will face death. Death will occur sooner in this exposure rate category, 20-35 days. Survivors' period of convalescence will be several months to years. Survivors of this dosage level will require all the previously described treatments in addition to bone marrow transplant, if possible.

600-800 Rads - Those exposed to this dosage of radiation will experience the symptoms listed for all the previous exposure rates, but in addition the circulatory system and parts of the central nervous system will quickly begin to malfunction. All those exposed to this level of radiation will begin to vomit soon after exposure due to the major damage that radiation has inflicted upon the epithelial cells of the gastrointestinal tract. Unless medicated, this gastric distress can continue for several days or until death. The typical symptoms here are vertigo, headache, nausea, cramps, vomiting and uncontrolled diarrhea which eventually becomes bloody. Medications can control some of the negative symptoms and make the patient more comfortable but at this level of exposure, it won't change the inevitable outcome.

About 75 to 99 percent of the people exposed to this level of radiation will face death quickly. Bone marrow transplants and advanced life support will most likely be necessary for gaining any survivors from this group. Consequently the period of convalescence or recovery will be years. Bone marrow transplants and advanced life support will probably not be available during a nuclear war. Consequently, treatment would mainly be emotional support because of the probability of death.

800-5,000 plus Rads - Rapid deterioration. Vomiting, falling blood cell count, diarrhea, fatigue, internal bleeding, internal organ failure, nervous system collapse, heart failure, coma and death. 100 percent death rate within hours. Treatment is purely supportive. Humans cannot survive this level of massive radiation exposure.

5,000 plus Rads - Cells of the central nervous system and the tissues of the heart are affected. This produces disorientation, shock, seizures and neurological problems. Death results within a few days due to the collapse of the central

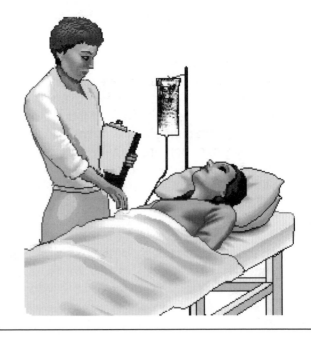

88

nervous and the cardiovascular systems.

Selective Uptake

Selective uptake is an important concept in nutritional protection against radiation. It means that sufficient levels of proper nutrients can effectively block the uptake of many types of radioactive pollution.

1. Potassium Iodide

Iodine is essential for the production of the key metabolism regulating hormone (thyroxin) in the thyroid gland. Radioactive forms of iodine are contained in fallout and are especially dangerous because they are taken up by the thyroid and damage it. About 100 milligrams of potassium iodide per day will block much of the uptake of radioactive iodine and should be taken for about 100 days after exposure to fallout. Potassium iodide supplements should be your first line of defense against radiation damage.

Thyroxin supplements are also recommended by some as an even more effective way to block the uptake of radioactive iodine. The thyroxin level sets your basic metabolic rate and it is very important that this be correct for proper health and well being. If you have plenty of energy and your weight is about right, your thyroxin level may be OK. If in doubt, it is inexpensive to have measured the next time you have a checkup.

2. Calcium

Calcium blocks the uptake of a common constituent of fallout, strontium-90. Calcium also has a calming effect. Good sources of calcium include whole grains, most green vegetables, soy products, many sea weeds, some nuts, fish and milk. Milk is tricky because it is high in phosphorous, which blocks calcium uptake, and is low in magnesium which should be taken in balance with cal-cium. Good absorbable calcium supplements which are balanced in magnesium are excellent to use on a daily basis and essential to store.

3. Iron

Iron blocks the uptake of plutonium from fallout. All grains are good sources and it should be noted that millet and barley are especially good. Meat is a rich source while beans, fish and some nuts are good.

4. Potassium

Potassium reduces the uptake of cesium-137 and helps balance the body. Good sources are vegetables, beans and sea weeds. Potassium tends to leak into cooking water so the water should be consumed.

5. Zinc

Zinc reduces the uptake of radioactive zinc-65. It is good for mental equilibrium, fighting viruses and as a general immune stimulant. Whole grains, green vegetables, sea weed and some nuts are good sources. For some reason the most effective way to take zinc supplements is in lozenges which are dissolved in the mouth. Zinc is very useful as an immune stimulant if you have a virus infection.

6. Vitamin B-12

This vitamin suppresses the uptake of cobalt-60 which is a very dangerous radioactive pollutant. B-12 is helpful to oppose anemia and to mellow out the nervous system. All B-vitamins are essential for good health. A supplement should be stored for use on a daily basis and should be increased in times of stress.

The Decay of Radioactivity

There are many popular myths propagated by those advocating unilateral disarmament and, to a certain extent, the media regarding the effects of nuclear war. These myths include ideas like: "In the event of a nuclear war, the earth will be uninhabitable for thousands of years" or "There will be no use trying to survive a nuclear war because when you come out of your shelter, the earth will be totally devastated". These statements are untrue. These false concepts are intentional distortions, fabricated to elicit an emotional response from uninformed people.

First of all, the radioactive fallout from the detonation of a nuclear weapon decays extremely fast. Within two weeks (14 days) after the detonation of a nuclear weapon, occupants of shelters could start going outside for an increasing number of hours each day. Generally speaking, 28 days after a nuclear detonation, radiation levels in the area affected by that detonation would have decayed enough that shelter occupants could come out of the shelter permanently. The exception would be areas that are downwind from targets which were attacked by multiple weapons and or multiple attacks over a period of time. Consequently these areas will receive heavier concentrations of fallout and the fallout from the detonation of subsequent or secondary attacks will produce an overlapping radiation decay rate. The only foolproof way to tell when it is safe to go outside your shelter is to take a reading with a reliable survey meter.

Secondly, the physical destruction caused by the detonation of a nuclear weapon is limited to a relatively small radius. For instance, the detonation of a one-megaton nuclear weapon at a surface burst altitude would cause fires and physical destruction for a radius of up to 5 or 6 miles. Anyone living seven to ten miles away wouldn't be affected by the heat, blast winds and overpressure generated by the nuclear detonation.

Where this myth gets its origin is nuclear power plants. Nuclear power plants use a different type of radioactive material which can take a thousand plus years to decay. Chernobyl was a real life environmental nightmare and the vast areas contaminated will remain so for many generations to come. This may be a genuine concern if you happen to live downwind of a nuclear power plant. In the event of an accident or melt down, radioactive isotopes would be carried by the wind and deposited over a wide area. These isotopes have a very long half life, thousands of years, whereas the isotopes found in the fallout from a nuclear weapon detonation decay rapidly.

Chapter 9
Shelter Types

When beginning the task of designing an underground shelter, the prospective builder is faced with a number of different options. The builder must decide first which particular type of shelter envelope system or structural shell is the best suited to his needs. Once this important decision is made, the rest of the peripheral design considerations will start falling into place.

The structural shell options include new or used steel tanks, steel culvert, steel quonset, molded fiberglass, conventional steel reinforced concrete, thin shell concrete domes and concrete arch systems, basement shelters and expedient shelters. Each one of these shelter envelope systems has its pros and cons. No one system is going to be the best system in all environments, under all circumstances and considering all budgets. In fact, a shelter can contain combinations of a number of these options. The following section will examine the most basic shelter

USED FUEL TANK

types with the intent of educating the reader so he can make his own decision.

Fuel Tank Shelter

Advantages

If you are an accomplished welder, (or have someone in you shelter group who is) a steel tank shelter may be your best option. The advantage of fuel tank shelters is that used steel fuel tanks are relatively inexpensive to purchase. They can, to a limited extent, be internally outfitted in a shop before actual burial. Also, steel tanks can provide E.M.P. (electromagnetic pulse), shielding. Steel tanks are generally watertight and can be used in areas where high water tables exist. It is much easier to weld

BASIC TANK SHELTER

WELDING - TANK SHELTER

and join together steel tanks than culvert.

If you are not an experienced welder and you intend to build a tank shelter, I suggest you research and purchase a good wire feed welder. Wire feed welders are much easier for novices to work with than are arc welders. A new Hobart wire feed welder costs about $1,200 new.

Disadvantages

There are certain disadvantages of steel fuel tank shelters to consider. They are a confining space, generally limited to 10'6" diameter and require a floor system. Steel tanks have very little integral strength and are dependent on earth-arching to remain struc-

turally intact. Backfilling has to be done very carefully with optimum compaction. **Please note that cutting into a used fuel tank is an extremely dangerous operation.** (Full precautions are given in the next chapter.) It can also be very difficult to clean the inside of a used diesel tank. Used fuel tanks can be rusted through in spots. New tanks are quite expensive.

Backfill & Compaction

A large shelter project, which I am personally involved in, successfully buried eleven steel tanks. I have friends who built tank shelter complexes which suffered severe deflection and shape distortion due to settling. This is usually a result of poor compaction during backfilling. The backfilling of steel tank shelters has to be done with optimum compaction of carefully selected or screened fill. Also, compaction cannot be accomplished when temperatures are freezing. When air temperatures are freezing, the application of pressure on the fill material does not produce compaction but instead produces ice crystals in the soil. This may produce the false

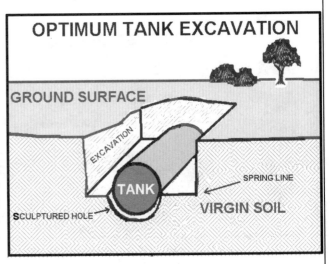

OPTIMUM TANK EXCAVATION

GROUND SURFACE

EXCAVATION

SCULPTURED HOLE

TANK

SPRING LINE

VIRGIN SOIL

appearance of compaction, but when the spring thaw comes, further settling and deflection of the tank or culvert will result.

Steel tanks can also suffer severe deflection and shape distortion if someone drives a heavy loader, backhoe or dozer over, or too close to, a partially buried tank or culvert.

Any rock bigger than a softball, that is buried within two feet of the tank, may eventually be forced by settling pressure directly against the tank and produce a deforming dent. Also large rocks, chunks of concrete, wood, or other debris buried next to the shelter when it is backfilled will cause differences in ground structure compression. This is a specific concern if the shelter is a blast shelter which is designed to withstand extreme overpressure.

The ideal way to control compaction and prevent deforming is to take some special steps during excavation. The hole for the tank should be only dug down to the spring line or equator point of the tank. At this point the remainder of the hole should be shaped to accommodate the rounded bottom of the fuel tank. The tank should then be placed in the hole and 3/4 inch minus gravel can be vibrated down around the sides into any unfilled voids. Instead of 3/4" minus gravel, a mud mixture can be slurried down around the underside of the tank. Such a slurry should be high in sand content and low in clay and the ground underneath needs to have good percolation. The danger with such a mud slurry is that the tank can float up out of the hole.

If you place a tank in a flat-bottomed hole, which has been dug to its total intended depth, it will be difficult or impossible to get adequate compaction on the underside of the tank.

If you are intending on internally outfitting a fuel tank before it is installed in the ground and backfilled be aware that the tanks tend to flatten out on the bottom once the backfill

settles. This is more pronounced with 12 ft. diameter tanks. The ramifications are that any braces used to support the floor or partition walls will bend and end up not being at their originally intended level or position. Also, pre-framed doors will end up out of square and not closing. The curved side walls of the tank, due to overhead pressure, will tend to expand away from pre-fitted partition walls leaving a gap. This same settling effect will put pressure on top of any pre-fitted partition walls and bow out the surface of the wall. Using great care during the backfill process will minimize these problems but not eliminate them.

Cleaning

As I previously mentioned, cleaning the inside of a used diesel tank can be a big and messy job. Used diesel tanks tend to have a fairly thick coating of hard, pasty residue on the inside walls and puddles of liquid diesel on the floor. We resorted to scraping the inside walls with a flat shovel to get this accumulated coating off. The liquid diesel can be easily soaked up with sawdust and then shoveled into old feed sacks for disposal. Make sure you dispose of the material you clean out of the tank in an environmentally safe manner. Used gas tanks are generally fairly clean on the inside, but have a greater tendency to be rusted.

Arthur Robinson is the preeminent authority on shelters made from fuel tanks. He is the author of "The Fighting Chance" Newsletter. Plans can be purchased from him for fuel tank shelters. Contact the Oregon Institute of Science and Medicine, P.O. Box 1279, Cave Junction, Oregon 97523.

Sharon Packer, with the Civil Defense Volunteers of Utah has a 160 page booklet on aspects of shelter building which includes an excellent section on steel tank and culvert shelters. This booklet is $20 post paid and can be obtained by contacting her at 1950 Forest Creek Lane, Salt Lake City, UT 84121.

Shipping Containers

I have been questioned about the feasibility of using shipping containers for a shelter. I'm sure you have seen these rectangular cargo containers single and double stacked on railroad flat cars moving down the railroad tracks from time to time. These are referred to as sea containers. They are used for transporting freight on ships. These containers are many times stacked seven high on the decks of ships loaded with cargo. If you live near a major seaport like Seattle, Long Beach, Los Angeles, New York, Boston, etc. you can find these containers which have been retired from service for various reasons.

The containers themselves are 8 ft. high and 8 ft. wide and come in lengths of either 30 or 40 feet. If they are in good condition, they are watertight. Even though sea containers can be stacked seven containers high, one on top of the other, they have very little overall strength. All of a sea container's strength is in its heavily reinforced corners.

The author has been involved in burying numerous sea containers and witnessed a number of other people attempting to do the same. The key to success when burying sea containers is reinforcing.

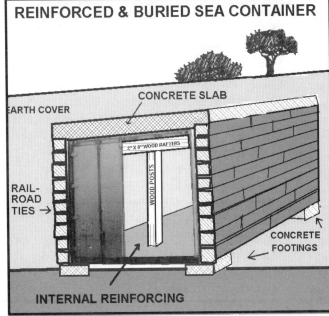

REINFORCED & BURIED SEA CONTAINER

CONCRETE SLAB

EARTH COVER

2" X 8" WOOD RAFTERS

WOOD POSTS

RAIL-ROAD TIES →

CONCRETE FOOTINGS

INTERNAL REINFORCING

Experience has shown that unreinforced sea containers buckle and collapse if dirt is loaded on top of them. I have also seen sides buckle inward from the weight of backfilled dirt pressing sideways against the walls.

The sea containers we buried did not buckle or collapse because we went to considerable effort to reinforce them. First, we reinforced the containers from the inside. This was done by installing vertical posts which supported horizontal cross pieces. This is similar to the way conventional mining operations shore up a tunnel. Then we built railroad tie walls, in a Lincoln log fashion, all the way around the exterior walls of the sea container and snug to them. These railroad tie walls went from the bottom of the container wall up to the roof. Our next step was to pour a reinforced concrete slab over the top of the container's roof. This slab overlapped onto the railroad ties which formed a perimeter around the edges of the sea container's roof. We then waterproofed the slab and covered it over with dirt.

In our case, we are using the containers, not for shelters, but for underground storage. One other thing we did which added to our success was to place the four corners of the containers on concrete footings. The logic behind this is that without footings the containers would settle once backfilled. As a result they would tend to tweak or twist slightly out of shape which would make the doors not open or close properly.

Buyer be wary. Personally inspect any container you are thinking of purchasing. Make sure they are not rusted out and make sure the doors work.

Culvert Shelters

Advantages

Shelters made from steel culvert can usually be installed and backfilled quicker than with concrete shelters. In other words, the excavation hole is open for a shorter time. This can be an advantage where confidentiality is a concern. Culvert is stronger than steel fuel tanks, and if the entry ways and penetrations are properly configured, culvert will shield against electromagnetic pulse. It is a somewhat flexible system which can be extended with additional lengths as more capacity is needed.

Disadvantages

Culvert shelters are a confining space, limited in width to 13' diameter. Culvert has minimal integral strength and is dependent on earth-arching. However, culvert is stronger than a steel fuel tank even though it is made of a thinner gauge steel. Culverts corrugated ribs greatly enhance its structural characteristics. Culvert, due to its spiral ribs and galvanized coating, is not as easy to weld and join together as steel tanks.

The initial cost of purchasing steel culvert is much greater than picking up a used fuel tank, but once you consider the cost and hazard of cutting into, cleaning, painting and recoating the outside of used tanks, new culvert ends up comparable in price.

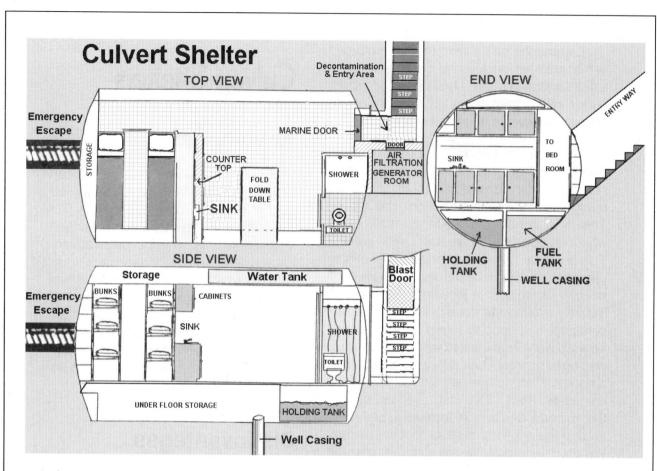

Culvert Shelter

TOP VIEW

Decontamination & Entry Area

STEP
STEP
STEP

Emergency Escape

STORAGE

COUNTER TOP

SINK

FOLD DOWN TABLE

MARINE DOOR

DOOR

SHOWER

AIR FILTRATION GENERATOR ROOM

TOILET

END VIEW

ENTRY WAY

SINK

TO BED ROOM

HOLDING TANK

FUEL TANK

WELL CASING

SIDE VIEW

Storage

Water Tank

Blast Door

Emergency Escape

BUNKS

BUNKS

CABINETS

SINK

SHOWER

TOILET

STEP
STEP
STEP
STEP

UNDER FLOOR STORAGE

HOLDING TANK

Well Casing

Culvert Shelter

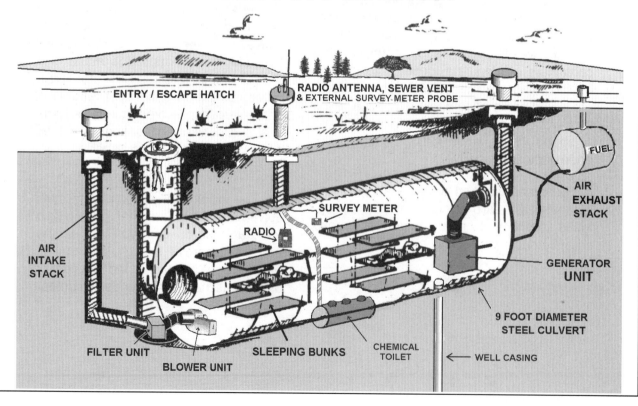

ENTRY / ESCAPE HATCH

RADIO ANTENNA, SEWER VENT & EXTERNAL SURVEY METER PROBE

FUEL

AIR EXHAUST STACK

SURVEY METER

RADIO

AIR INTAKE STACK

GENERATOR UNIT

9 FOOT DIAMETER STEEL CULVERT

FILTER UNIT

BLOWER UNIT

SLEEPING BUNKS

CHEMICAL TOILET

WELL CASING

Backfill

Rocks backfilled too close to the outside surface of a buried fuel tank produce large dents on the inside and hitting them with the bucket of a backhoe also produces large dents. This is not the case with culvert. None of the culvert shelters I have seen have shown any signs of point loading dents caused by rocks being backfilled too close to the outside surface of the culvert. Also, experience has shown that culvert does not buckle and dent when bumped by a backhoe bucket.

Backfilling has to be done very carefully with optimum compaction. There have been some serious instances of deflection as a result of the settling of the earth which was backfilled over the culvert. As we previously discussed concerning steel fuel tanks, compaction cannot be accomplished when temperatures are freezing. The ideal way to control compaction and prevent deforming is to take some special steps during excavation. Please see the previous discussion at the beginning of this chapter under Tank Shelters, Backfill and Compaction.

If you place a culvert in a flat-bottomed hole, which has been dug to its total intended depth, it will be difficult or impossible to get adequate compaction on the underside of the culvert.

Even with the best compaction and backfill, culvert manufactures say you should expect about 10% deflection. This means that any internal doorways and wall structures which are fitted tightly to the inside curve of the culvert will eventually become compressed. Cabinets fitted to the original inside curved surface of the culvert will displace themselves from the walls when the culvert squats due to settling and changes its internal curved shape. The punch line is either wait a year or two after backfill and subsequent settling to internally outfit your culvert shelter, (I do not recommend such a delay), internally outfit the shelter without wall and ceiling attachments which will be effected by the settling or build a steel reinforced concrete shelter.

Steel Quonset Shelters

Advantages

Steel quonset shelters allow for larger spans than steel culvert and have more usable internal space. The military has been using these for munitions bunkers for many years.

STEEL QUONSET

ABOVE GROUND QUONSET WITH EARTH COVER

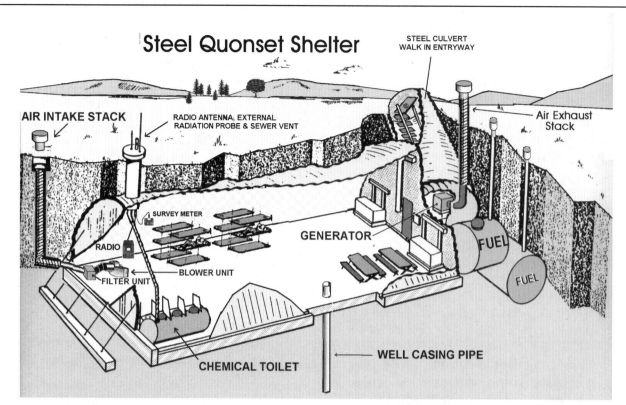

Steel Quonset Shelter

STEEL CULVERT
WALK IN ENTRYWAY

AIR INTAKE STACK

RADIO ANTENNA, EXTERNAL
RADIATION PROBE & SEWER VENT

Air Exhaust
Stack

SURVEY METER

GENERATOR

RADIO

FUEL

FILTER UNIT

BLOWER UNIT

FUEL

CHEMICAL TOILET

WELL CASING PIPE

Disadvantages

Steel quonset shelters have very little integral strength and are dependent on earth-arching. Backfilling steel quonset has to be done very carefully with optimum compaction. End walls are made of steel reinforced concrete or masonry block. In the event of an overpressure, there is a possibility of point loading on the structure where the flexing steel meets the rigid wall. Such point loading can result in buckling and structural failure.

In the past the military has had considerable water leaking problems with its steel quonset ammunition bunkers. The problem was so bad that in instances the Army made P.V.C. tents inside the quonsets to keep the leaking moisture off the munitions. This problem has now been resolved by spraying the exterior of the quonset with polyurethane foam. Steel quonsets cannot be used in areas with high water tables.

Deep Core is a steel quonset structure made by Syro Steel, For more information

contact Syro Steel Company, P.O. Box 99, Centerville, UT 84014. (800) 772-7976 or (801) 292-4461. A similar product is made by Arch Technology Corporation, P.O. Box 6, Plato Center, IL 60170. (312) 464-5656

Pre-made Fiberglass

Advantages

The great feature about this system is that it comes completely internally outfitted from the factory. All of the life support systems are premade at the factory. This includes air filtration, water storage, entry way, sewage system, power and observation tower. They can be used in areas where a high water table exists, and often can be ordered, delivered and installed faster than building a comparable shelter from scratch. The advantages to the single family shelter are that less people know about the

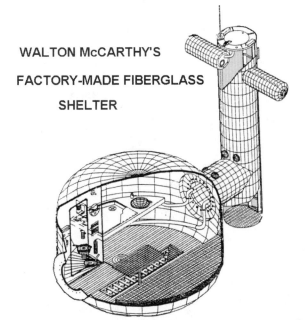

WALTON McCARTHY'S
FACTORY-MADE FIBERGLASS
SHELTER

shelter and ideally the project gets done faster. Another advantage to the single family approach is that management and decision making are more simplified.

Disadvantages

This system is expensive. It is a confining space and it is generally not conducive to community application. Molded fiberglass shelters have very little integral strength and are dependent on earth-arching. Like steel tanks and culverts, backfilling has to be done very carefully with optimum compaction. It does not shield E.M.P.

Like any other one room shelter, this system has minimal personal privacy. These shelters have only one entry way. The advantages of two entry ways are covered in the next chapter.

Walton McCarthy is the owner of Radius Defense Inc. which makes several different sized models of fiberglass shelters. He is also the author of several books on this subject including *Sheltered, Nuclear Weapon Fundamentals* and *Blast Shelter Design Standards.* This book contains some information for the

shelter builder. He can be contacted at 222 Blakes Hill Road, Northwood, NH 03261, (603) 942-5040.

Conventional Steel Reinforced Concrete Shelters

This type of shelter is basically a concrete box. It is formed up with standard basement wall forms. The roof is made of a thick steel reinforced slab.

Advantages

Most local masonry contractors are familiar with the concepts involved in building a below-ground conventional steel reinforced concrete structure. This system does not derive its strength from earth-arching and the backfilling can be done without any particular care.

Disadvantages

Conventional steel reinforced concrete shelters require massive steel reinforced concrete slabs to span ceilings and support the earth cover on top of the shelter. Concrete slab ceilings can only free-span limited distances without the addition of vertical supports. Concrete shelters do not provide E.M.P. shielding and are not usable in areas with high water tables.

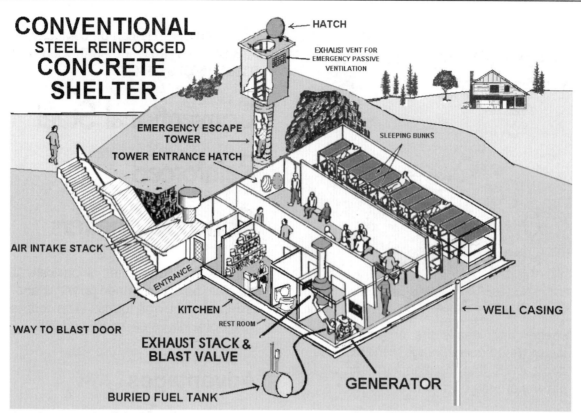

CONVENTIONAL STEEL REINFORCED **CONCRETE SHELTER**

HATCH

EXHAUST VENT FOR EMERGENCY PASSIVE VENTILATION

EMERGENCY ESCAPE TOWER

TOWER ENTRANCE HATCH

SLEEPING BUNKS

AIR INTAKE STACK

ENTRANCE

WAY TO BLAST DOOR

KITCHEN

REST ROOM

EXHAUST STACK & BLAST VALVE

BURIED FUEL TANK

GENERATOR

WELL CASING

Thin Shell Concrete Technology

This category of shelters includes steel reinforced concrete domes, barrel vaults and donut shaped structures. All of these structures are different shapes of the same technology. The author has considerable experience with this type of shelter system. I organized a project in which our group built a barrel vault which was 32 feet wide and 122 feet long with three floor levels. As far as we know, this is the largest underground concrete dome-type shelter ever built.

Advantages

The advantages of thin shell concrete domes and arches are that they facilitate large

structures with wide spans. This system capitalizes on the natural geometric strength of an arch. The resulting structure is a very strong shell which is not dependent on earth-arching for its strength. Thus, no particular care is needed when backfilling these shelters.

Disadvantages

The disadvantages of this particular system are that it does not shield E.M.P. and it is not usable for areas with high water tables. Concrete is applied with shotcrete equipment. Shotcreting is expensive and takes substantial amounts of manpower to accomplish. The kits for the shell of these structures are fairly expensive.

The large concrete barrel vault shelter shown in the pictures was built by the author and his friends. The shell is made from a kit supplied by E.S.I., Earth Sys-

FRAMEWORK FOR CONCRETE DOME

FABRIC COVER BEING APPLIED

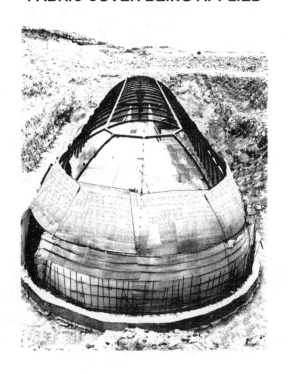

tems, Inc., P.O. Box 3270, Durango, Colorado, 81301.

There are at least two other companies in the United States which make steel reinforced concrete dome kits. Both of these companies, Monolithic and Terra Dome, make a dome kit which utilizes an inflatable form.

Terra-Dome

The Terra Dome is another steel reinforced concrete dome-like system. The Terra-Dome system uses aluminum vertical wall forms and a fiberglass roof form. Waterproofing for this system is applied to the outer concrete surface.

101

FABRIC AND REINFORCING STEEL BAR COMPLETE

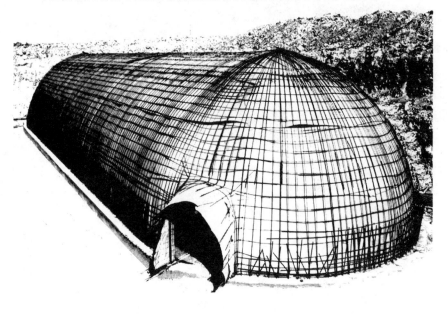

Shotcreting

Finishing Touches

Terra-Dome Modules During Construction

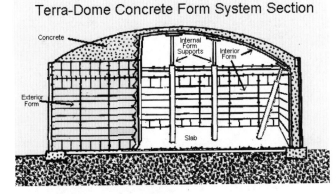

Vent Pipes

Entrance Way

Footing

Basic Terra-Dome Unit

Terra-Dome Concrete Form System Section

Concrete

Internal Form Supports

Interior Form

Exterior Form

Slab

Advantages

This system has the same advantages as other thin-shell concrete domes. There are several other advantages. The shell form system can be reused (this is not necessarily an advantage because it belongs to and goes back to Terra-Dome.) Terra-Dome is constructed on-site by Terra-Dome employees who know what they are doing and can complete the shell of the structure in three days, once all site preparation is completed. This is the big advantage to Terra-Dome, if you have the money. Terra-Dome is potentially the quickest one you can get built. This is a definite option for those who do not want to do the work themselves.

Disadvantages

One minor disadvantage is that this system, like the E.S.I. system, needs to be waterproofed. The Terra Dome system is not insulated, which is usually important for underground shelters. Terra-Dome is not a do-it-yourself kit system and thus is relatively expensive. This system is not a pure dome shape and thus does not have the same geometric strength advantages. Thus the Terra-Dome would not be usable for a blast type shelter without structural reengineering and reinforcing.

For Terra-Dome information and prices write to Ben Thorhal, Terra-Dome, 8908 S. Shrout Rd., Grain Valley, Missouri 64029, or call 1(800) 481-3663.

Monolithic System

The Monolithic system involves inflating a balloon form and then spraying urethane foam on the inside surface of this form. Steel rebar is then attached to the resulting inside surface and concrete is sprayed over the rebar, again from the inside of the balloon form.

The resulting product is a steel reinforced dome, with foam insulation and the original form balloon becomes a fixed waterproofing barrier.

Advantages

This system has the same advantages as other thin-shell concrete domes. Another advantage to the

Monolithic system is that it has built-in water-proofing.

Disadvantages

The disadvantages of this particular system are that the kit is relatively expensive and the expensive balloon form can't be reused for another project. Another potential disadvantage of this system is the foam insulation. The foam insulation is probably an asset for a sub-earth home or an above ground structure. If, on the other hand you are building an underground shelter, the foam insulation will restrict the conduction of heat from the inside of the shelter to the surrounding earth cover. Heat build-up is a potentially serious problem in crowded underground shelters as we will discuss in the Chapter on Air Supply .

Prices and information on the Monolithic Dome system can be obtained by writing to Monolithic, P.O. Box 479, Italy, TX 76651 or calling (214) 483-6662.

Home Grown Versions

The final option in steel reinforced concrete domes and barrel vaults are the home-grown versions. This option includes systems which involve making your own form and thus eliminating the expensive kit cost.

Grenier - Barbier System

The first system in the home brewed category is what I call the Grenier - Barbier system. Mark Grenier is a close friend and a master builder. He built a donut arch system designed by Marcel Barbier, who is a well known engineer in the civil defense field. The form was made of redwood boards which were steamed and bent over a wooden rib system. The outside surface had rebar tied

HOME MADE BARREL VAULT DOWN UNDER
(AUSTRALIA)

over it and concrete applied. The resulting structure was then waterproofed and backfilled.

Advantages

This system has the same advantages as other thin-shell concrete domes. Other advantages to this system are that there are no kit costs and the interior surface is beautifully finished redwood. For a larger shelter, the donut shape has optimum geometric strength.

Disadvantages

The disadvantage with this system is that the process of steaming and bending redwood boards is very labor intensive and the form is not reusable.

Newhouse System

Another home brewed system is what I call the Newhouse system. This system was devised by a friend of mine named Dave Newhouse. Dave built tract homes for 30 or so years before he started building underground shelters.

BASEMENT / FOUNDATION FOR GRENIER - BARBIER SHELTER

REDWOOD BOARDS ON GRENIER - BARBIER SHELTER

INSIDE THE DONUT

After building concrete dome systems from expensive E.S.I. kits, Dave and I started contemplating how we would do it if we had to do it over again. The whole idea was to get away from the expensive kit and the labor intensive shotcreting. Dave came up with a new system which involved making your own form system with plywood and two-by-fours.

The forms for this system can be reused over and over again and shotcrete equipment is not needed for the Newhouse system. The concrete can be applied to the structure by a concrete pump truck. This reduces the labor by about two thirds, and the pump truck costs a fraction of what shotcrete application costs. As I mentioned earlier, for an E.S.I. system you need to apply the concrete with a shotcrete equipment. Shotcreting is expensive and takes massive amounts of manpower to accomplish.

Disadvantages

The disadvantages of the Newhouse system are that it does not produce a finished interior surface, and the home brewed systems in general are not as easily configured for arches larger than 33 feet in diameter.

For Newhouse Shelter plans contact Yellowstone River Publishing, P.O. Box 206, Emigrant, MT 59027. (800) 327-7656

Last but not least, I know some people in Australia who made their own steel frame kit and successfully built a concrete dome very similar to the E.S.I. system. This may be an option if you have

familiarity with and access to the heavy steel bending industry.

Basement Shelter

Probably most civilian shelters are in basements.

Advantages

Basement-type shelters can be more easily concealed, they make use of an existing structure and in some cases, the sewer, water and gas. A basement shelter is probably the most inexpensive alternative. If you already have a basement and are not in close proximity to a target, this may be your best option. It is much easier to make effective dual-use out of a basement shelter such as doubling as a family room, rec room or storage. Your basement shelter is so close that it is much easier to get yourself and all necessary supplies in there on short notice.

Disadvantages

Basement shelters do not have blast protection. In the event of overpressure, the house may be reduced to rubble, covering up or destroying the shelter. In the event that the shelter didn't have a dedicated subterranean exit leading away from the house, the occupants would be trapped. If the blast-generated rubble ignites, high temperatures will be produced right on top of the shelter. Heavier than air gases like carbon monoxide would penetrate non-airtight shel-

NEWHOUSE SYSTEM - SECTION DRAWING
SHOWING PRE-FAB FORMS IN PLACE

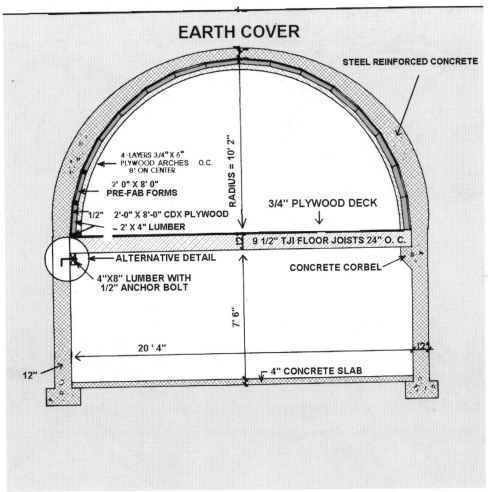

EARTH COVER

STEEL REINFORCED CONCRETE

4-LAYERS 3/4" X 6"
PLYWOOD ARCHES   O.C.
8' ON CENTER

2' 0" X 8' 0"
PRE-FAB FORMS

1/2"   2'-0" X 8'-0" CDX PLYWOOD

2" X 4" LUMBER

RADIUS = 10' 2"

3/4" PLYWOOD DECK

9 1/2" TJI FLOOR JOISTS 24" O. C.

ALTERNATIVE DETAIL

4"X8" LUMBER WITH
1/2" ANCHOR BOLT

CONCRETE CORBEL

7' 6"

20 ' 4"

12"

4" CONCRETE SLAB

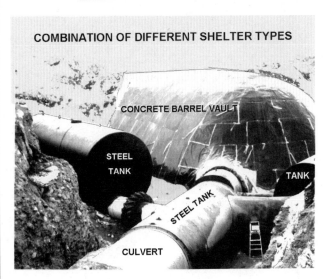

COMBINATION OF DIFFERENT SHELTER TYPES

CONCRETE BARREL VAULT

STEEL TANK

TANK

STEEL TANK

CULVERT

ters, killing the occupants. Carbon monoxide poisoning killed most of the people who died in basement shelters during World War II. Home basement shelters are generally not applicable for community shelters because they are usually small spaces. Basement shelters don't offer the same protection as buried shelters when it comes to earthquakes and high winds.

NEWHOUSE SYSTEM
CONCRETE FORM DETAIL

PLYWOOD ARCH

COMPRISED OF 4 LAYERS OF 3/4" PLYWOOD

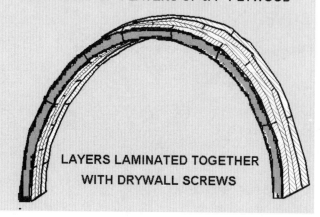

LAYERS LAMINATED TOGETHER
WITH DRYWALL SCREWS

PLYWOOD ARCHES SET 8' APART

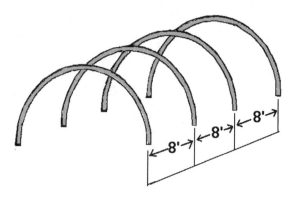

8' — 8' — 8'

2' X 8' FORM

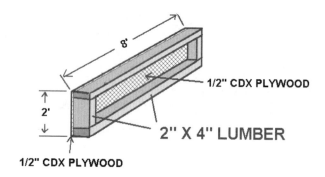

8'

2'

1/2" CDX PLYWOOD

2" X 4" LUMBER

1/2" CDX PLYWOOD

DETAIL OF FORM

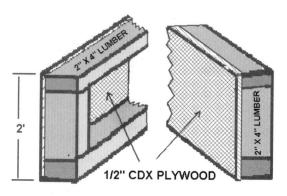

2" X 4" LUMBER

2'

1/2" CDX PLYWOOD

2" X 4" LUMBER

PLACING THE FORMS BETWEEN THE ARCHES

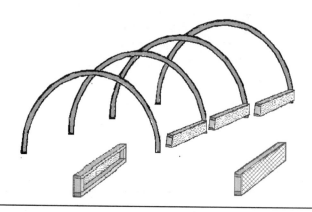

BARREL VAULT FORM SYSTEM

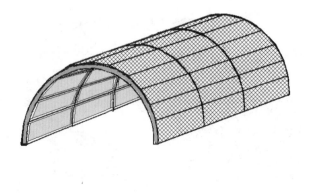

108

TYPICAL BASEMENT SHELTER

Expedient Shelters

Nuclear War Survival Skills, by Cresson Kearny is the authoritative book on this subject. This book covers all aspects of improvising a fallout and light blast shelter. If you are intending to utilize an expedient fallout shelter, his book is a necessity. It's a how-to book and it contains all the facts you need to know to protect yourself from the effects of a nuclear weapon. To obtain this book contact the Oregon Institute of Science and Medicine, P.O. Box 1279, Cave Junction, Oregon 97523.

Advantages

The advantage of expedient shelters is they are dirt cheap! They can work for fallout protection if built correctly and no particular care is needed in backfilling. In all fairness, they are certainly better than nothing.

Disadvantages

The disadvantages are many. Expedient shelters will work only if the occupants get enough advance warning to

have time to construct them. If someone constructs an expedient shelter in advance of crisis, this problem would be solved. An additional concern is that expedient shelters become victims of moisture and rodent damage. Most expedient shelters tend to be small and confining and they usually have limited, power, light, and sanitation. Any extended stay in an expedient shelter would tend to be a very trying experience at best, but discomfort is preferable to death.

Public Shelters

What about public shelters? You know those designated public civil defense shelters, usually located in basements of schools, banks, etc. Unfortunately, these so-called shelters will tend to be a place where the average American family will find lots of death and misery for company. Very few public shelters have any emergency stores of food, water, medical supplies or back-up power. None have air filtration capabilities and most will not have working sanitation facilities. When the state civil defense agencies started surveying buildings for use as potential public shelters, they found that there weren't enough structures to accommodate all the people who needed to be housed. The solution was to lower the minimum protection factor so more buildings would qualify.

The United States Civil Defense program, for all intents and purposes, exists only on paper. The bulk of the program involves the mass relocation of the metropolitan populations to rural areas. The plan has obvious deficiencies. First of all a one-week advance warning is needed to implement a relocation, secondly people would be moved to areas with inadequate shelter, food, and sanitation facilities and finally most of these relocation areas would be subject to contamination by airborne radioactive fallout.

Shelters for the Elite

The U.S. government has spent billions on secret shelters for important federal, state and corporate officials. Mount Weather is a virtual underground city located near Bluemount Virginia, 46 miles from Washington D.C. and is officially designated the Western Virginia **Office of Controlled Conflict Operations.** The Mount Weather staff includes a working duplicate of the Executive Branch of the Federal government and is the central command center for FEMA. It is interesting to note that a privately owned profit corporation, the Federal Reserve, also has an office in Mount Weather. In the event of war, the declaration of martial law or other national emergencies (potentially including serious financial or political crises), the President, the cabinet and the rest of the Executive Branch would be relocated to Mount Weather. Believe it or not, Congress has little knowledge of this facility and absolutely no budgetary oversight.

In all fairness we shouldn't be too hard on our elite bureaucrats for trying to save their own necks. After all, the Soviets built blast hardened facilities for the KGB, their military command and top party officials; (they obviously are not expecting a nuclear war). Most recently this includes the huge Yamantau complex in the Southern Ural Mountains near Beloretsk. This complex is purported to be as big as the area inside the Washington beltway. Yamantau is not a relic left over from the cold war, but was constructed under Yeltsin's direction. Not bad for a bankrupt country that needs aid from the U.S. to pay for its own compliance with disarmament treaties. The one thing we can say for the Russian elite is they seem to value their civilian cattle a little more than their western counterparts. The Russian leadership built shelters for 70 percent of Russia's civilian population.

Public Shelter

Fallout Shelter

Small Versus Large Shelters

Shelter size and capacity should be carefully considered.

Smaller shelters are easier to run in terms of the management and the decision-making process, because fewer people are involved. It is easier to build a smaller shelter secretly while it is virtually impossible to build a larger shelter without people knowing about it. The down side of smaller shelters is that they are usually minimally equipped, are more expensive per person and are potentially harder to protect. Doing one's own thing is a limited option and the overall potential for survival is greatly reduced. Confinement in a small shelter over a long period of time with limited interaction options can produce adverse effects.

Larger shelters provide a greater resource base in terms of personal skills and able-bodied people. There are defi-

CROSS SECTION OF ACTUAL SHELTER

3 LEVEL CONCRETE BARREL VAULT

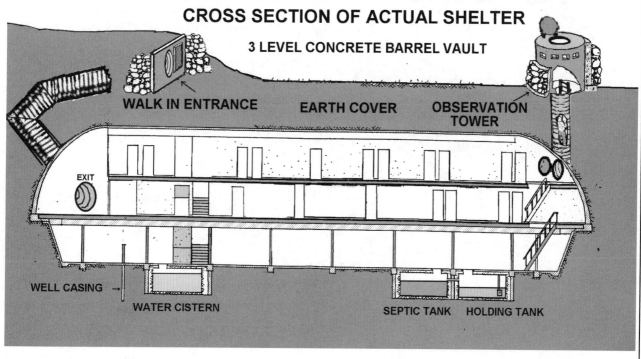

WALK IN ENTRANCE   EARTH COVER   OBSERVATION TOWER

EXIT

WELL CASING →

WATER CISTERN

SEPTIC TANK   HOLDING TANK

nite strengths in union and in numbers. Building larger shelters with more occupants makes more effective use of financial resources and allows for better systems and facilities. Larger shelters, generally speaking, provide more potential for privacy and are potentially easier to protect. On the down side, larger shelters may take longer to build and the systems are more complex.

Sub-Earth Home With Dual Use Shelter

The final shelter option to be covered in this chapter is a sub-earth home which has a section of the living quarters shielded. On one of the following pages is a floor plan of an actual sub-earth home. In this house the master bedroom, master bath, pantry and a loft over the bedroom, are situated into the back of the house. This area is separated

CONCEPTUAL BARREL VAULT SHELTER

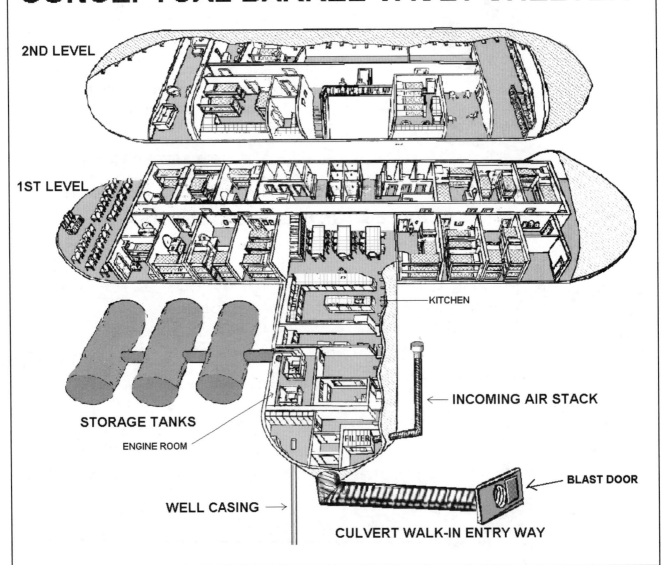

2ND LEVEL

1ST LEVEL

KITCHEN

INCOMING AIR STACK

STORAGE TANKS

ENGINE ROOM

FILTER

BLAST DOOR

WELL CASING →

CULVERT WALK-IN ENTRY WAY

Arches Up

Form Panels Being Attached To Arches

Steel Reinforced Bar In Place

Concrete On The ROOF

The Finished House

114

Sub-Earth Home With Dual Use Shelter Area

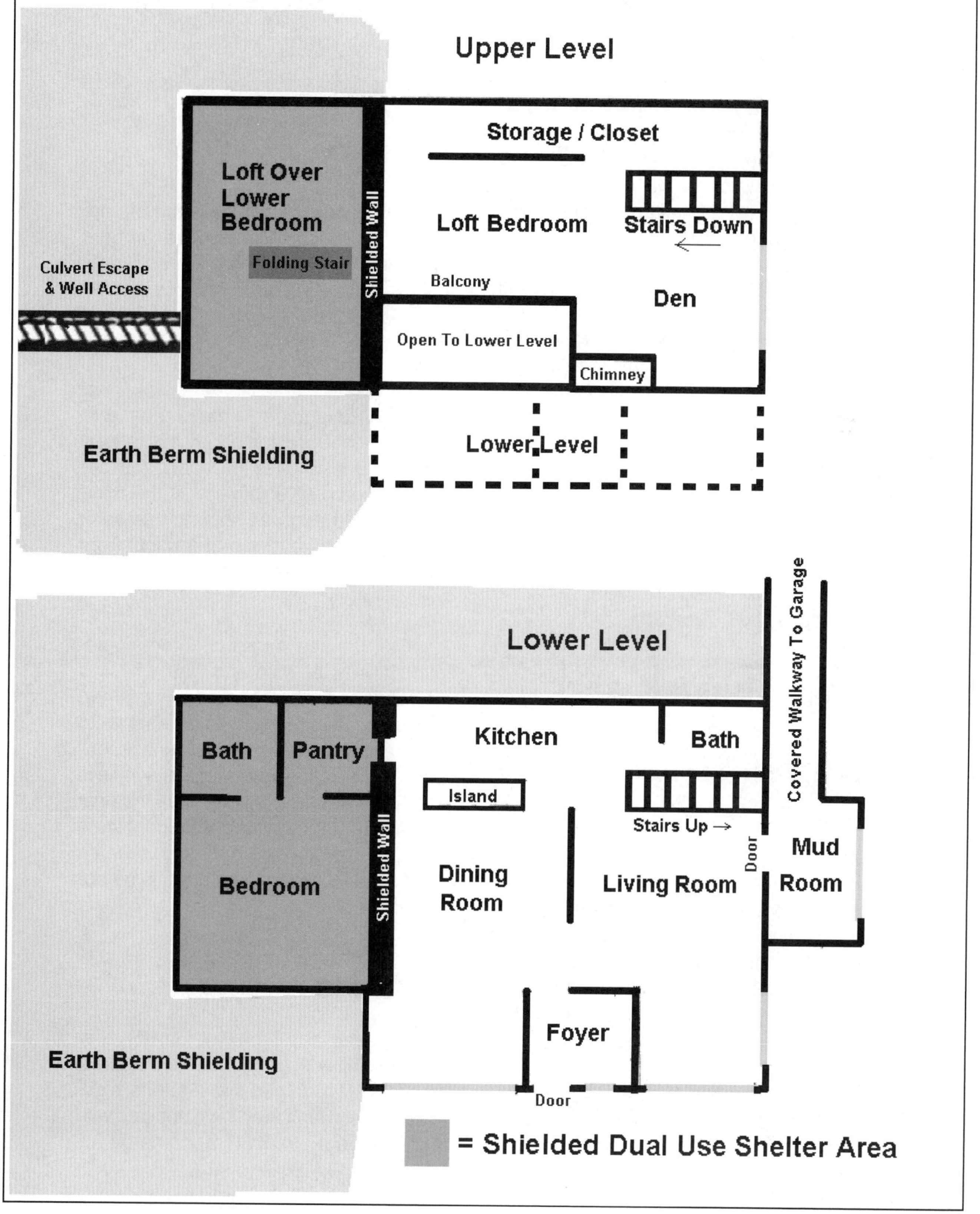

Upper Level

Storage / Closet

Loft Over Lower Bedroom

Folding Stair

Loft Bedroom

Shielded Wall

Stairs Down ←

Den

Balcony

Open To Lower Level

Chimney

Culvert Escape & Well Access

Earth Berm Shielding

Lower Level

Lower Level

Covered Walkway To Garage

Bath | **Pantry**

Kitchen

Bath

Island

Shielded Wall

Stairs Up →

Door

Mud Room

Bedroom

Dining Room

Living Room

Foyer

Door

Earth Berm Shielding

= Shielded Dual Use Shelter Area

from the rest of the house by a shielding wall and the roof and side walls are shielded with earth.

In the event of a disaster, the owners would quickly move into the shelter portion of the house and have a living space, bathroom and also the loft space for friends and relatives. The kitchen operations would move into the pantry, which has its own sink. The house also has a low voltage system with battery storage to provide power.

This particular house is built using the Newhouse shelter system. You can view the progressive steps involved in building a Newhouse type steel reinforced concrete barrel vault in the pictures shown on this and on the next page.

One excellent book on building sub-earth homes is *Passive Annual Heat Storage, Improving The Design of Earth Shelters*, by John Hait. This book and other literature can be obtained from The Rocky Mountain Research Center, PO Box 4694, Missoula, Montana 59806, phone (406) 728-5951.

Mines, Tunnels and Caverns

Mines

If you happen to be so fortunate to own a mining claim with a hard rock mine, it is very possible that you can modify it to become an effective shelter. The three big considerations are air supply, safety and moisture control.

Most heavily worked mines have air supply shafts. If the mine does not, it is possible, if access facilitates, to get a well driller to bore

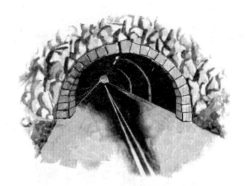

a shaft down to the tunnel from on top. Air supply ducts can also be brought back into the tunnel from the entrance, but it is better if you can draw your supply air from the entrance and use natural convective air flow and exhaust air vertically.

The advantage to using a mine or cavern is that you can drastically reduce building and excavation costs and secure unheard of shielding. If the mine tunnel is old and has shoring braces in it, you should have a professional from the mining industry inspect the mine to insure safety.

Old, unused mines can contain accumulations of methane gas which is colorless, odorless and deadly. When surveying the mine, it is advisable to bring along a methane gas detector. Another potential disadvantage of a mine is that any actual excavation of material can bring the scrutiny of the local state mine safety agency and some states may actually require that permits be obtained.

Tunnels

Railroad and highway tunnels offer the possibility of large scale emergency shelters. The Swiss have a long highway tunnel which they have outfitted with a blast door, ventilation, emergency power generation, sanitation and stored beds and supplies to accommodate 7,000 people. In most cases utilizing a highway or rail tunnel will be limited to a government organized project. The local rail company and the state highway department might not be too excited about you and your friends commandeering a tunnel, especially for a drill, but it is not inconceivable that a group of citizens could make use of a tunnel in an emergency if they did a little preplanning.

If you were to drive in a water truck, a trailer full of porta-potties, and a truck full of food from your local grocer or other sources, you would have all the basics for an expedient tunnel shelter. The only other essential ingredients would be a backhoe, sand bags (or feed sacks), and shovels. The backhoe would break up the compacted earth, and there shouldn't be any shortage of volunteers to man the shovels, and fill and stack sand bags. If an 8 to 10 foot high by at least 4 foot deep sand bag barrier was constructed on either end of the tunnel, you could obtain reasonable shielding against radioactive fallout. People can shovel a lot of dirt in a real short time when their lives depend on it.

The open space between the top of the sand bag barrier and the roof of the tunnel on either end would facilitate ventilation and minimal lighting during the day. Have everyone bring their sleeping bags and camping gear, and have a two to three week campout.

Caverns

There are many extensive complexes of underground caverns in various parts of the United States. Most of these are government owned and access is controlled. This fact would tend to prohibit modification and stockpiling of equipment. But if individuals were to pre-plan and be organized, they could probably make use of such caverns during a crisis. Of course each cavern would have to be evaluated for its natural ventilation capabilities and the availability of water.

To find out more about caverns in your area, contact the local cave exploring group. Get certified training from these people and then go tour the local caves.

A last minute shelter builder getting ready for a two week stay under the basement workbench.

(More advanced thought and planning will produce happier campers.)

Chapter 10
Shelter Design and Construction Considerations

A shelter project, like any other project, can seem overwhelming at the onset. A large, intimidating shelter project can be tamed by segmenting the project into smaller manageable units such as the shell, air system, power system, lighting system, water supply, etc. Deal with the project a segment at a time. And remember, what man can conceive and believe, man can achieve.

Space Requirements

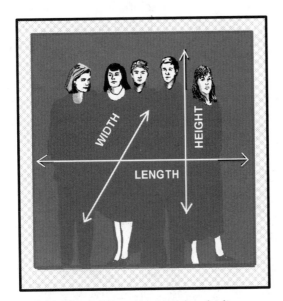

Space requirements for shelter occupants should be considered from both a basis of square footage per person and cubic feet per person. The U.S. Government publications recommend a minimum of net 10 square feet per person. Net square feet is floor space that can also include walls, bunks, storage and fixed equipment. Try living in 10 square feet for two weeks, especially with children. What if it ends up being 60 or 90 days? Some of the larger shelters I am aware of have up to 50 net square feet per person.

Design Specification Questions

The following are questions, which when answered, can stimulate creative thinking and bring to light considerations that the first time shelter builder may have overlooked.

1. How many people is the project going to accommodate?

2. How many family groups does the above number of people represent and what sizes are these groups?

3. Do you want private rooms for family groups?

4. Where generally is the shelter going to be located?

5. Do you want blast protection or do you want fallout protection?

6. How long do you want to be able to remain in the shelter without having to come out? 1 month, 2 months, 3 months, 9 months, or more?

7. How much of a food supply do you need to stockpile in or near the shelter? 1 month supply, 2 months, 6 months, 1 year, or more?

8. Do you want a separate clinic room in the facility?

9. Do you want to try to shield your sensitive electronic equipment from electromagnetic pulse?

10. Do you intend on having communication equipment, (i.e. shortwave, ham, etc.)

11. Do you intend on having a decontamination area?

12. Do you plan on having a crawl-in or a walk-in entry way?

13. How deep are the wells in the area where the shelter is going to be built?

14. Do you want a pressurized water system with an electric well pump or do you want a hand pump or both?

15. Do you want hot water and flush toilets?

16. Do you want both a men's and a women's bathroom, or one bathroom?

17. Do you plan on having a water storage cistern?

18. How big of a cistern, in terms of balloon capacity, do you want or need?

19. How long do you want to be able to run your generators, (i.e. how long do you want the fuel to last)?

20. Do you want to have an extra generator for back up?

21. Do you want to incorporate the ability to run on battery power into the shelter's power system?

22. Do you want battery power as an emergency reserve or as a primary operating source?

23. How long do you want to be able to run on battery power without recharging the batteries?

24. Is the site for the shelter such that the shelter can be built into the side or into the top of a slope?

25. Are you concerned about the security of your shelter in the event of an emergency occupation?

26. Do you want to have an observation tower incorporated into your shelter which would enable you to have some control of the perimeter around the shelter site?

27. Is the shelter going to be connected to or part of any above ground structure?

You probably don't have the answers to many of these questions but hopefully they have caused you to consider some important aspects of shelter building that you hadn't thought of before.

Site Location Considerations

Proximity to Target Areas

When standing on railroad tracks, if you observe an oncoming train, the logical thing is to get out of the way. If you believe in the possibility of an oncoming nuclear war or natural disaster, the logical thing to do is to relocate out of any target areas or any potential area that could be drastically affected by earth changes. If you live in the crime dominated inner city, what would be a better investment, buying bullet proof vests for your family or moving to a better neighborhood? In the same fashion building a blast shelter, because you live near a known target area, might not be as smart as moving to a non-target area (perhaps as little as ten miles away) and building a less expensive fallout shelter. When you see a locomotive coming down the tracks toward you, the most cost-effective thing you can do is get out of the way.

In order to be effective, shelters in target areas must be blast hardened; built to withstand the direct effects of a nuclear weapon. Even if you have a blast shelter, your first indication of a threat might be the flash of a strike. At that point, it is too late to run to your shelter.

It is my personal opinion, that if you determine that you need blast protection you should seriously consider relocating. Distancing yourself from target areas and areas of vulnerability has both practical and financial advantages. It is practical in the sense of avoiding the direct effects of a nuclear weapon and financial because building a fall-out shelter is less expensive than building a blast shelter. Living close to a target area greatly reduces your chance of survival and greatly increases the cost of a shelter.

The best place to relocate to is a remote area, at least ten miles from any target area, and preferably, thirty miles away from any areas of major population. It would be ideal if this area was in proximity to other like-minded people.

Once a decision is made on a general location of a shelter there are certain factors which should be considered for specific locations.

Available Water

You can't live long without water. You should investigate how deep the wells are in the areas you are contemplating for a shelter and how many gallons per minute they produce. Water capability is an important issue. Drilling a 600 or 800 ft. well is expensive and in the event of well pump failure, it is hard to pull up and service the pump. It is impossible to pump water with a hand pump from a well deeper than 250 feet.

I would highly recommend having a well for your shelter as water is very difficult to store over a long period of time. This subject is covered extensively in the Food and Water chapter.

Drainage

Drainage is very important. Consideration should be given not only for a high water table and its obvious complications, but also for the potential impact of unusual surface water runoff which could swamp the shelter.

Depth of Bedrock

It is wise to drill or dig a test hole to determine if there is any underlying bedrock that would obstruct the excavation necessary to accommodate the shelter. I know one shelter project in particular which spent considerable sums of money having to blast through unexpected bedrock.

Tactical Problems

It is ideal if the terrain around a potential shelter site enhances the security of the shelter's perimeter. In other words, it would be best to locate the shelter on a relatively high point which would give its observation tower unobstructed tactical view of the area surrounding the shelter.

Access

Access is a two-edged sword. Too easy an access is not good because you don't want to be constructing your shelter under everyone's gaze and inspection. And you don't want to make it easy for someone not invited to make their way to your shelter. On the other hand, you want to be able to get to it in the winter if needs be. It won't be much good having a shelter you can't get to in time, even if it is an ideal tactical location. Another consideration relating to access is getting equipment and material to the site during the construction phase.

Earthquake Considerations

Buried structures have the advantage of being earth-integrated. In an earthquake, the ground accelerates causing structures above the ground to move back and forth in a rocking motion. On the other hand, structures buried in the earth are moved only slightly and then in sync with the surrounding ground. This can be compared to the way a ship tosses and rolls in a storm on the water's surface and a submarine below the surface moves stably within the sea.

You obviously do not want to build a shelter in close proximity to a known fault, but in general, a buried shelter has the best chance of surviving an earthquake, far better than any type of above ground structure. Nevertheless, shaking will occur. Supplies and equipment on shelves, generators, and water and fuel storage tanks should be properly secured.

Other Shelter Building Considerations

Cutting Into A Used Fuel Tank

DANGER

Extreme caution should be exercised when cutting into a used fuel tank. Over the years numerous people have been killed by explosions resulting from trying to cut a hole in a used fuel tank. Tanks which contained diesel fuel are less susceptible to explosion than tanks which held gasoline. Old gasoline tanks are extremely dangerous to cut into. Regardless if it is an old diesel tank or an old gasoline tank, treat both kinds as if they were gas tanks. You can never be sure exactly what was in the tank.

A fuel tank does not need to have fuel in the bottom of it to be dangerous. The fuel vapors or fumes, mixed with air, will be ignited by a spark from a metal cutting saw or torch. The intensity of such an explosion can rarely be survived.

There are three ways to prevent fuel tank explosions when cutting into a tank. The first is to use dry ice, the second is to use compressed CO_2 and the third is to use exhaust off the tail pipe of a truck. The whole concept is to displace the vapor fumes out of the tank with CO_2 which is a heavy gas. Begin by making sure the tank is standing upright in its natural position. Then open one of the plumbing access fittings on the top of the tank.

If using dry ice, put 20 lb. of the dry ice in the tank for each 1000 gallons of tank capacity. The temperature has to be warm enough to facilitate the melting of the dry ice in order to produce the CO_2 gas. Wait until the dry ice has melted completely before cutting into the tank.

If using CO_2 compressed gas, insert a discharge tube from the tank of CO_2 into the access hole in the fuel tank and slowly let the gas fill the tank. Use 150 cu. ft. of compressed CO_2 gas for every 1000 gallons of tank capacity.

When using the exhaust from a truck you need a connecting piece of black flexible PVC waterline slightly larger than the exhaust pipe. Slip this waterline over the exhaust pipe and secure it with duct tape or a hose clamp so that no exhaust leaks out. Put the other end

into the access hole in the top of the fuel tank. Turn on the engine and let the truck idle. Check the end of the water line to make sure the exhaust is coming out. Don't use a truck with holes in the muffler. The exhaust contains CO_2 gas which will displace the gas vapors out the top of the access hole so be sure not to tape this connection shut. A truck with a conventional 350 Chevy engine needs to idle 10 minutes for every 1000 gallons of tank capacity. Make sure the truck is tuned up and not running rich or having choke problems. You don't want to be putting unburned fuel and CO into the tank. CO can actually combust. A diesel engine would be better to use for this application if it is available.

As an added precaution have someone duct tape the cut behind the cutting torch to prevent air from entering the tank during the actual process of cutting the hole. Any fuel you find in the bottom of the tank, once you've cut the hole, can be absorbed with sawdust, shoveled into feed sacks and properly disposed of.

NOTICE

Unless you are knowledgeable and experienced in this area, you should purchase a tank which has already been cut open by a licensed fuel tank remover or purchase a new tank. This information is not intended to be encouragement or advice from the author in regards to cutting into a tank. If you decide to engage in such a dangerous and unpredictable activity, you will have to do it at your own risk!

Waterproofing

Putting on Paraseal Waterproofing

When constructing sub-earth homes and underground shelters which have shells comprised of porous materials, such as concrete or wood, make sure not to skimp on the waterproofing. There is nothing much worse than to backfill and landscape an underground shelter and then discover a leak. The only real solution is to call the backhoe or excavator back, uncover the shelter by removing the dirt, and re-do the waterproofing.

For waterproofing concrete shelters, a product called Para Seal is probably the best commercial product on the market today. Para Seal is a heavy mil P.V.C. material with a bentonite backing. It is a self healing waterproofing system. If a hole is punctured through the P.V.C. and water seeps in, the bentonite swells up and stops the leak. Paraseal is made by Paramount Technical Products, 2600 Paramount Drive, P.O. Box 1042, Spearfish, S.D. 57783, (605) 642-4787 or (800) 658-5500. Another fairly good product is called ADF which is made by Tec Coatings, located in San Antonia, Texas.

For fuel tanks, the best material for waterproofing is epoxy coal tar. You can locate this coating material by contacting the nearest underground tank manufacturer or by ordering it through a commercial paint store.

For culvert, a tar based foundation coat is adequate since the shell is galvanized material. Caution should be taken to make sure that the joints between sections and welded intersection joints are properly sealed to prevent leaking.

The steel quonset structures contain many bolt connections and tend to leak if not properly waterproofed. Polyurethane foam has been used successfully to waterproof this type of structure.

The newer generation of fiberglass shelters like the Subtec ES10 are considered to be watertight, but the old Theta modules have a tendency to leak.

Backfilling

Backfilling the Shelter

Backfilling is the most critical construction phase of any structure system which relies on earth-arching for its structural strength. This includes fuel tanks, culvert, fiberglass pods, and steel quonset. The potential for serious problems with settling and deforming due to improper compaction should not be underestimated! Generally the backfilling material should be screened in such a manner that nothing larger than a softball is placed within two feet of the surface of the tank or culvert. Larger rocks, due to the compaction pressure of the material above, tend to gradually migrate down onto the surface of the tank or culvert. This produces dents and point loads which can potentially compromise the structural integrity of the system.

Earth-Arching

Earth-arching is a phenomena whereby the earth covering a "fully buried" structure, when subjected to an overpressure, acts as one integral unit, absorbs the pressure and as pressure penetrates down into the earth, the earth reduces and dissipates the pressure. The effect is that "fully buried" structures, without significant strength, only receive a fraction of the initial pressure which was applied to the surface of the ground, and survive without collapse. The term "fully buried" means the structure in question, be it a fiberglass dome, tank or culvert, is covered by at least a depth of earth which is equal to or greater than the diameter of the structure. In other words, earth-arching will not work if the shelter is not buried at a depth which provides an earth cover over the top of the shelter which is equal to or greater than the diameter or width of the tank, culvert or arch that it is covering.

Compaction

Another significant factor in relation to earth-arching effects is compaction. The earth-arching effect will not function to its

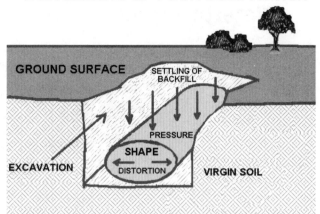

THE RESULTS OF POOR COMPACTION

GROUND SURFACE · SETTLING OF BACKFILL · PRESSURE · SHAPE · DISTORTION · EXCAVATION · VIRGIN SOIL

capacity if the earth cover over the shelter is not uniformly compacted to a percentage of 95% Proctor. Compaction is not only important from the standpoint of achieving the earth-arching effect to protect the shelter in the event of an overpressure, but it is also important in terms of keeping the tank or culvert from deforming and deflecting during the settling process.

A simple way to test for this 95% Proctor density is to jam your heel into compacted earth and if it barely leaves an imprint you have approximately 95% Proctor density.

Compaction can be thoroughly accomplished with a wacker tamper applied to each 8 to 12 inch layer of earth cover. The flat side of the bucket of a large excavator can be used as long as a repeated pounding action is applied.

The final aspect of achieving the earth-arching effect is the aggregate nature of the soil. Silty and clay soils will not, even when properly compacted, produce the earth-arching effect. The overall material must not contain any more than 15% silty fines or clay. On the other extreme, if the compacted backfill is comprised of significant amounts of material which is larger than 12 inches in diameter, the earth-arching effect will be compromised. Both coarse sand and gravel work well and facilitate earth-arching.

Landscaping Considerations

Efforts should be made to scrape off, stockpile and reserve any topsoil in the area on top of and around the shelter site. Be careful that this soil does not get used for backfilling, bedding pipes or other activities where earth with minimum rocks is needed. The topsoil may not seem all that valuable at the start of the project, but when the bulk of the project is done and it is time to landscape, the topsoil is extremely valuable. Not only should the soil be scraped off areas where excavations are going to occur, but also areas where subsoil from the excavation is going to be piled. The topsoil under these piles usually gets dug up and mixed in with the sub soil and thus wasted during the backfilling operations.

The landscape should also make consideration for security. Bushes, earth mounds, and large rocks should not be located on the grounds in such a way as to create concealment and cover for uninvited guests and intruders.

Doors

Shelter entry doors and hatches have three primary requirements: airtightness, security, and heat and overpressure resistance. An airtight seal is the primary requirement of any shelter door. This can be especially difficult to achieve with a homemade blast door. Doors built to resist heat, overpressure and thwart security threats are by their nature massive and difficult to get airtight seals on. One solution is to have a primary outer door which provides heat, overpressure and secu-

Shelter Entrance Blast Door

rity protection and a secondary inner marine door which facilitates an excellent airtight seal.

Most shelters have doors which swing open to the outside (in contrast to doors that swing inward). The reason for this is that the door is easier to build that way. In the event of an overpressure, the outside pressure presses the door onto the door frame and the door is supported by the frame on its entire periphery. The disadvantage is that rubble resulting from a blast can prevent the door from being opened and if the shelter doesn't have an unblocked alternative entrance, the occupants must be freed by rescue teams coming from outside.

Doors opening to the inside allow occupants to open them even when blocked by accumulated rubble. These style doors are more difficult to construct because the overpressure is carried by dogs or pins which extend from the door into the door frame. These dogs transfer the overpressure to the door frame. Therefore inwardly opening doors need to be small.

Marine Door

The best solution to being trapped inside a shelter by rubble is to have an alternative entrance/exit with a different profile, exposure and elevation. The idea here is that if one door is a horizontal surface walk-in type then the other entrance/exit would be a vertical hatch coming up to the surface ideally at a higher elevation. The problem of rubble blocking a vertical type hatch can be reduced by elevating the concrete reinforced hatch tube (see following pictures) several feet above the ground level and have the hatch pivot horizontally to open as opposed to flipping open vertically. One design which seems effective is the jack pivot type (as shown in the upper right drawing) which would allow occupants to lift the weight of any debris accumulated on top of the hatch.

Door frames for exterior entry doors should be structurally substantial. In the case of a blast shelter, the door or hatch frame should be constructed of steel reinforced concrete and have enough surface area and mass to disperse the force of overpressure into the ground. This is important because if the overpressure is not transferred to the ground, it will be transferred directly to the

entry tube which will be either crushed or driven into the shelter.

Doors should be outfitted on the inside with at least two chain binder arrangements which would tightly secure the door against the effects of negative overpressure and attempts by hostile individuals to make an unauthorized entry.

Another important aspect of doors are seals. Usually these are made of rubber. Rubber burns in case of intense surface fires. Ideally fire-resistant fiber gaskets made of kevlar (which would retain integrity at high temperatures) should be used.

To protect the door from the effects of large calibre firearms and cutting torches, steel doors should be lined internally with at least 1 inch of steel reinforced concrete.

Pivoting Hatch Door Vertical Entrance

Section of Vertical Entry Hatch

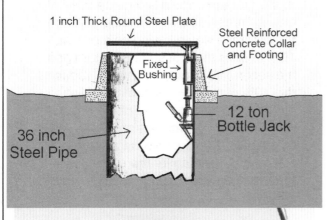

1 inch Thick Round Steel Plate

Steel Reinforced Concrete Collar and Footing

Fixed Bushing

36 inch Steel Pipe

12 ton Bottle Jack

Vertical Entry/Exit Looking Up from Inside Shelter

Entry Ways

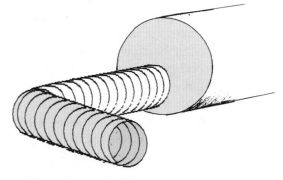

90 Degree Turn on Culvert Entry Way

When designing shelter entry ways, the two factors which must be considered are attenuation of radiation and ease of access. The diameter, length and number of turns in an entry way will determine how much outside radiation comes into the shelter through the entrance way. Most shelter doors do not attenuate a significant amount of radiation. Shelter entrance way tunnels need to be configured so that they attenuate radiation, both gamma and neutron, and provide adequate access and escape.

Radiation attenuation is usually accomplished by extending the length of the entrance tunnel, incorporating turns in the entrance way tunnel, reducing the diameter of the entrance way tunnel, creating a barrier inside the tunnel, or any combination of these four.

Neutron radiation is only a problem in proximity to areas affected by the blast. Neutron radiation is not attenuated as easily as gamma radiation. Much more attenuation is needed to protect the occupants of a shelter from the initial deadly pulse of neutron radiation which will attempt to penetrate into the shelter through the entrance way in the areas affected by the blast.

You should never be able to open the door to a shelter and see directly into the shelter. If you can see into it, radiation can penetrate into it. One good test which will indicate how well the entrance way will reduce radiation is to open the hatch-ways and doors to your shelter and go inside to where the entrance way actually meets the shelter. Make sure no lights are on inside the shelter. Turn around and look back out the entrance way toward the outside. Observe how much sunlight, if any, is being reflected down the tunnel way from the outside.

The presence or absence of outside light being reflected down into the interior of the shelter via the entrance way tunnel is a good general indication of how adequately the tunnel will shield the occupants from the penetration of neutron radiation. Neutron radiation is much like light in its reflecting capabilities.

Neutron radiation has a scattering effect which enables it to effectively penetrate around corners. A 90 degree turn in a entrance way tunnel will only reduce the penetration of neutron radiation by 20%. In effect, 80% of the neutron radiation will get around the 90 degree turn in the entrance tunnel. A second 90 degree turn in the entry way tunnel will reduce the 80% that got around the first turn in the tunnel by a factor of 20%. In essence, if you had 9,000 rems of neutron radiation pulsing into the shelter entry way, the first 90 degree turn would reduce the neutron radiation to 7,200 rems and the second 90 degree turn in the tunnel would reduce this 7,200 rems to 5,760 rems. It is going to take many 90 degree turns and/or the addition of tunnel length and barriers to reduce the neutron radiation to a survivable level.

Barrier shielding is a practical solution to the penetration of neutron radiation. Substances with high hydrogen contents provide the best shielding and attenuation of neutron radiation. Stacking full water containers in the entrance way is one solution. Sandbags full

of sawdust is another. The ideal neutron shielding barrier would be a massive, hinged interior door which contained an interior water bladder. Water is an excellent substance for shielding neutron radiation.

Gamma radiation is a product of airborne fallout dust particles and its presence will extend far beyond the area affected by blast. However, gamma radiation does not have the same scattering effects as neutron radiation. Generally speaking, one 90 degree turn in an entrance way will attenuate 90% of any gamma radiation coming in through the entrance way. If you had an entrance way with two 90 degree turns in it, the first turn would attenuate 90% of any gamma radiation coming through the entrance way allowing no more than 10% to pass the first 90 degree turn and the second 90 degree turn would reduce this 10% of the original gamma radiation by 90% allowing only 1% of the original radiation through the entrance way into the shelter. See the following chapter on Radiation Shielding for more detailed information on attenuating radiation in entry ways.

This does not take into account any attenuation of gamma radiation as a result of the length of the entrance way, its diameter or any shielding which the doors may provide.

You can start to see how much trouble and expense can be avoided by locating oneself and one's shelter away from known target areas which could experience the effects of blast.

Walk-in type entrances have certain advantages and disadvantages. They can be a blessing when it comes to getting people and equipment in and out of the shelter quickly. If you have elderly people in your shelter group, a walk-in entry may well be a necessity. Having a crawl-in or climb-down type entrance way can make movement of people and equipment in and out of the shelter slow or in the case of larger equipment, impossible. Walk-in entry ways can provide the needed space for decontamination facilities and extra storage.

The disadvantage to a walk-in entry is that its nature involves utilizing a larger diameter pipe or culvert. This larger diameter dictates the need for more 90 degree turns and more tunnel length to effectively attenuate the increased radiation coming into the shelter through the larger entry way opening.

It is important to have more than one entrance way in the event that one becomes blocked by debris or compromised due to a security problem. In consideration of the latter situation, it would be good to make the alternate entrance-exit hidden on the surface, if possible.

Fire Suppression

The threat of fire should not be taken lightly. A fire in a confined area can lead to a rapid build-up of heat, hot gasses and smoke. The last thing you want is to have a fire that drives you out of a shelter at a time when outside conditions have driven you into the shelter. **Install smoke detectors**, have an ABC fire extinguisher, and if you have running water, have garden hoses in the shelter. Sur-viving a shelter fire will mean quick detection and quick suppression. Also, make sure flammable liquids are kept isolated in metal cabinets or behind airtight bulkheads away from living areas.

Sewage Disposal and Sanitation

Toilets and sewage disposal are an obvious necessity. If you have a smaller shelter it may mean a chemical toilet and some sort of holding tank. In a larger shelter where you have running water, flush toilets are the way to go. This poses a problem with the disposal of the resulting sewage. The ideal is to have your shelter situated in the slope of a hill, enabling you to utilize gravity flow for removing waste water to a drain field. If you are looking for blast protection, you should consider that conventional septic tanks have little or no resistance to overpressure. This can be solved by building or purchasing a special blastproof tank or placing the septic tank under some portion of the shelter or its entryway where it will enjoy existing blast protection.

If you are not located in some sort of slope and cannot use gravity flow, a holding tank and pumping station must be used. This equipment is commercially available off the shelves at plumbing supply stores.

A popular subject when shopping for shelter space or designing shelters is per capita of persons per toilets. Ideally the shelter would have no more than 20 persons per toilet.

Being able to shower is an extremely important issue. Sanitation and cleanliness are not only important to prevent the spread of sickness and disease but also have psychological benefits. Showers, unlike toilets, can be scheduled. One shower for every 25 persons should be adequate.

Government Regulations

All states have requirements for permits when electrical work is being done. Following electrical codes will greatly enhance the safety and performance of the shelter wiring.

Most counties have regulations regarding septic systems. If the shelter is a single purpose shelter you may be able to get an exemption as long as the sewer system is not being used.

Regulations in regard to plumbing permits vary from state to state and county to county.

Building permits are required in most areas. The only salvation for the shelter builder in this regard is that there is an appendix in the back of the U.B.C. codebook for Fallout Shelters. It is as liberal as can be and is the one thing you should politely bring to the attention of the building, plumbing, and electrical inspectors, the county sanitarian and state fire marshal, if they happen to come around.

Unfortunately there are enough laws that if you get on the wrong side of the government, they can make your life miserable with arbitrary enforcement.

Most people who build shelters want confidentiality, and for good reasons. Unfortunately the government makes this very difficult.

The following is a reprint of the U.B.C. Appendix on Fallout Shelters.

1991 UNIFORM BUILDING CODE

APPENDIX

Chapter 57 (Pages 991 & 992)

REGULATIONS GOVERNING FALLOUT SHELTERS

Purpose

Sec. 5701. The purpose of this appendix is to establish minimum criteria which must be met before a building or building space can be constructed, occupied, used or designated a fallout shelter.

Scope

Sec. 5702. The scope of this appendix extends to building spaces designated for use as fallout shelters, including periods of drill and instruction for this purpose.

Definitions

Sec. 5703. FALLOUT SHELTER is any room, structure or space designated as such and providing its occupants with protection at a minimum protection factor of 40 from gamma radiation from fallout from a nuclear explosion as determined by a qualified fallout shelter analyst certified by the Office of Civil Defense. Area used for storage of shelter supplies need not have a protection factor of 40.

DUAL-USE FALLOUT SHELTER is a fallout shelter having a normal, routine use and occupancy as well as an emergency use as a fallout shelter.

SINGLE-PURPOSE FALLOUT SHELTER is a fallout shelter having no use or occupancy except as a fallout shelter.

PROTECTION FACTOR is a factor used to express the relation between the amount of fallout gamma radiation that would be received by an unprotected person and the amount that would be received by one in a shelter.

UNIT OF EGRESS WIDTH is 22 inches.

Occupancy Requirements

Sec. 5704. (a) **General**. Nothing in these regulations shall be construed as preventing the dual use or multiple use of normal occupancy space as fallout shelter space, providing the minimum requirements for each use are met.

(b) **Mixed Occupancy**. The occupancy classification shall be determined by the normal use of the building. When a normal-use space is designed to have an emergency use as a fallout shelter in addition to the normal use, the most restrictive requirements for all such uses shall be met.

(c) **Occupancy Separation.** No occupancy separation is required between that portion designated as a fallout shelter and the remainder of the building.

(d) **Space and Ventilation.** A minimum of 10 square feet of net floor area shall be provided per shelter occupant. Partitions, columns and area for storage of federal shelter supplies also may be included in net area. A minimum of 65 cubic feet of volume shall be provided per shelter occupant. A minimum of 3 cubic feet of fresh air per minute per person shall be provided.

In addition, the shelter shall have a ventilating rate sufficient to maintain a daily average effective temperature of not more than 82°F. for at least 90 percent of the days of the year.

(e) **Illumination.** No special lighting levels are required.

(f) **Hazards.** Hazardous utility lines such as steam, gas and oil shall not be located in or near the shelter unless provision is made to control such lines by valving or other approved means.

Exits

Sec. 5705. There shall be no fewer than two widely spaced exits from a fallout shelter, leading directly to other spaces of the building or outdoors. Exits from the fallout shelter shall aggregate at least one unit of egress width for every 200 shelter occupants. In no case shall a single exit be less than 24 inches wide.

Flame-spread Index of Interior Surfaces

Sec. 5706. Interior surfaces of single-purpose fallout shelters shall have a flamespread index not exceeding 200.

Minimum Design Loads

Sec. 5707. (a) **Dual-use Fallout Shelters.** In the case of dual-use fallout shelters, design live load required for the normal use shall govern, except that concentrated loads shall be considered.

(b) **Single-purpose Fallout Shelters.** Minimum live loads for floor design in single-purpose fallout shelters shall be 40 pounds per square foot except that concentrated loads shall be considered.

Sanitation

Sec. 5708. Toilets, either flush-type operating from the normal water supply system, or chemical or other types, shall be provided on the basis of one toilet per 50 fallout shelter occupants. Fifty percent of the toilets may be provided outside the fallout shelter area. Empty water containers may be considered as fulfilling this requirement.

Chapter 11
Radiation
Shielding

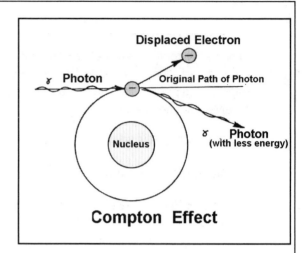

Compton Effect

Shielding is the use of mass, such as dirt, sand, concrete, lead, etc., to protect yourself from the effects of a disaster. These effects could involve intense heat, radiation, over-pressure, and high winds.

Attenuation - Ionization

Radiation shielding is the primary concern of any fallout shelter. This is accomplished by placing mass between people and radioactive fallout particles. Nuclear radiation is attenuated or weakened by the ionization process. Each time an alpha or beta particle or a gamma photon come in contact with an atom, they cause an electron to leave the atom's orbit, either through attraction, repulsion or propulsion. Each time this electron interaction process occurs, the radiation loses a little of its energy. This happens whenever radiation comes in contact with, and attempts to pass through, any type of mass. The amount of radiation attenuation which occurs depends on the type and thickness of the mass and the particular type of radiation which is attempting to penetrate the mass.

Alpha and beta particles are easily attenuated and are not a major concern when it comes to shielding. When considering fallout shelters, gamma radiation is the major concern. Gamma radiation is not readily attenuated and is biologically destructive. Gamma radiation is a wave which transmits energy through space, a

stream of tiny bundles of energy. These tiny bundles of energy are called photons.

There are three ways that gamma radiation is attenuated, but only one of these, the Compton effect, is of any relevance to fallout shelter shielding. The essence of the Compton effect is that when the gamma photon contacts an electron in an atom of matter, be it air, dirt, concrete or living tissue, it transfers some of its energy to the electron. The electron is then propelled at an angle out of its orbit. The remaining photon, with less energy, also continues on at a different angle from its original path. This changing of angle direction is called scattering. Both the electron freed from its orbit and the reduced-energy photon will continue to strike other atoms and produce further ionizations until they completely lose their energy and are absorbed into another atom.

This scattering tendency (the Compton effect) enables gamma radiation to scatter around corners to a limited extent. This is why entry ways need a combination of multiple 90° turns, distance and/or removable mass barriers in order to attenuate gamma radiation. Radiation scattering can be compared to light entering a dark tunnel. Light gradually loses its penetrating effect as the tunnel twists around corners and extends deeper into the earth.

The penetrating and scattering effect of neutron radiation is significantly greater than that of gamma radiation. Neutron radiation is extremely biologically destructive and is hard to attenuate, much more so than gamma radiation. But since neutron radiation only affects areas in close proximity to the weapon detonation, the need for extensive neutron radiation shielding is limited to blast shelters.

There are four means of shielding nuclear radiation, which are mass, geometry, distance and time.

Mass Shielding

In this type of shielding, a barrier of earth, steel, concrete, or lead is placed between the shelter and the source of radiation.

Protection Factor

The protection factor of a shelter is used to express the relation between the amount of fallout gamma radiation that would be received by an unprotected person and the amount that would be received by one in a particular shelter. The protection factor represents a rating of the effectiveness of the shelter's mass, geometry and distance shielding.

Since most shelters are underground, radiation can enter only through the overhead mass or through the entrance ways.

A shelter with a protection factor of 20 indicates that a person in that shelter, would receive 1/20th the dose of radiation that a person outside the shelter would receive. For example, if the exposure outside was 1,500 rads per hour, the person inside the shelter would be receiving a dose of 75 rads per hour (1/20 of 1500). In this case, the person inside might want to look for a better shelter.

1 Half Value Layer Thickness or 1HVL

The thickness of a material that is required to stop one half of the radiation is referred to as 1 half value layer thickness or 1HVL. This varies with the material being used and the type of radiation. Each HVL added to the next reduces the radiation penetration by one half again.

Shown below is a comparison of different materials and the number of inches required of each material to achieve 1 HVL for both gamma and neu-

Half Value Layer Thickness (HVL)

Material	1HVL - Gamma	1HVL - Neutron
Lead	0.8 inches	3.5 inches
Steel	1.2	2.0
Concrete	3.9	4.7
Earth	5.5	4.7
Wood	11.8	3.9

Using Various Mass Thicknesses to Attenuate 1,000 Rads of Gamma Radiation

Concrete	/	Earth	=	# of HVLs	=	Protection Factor	=	Rads
19.5 in.		27.5 in.		5		32		31.25
50.7		71.5		13		8,200		.1221
85.8		121.0		22		4,200,000		.0002384

Possible Fallout Protection Factors
Shelter Locations Indicated By Dots

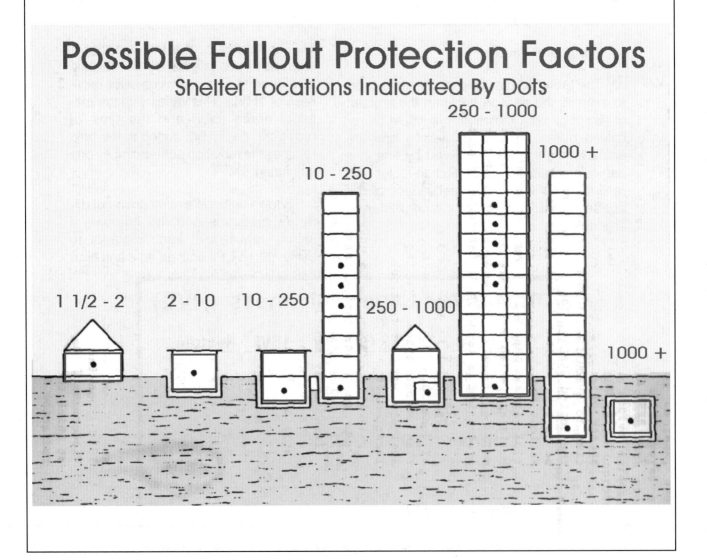

tron radiation. One HVL would not provide sufficient shielding against radiation.

Dense materials like lead, concrete and steel are the most effective substances for stopping gamma radiation. On the other hand, neutron radiation is not effectively stopped by dense material. Materials with high hydrogen contents, such as wood, water and polyurethane, provide the best shielding from neutron radiation. Hydrogen is the most effective substance for shielding neutron radiation.

Geometry Shielding

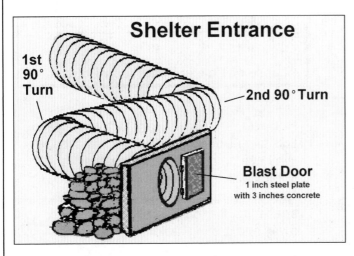

Shelter Entrance

1st 90° Turn

2nd 90° Turn

Blast Door
1 inch steel plate
with 3 inches concrete

In a fallout shelter, openings in the mass shielding are required for entry, exit and air supply. Radiation does not travel in a perfectly straight line and does, in fact, bend or scatter around corners. As it goes around a corner, it is significantly reduced. That is why entrances for shelters should have a 90 degree angle in the entryway. Another aspect of geometry shielding is that the radiation decreases in intensity with distance.

Let us look at an example of attenuation of radiation through geometry in a shelter entrance way. We will first establish that the outside fallout exposure rate is 1,500 rads per hour. We will also determine that we have a

blast door which is closed, composed of 1.2 inches of steel plate and 3.9 inches of concrete. The entry way itself has two 90° turns.

The steel plate on the door equals one half value thickness. Thus this steel plate would reduce the 1,500 rads to 750 rads. The 3.9 inches of concrete on the door equals another half value thickness. This concrete would reduce the 750 rads, which were able to pass through the steel, to 375 rads. The 1st 90° turn would then reduce this 375 rads by approximately 90% or to 37.5 rads. The 2nd 90° turn would then reduce this 37.5 rads to 3.75 rads per hour.

Additional shielding, such as another door, sandbags or buckets of food stacked in the entry way, would further reduce 3.75 rads of exposure per hour. This additional mass shielding can be placed in the entry way once everyone is inside the shelter. If you were to place a 12" barrier of sandbags at the lower end of the entry way, this 3.75 rads per hour would be reduced to about .9375 rads per hour. (12" of sand or earth has approx. 2 HVL; divide 3.75 in half two times.) This means anyone sitting for 24 hours where the entry way meets the shelter would receive an accumulated dose of (.9375 X 24) or 22.5 rads of gamma radiation. Please refer to the chapter on Radiological Monitoring for information on acceptable doses of radiation.

Openings in the shielding created by air supply and exhaust pipes are attenuated by the same means as entryways. See the chapter on air supply for detailed instructions and examples on configuring air supply and air exhaust penetrations.

Distance Shielding

The farther someone is from a source of radiation, the less negative effect or strength it will have on that person. This concept of distance shielding is more applicable to expedient shelters which do not have adequate mass, such as home basements and public buildings. Airborne fallout particles will accumulate on the roof of a structure and on the earth's surface around the structure. The harmful effects of radiation can be reduced by positioning oneself at a point in the building which is farthest from the accumulated radiation on the roof and on the ground adjacent to the building. The position safest from the harmful effects of radiation would be a spot in the center of the building on the lowest level. Distance shielding is maximizing the space between people and the accumulated outside fallout particles.

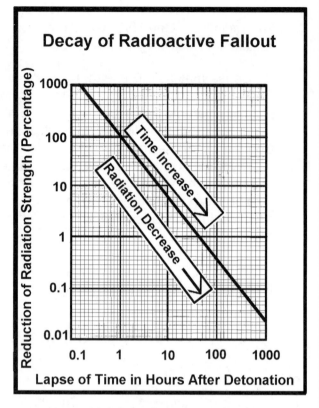

Decay of Radioactive Fallout

Reduction of Radiation Strength (Percentage)

Time Increase

Radiation Decrease

Lapse of Time in Hours After Detonation

Time Shielding

Time as a shielding factor may seem a little less tangible. When we talk about time as a shield, we are talking about staying away from any exposure to the effects of radiation until the radioactive elements have undergone their natural cycle of decay. Thus we are talking about using the passage of time as another barrier against the destructive effects of radiation.

Time shielding is the relationship of time to radioactive decay. Time is not a significant factor in relation to neutron radiation, because it is associated only with the initial burst of the weapon which lasts for a second or less. However, it is a significant factor in relation to the decay of gamma radiation. We have no control over the decay rate of radioactivity, although it is affected by weather conditions. Rain or snow will speed up the decay rate by washing away and diluting radioactive particles.

Seven - Ten Rule

In general, for every seven-fold increase in time after the detonation of a nuclear weapon, there is a ten-fold decrease in the strength of the radiation. For more detailed information on the decay rate of radioactive fallout, see the chapter on Radiological Monitoring.

Chapter 12
Air

Adequate air supply is as important a consideration as shielding against radiation. Without a continual fresh air supply, shelter occupants will quickly perish from CO2 poisoning.

If you want your shelter to have adequate air filtration, with an operational capacity of 60 days or more, power generation will be required. Pulling air through a good filter bank produces a pressure drop. It is not feasible to draw air through a filter with a hand crank blower. Besides the difficulty of continually turning the blower 24 hours a day, the increased level of activity results in more oxygen being consumed, more CO2 being produced, and more heat being generated.

Adequate amounts of air are required, not only for supplying the occupants with their chemical requirements (supplying oxygen and removing CO2), but also for environmental requirements (controlling temperature and humidity levels).

The average human body needs about 0.25 CFM, (cubic feet per minute), of oxygen to support life. In the process of breathing people consume oxygen and produce carbon dioxide proportionate to their level of activity:

The concern for adequate air supply is not limited to the issue of oxygen sup-

Chemical and Environmental Effects of Activity

Activity Level	Heat Released Btu/Hr	Oxygen Consumed Cu Ft/Hr	CO2 Produced Cu Ft/Hr	Breathing Air Cu Ft/Hr
Resting	300	.6	.5	15
Sedentary	400	.8	.67	20
Standing	600	1.2	1.0	30
Walking	1,000	2.0	1.67	50
Heavy Work	1,500	3.0	2.5	75

ply. It is also a matter of maintaining a bearable condition in the shelter. The three critical elements which must be conditioned or controlled in the shelter's interior atmosphere are dangerous levels of CO_2, heat build-up and high levels of humidity. This can be accomplished through mechanical air conditioning or, in some cases, simply ventilating the shelter with an adequate volume of outside air. This is fairly critical in smaller shelters which do not have any air conditioning - cooling equipment. Smaller shelters are many times reliant on earth cooling or mass transfer to the surrounding soil.

Ventilation

The U.S. Department of Defense recommended minimum shelter ventilation rate per person is 4 CFM. This is supposed to provide adequate oxygen, control dangerous levels of carbon dioxide and meet tolerance limits for heat, cold and humidity. The reality is that 4 CFM probably means keeping CO_2 at high "headache" levels, temperatures just under 100 degrees and humidity potentially as high as 100 percent. This low rate of ventilation is questionable in most locations in the United States. It is advisable to have at least a minimum of 5 CFM or 300 CFH, (cubic feet per hour) per person, and have some sort of air conditioning equipment.

The minimum rate of fresh air flow required to control heat and humidity will be greater than the fresh air required to control oxygen and carbon dioxide levels. The governing factor in determining the required volume of fresh air, or the ventilation capacity, for a shelter, then, is the control of heat and humidity. This being the case, the air volume must be based on meeting the ventilation requirements during hot weather conditions. People feel comfortable when the relative humidity is between 20 and 60 percent. The mean air temperature at which people feel comfortable is about 77 degrees F. The temperature and humidity that will develop in a shelter are determined by the heat and moisture balance.

Sources of heat which might be present are:

1. Heat losses of the occupants. (see Appendix A)

2. Heat in the ventilation air. (see Appendix B)

3. Heat from lights. (see Chapter on Lighting)

4. Heat from mechanical equipment. (See Appendix C and D)

5. Heat transfer to or from the surrounding earth or air.

6. Heat from combustion processes, such as open flames for cooking, candles, lamps, absorption refrigeration equipment

On the next page is a chart showing a breakdown of heat loss as a result of activity. If you are producing latent heat you are putting moisture into the air, (producing humidity).

Sources of moisture in the shelter might include:

1. Moisture loss from occupants. The moisture loss from people is a result of evaporative cooling, (latent heat) and breathing.

2. Moisture in the ventilation air.

3. Moisture from leaks in the structure

Heat Loss As a Result of Activity

Total Heat = Sensible Heat (dry) + Latent Heat (wet)

Activity	Total BTU/Hr	Sensible BTU/Hr	Latent BTU/Hr	Lbs. Moisture Per Day
Resting	300	183	**117**	.64
Sedentary	400	222	**178**	.98
Standing	600	300	**300**	1.66
Walking	1,000	375	**652**	3.47
Working	1,500	600	**900**	5.0

4. Evaporation from open containers of water, food, or from sanitation system

5. Moisture from combustion of hydrogen fuels used in cooking, lighting or refrigeration

6. Cooking and dishwashing, (1.5 lbs. per person per day)

7. Moisture from bathing or showers, (approximately 0.5 lbs. per shower).

8. Cleaning, (3 lbs. of moisture per 100 sq. ft.)

A simplified method of figuring ventilation requirements has been developed which is based on the fact that only a small percentage of the total metabolic heat generated in large shelters will be dissipated by heat transfer to the shelter walls during hot summer weather. This would be essentially true for most below-ground shelters after the first week of occupancy. The method neglects any heat loss or gain through the shelter surfaces and requires that all heat and moisture be removed by the ventilation air. The shelter is treated as an adiabatic (without loss or gain of heat) system, and the need for detailed information concerning the thermal characteristics of the shelter and its surroundings is eliminated. The resulting calculated figures for different geographic areas are shown on the map on the next page.

To use the above map it is necessary to determine the number of occupants of a shelter and multiply this by the ventilation rate as read from the map in order

Ventilation Rate Zones
(C.F.M. per Person)

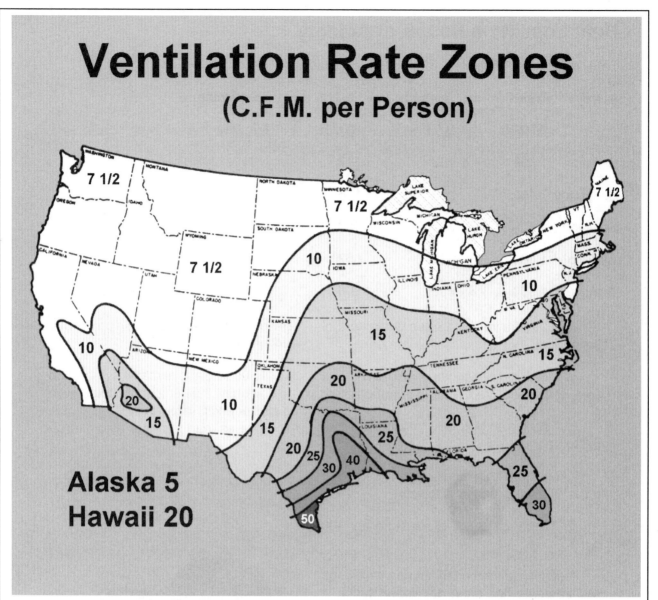

Alaska 5
Hawaii 20

to determine the total ventilation capacity required for the shelter. The method is based on the heat load of sedentary people. No other heat loads are considered. Therefore, if other heat loads are present, it is suggested that they be treated as additional occupants at the rate of one additional occupant for each 400 Btu per hour of additional heat load. Ventilation, inducing volumes of fresh outside air flow into and through the shelter, can ideally stabilize the chemical and environmental factors of the shelter's atmosphere.

The problem with using ventilation to control chemical and environmental factors is that the circumstances which would force you into your shelter may prohibit or restrict your ability to induce adequate volumes of ventilation air into the shelter. Contaminated outside air would necessitate that all fresh air induced into the shelter would have to go through a filtration process. In general, an adequate shelter air filtration system designed to remove radioactive contaminates and/or war gasses, by its nature, restricts or limits the ability to induce large volumes of ventilation air into the shelter for the purpose of controlling heat and moisture buildup.

The issue of heat and humidity is obviously less when outside air temperatures are cool or cold, but if the temperatures outside are 80 to 95 degrees you may have a life threatening situation on your hands. Heat build-up is a serious threat in a below-ground shelter where people may be closely confined.

Earth Cooling

Smaller shelters which do not have any form of air conditioning equipment are reliant on earth cooling or mass transfer to the surrounding soil. The earth in Montana, six feet down, maintains an average temperature of about 49 degrees F. The earth cover, in close proximity to the shell of the shelter, will absorb heat generated inside of the shelter. After several weeks of occupancy the earth cover surrounding a shelter will become saturated with heat and will no longer absorb the accumulated heat given off by the shelter occupants. Once this saturation point is reached, temperatures can rise to unbearable levels if other forms of cooling are not available.

The floor of an uninsulated concrete or metal shelter, with at least 8 feet of earth cover, should absorb about .08 Btu per hour per square foot. The walls of such a shelter should absorb about .05 Btu per hour per square foot and the ceiling should absorb about .09 Btu per hour per square foot.

Human Thermal Comfort

Heating and cooling are implemented as a means of maintaining human thermal comfort. Human thermal comfort is a phenomena which occurs when the human body's metabolic rate stabilizes. This happens at a certain range or zone of air temperature which varies according to the influence of certain contributing factors. These factors are: relative humidity, radiant heat, and air motion or velocity.

Human thermal comfort generally occurs in an air temperature range of about 70 to 90 degrees F. When temperatures rise above or below this comfort zone the body makes an effort to compensate for the change as a means of maintaining its deep body temperature at about 98.6 degrees F (this corresponds to a human skin surface temperature of 92 degrees F.) Brain damage occurs when body temperatures rise to 110 degrees F and death occurs when body temperatures lower to 60 or 70 degrees F.

Relative Humidity

One factor which causes the body to have a varied reaction to same air temperature is relative humidity. When air temperatures rise above the comfort zone, the human body attempts to dissipate heat through evaporation. The skin pores open and sweating occurs which dissipates heat through evaporative cooling on the skin surface. The effectiveness of evaporative cooling is varied depending on the relative humidity in the air. Relative humidity is the ratio of the quantity of water vapor actually present in the air to the greatest amount of water vapor that could possibly be present in the atmosphere at a given temperature. The maximum quantity of vapor which the atmosphere can contain varies according to the temperature. Relative humidity affects evaporation of water. A higher relative humidity means a lower rate of evaporation and a lower relative humidity means a higher rate of evaporation.

The human body can tolerate or compensate for higher air temperatures when the relative humidity is lower. This is because increased humidity inhibits the effectiveness of the body's evaporative cooling through sweating. At 20 percent relative humidity, a person will start feeling slightly warm at 79.6 degrees F. Higher levels of relative humidity make people start feeling warm at lower air temperatures. At 60 percent humidity the same person will start feeling slightly warm at about 77.1 degrees F.

Radiant Heat

Another factor which causes the body to have a varied reaction to same air temperature is the presence of radiant heat. Radiant heat, or infrared heat, is a form of heat energy which is transferred by wave motion, similar to light and radio waves. People, lights, and appliances produce radiant heat. Radiant heat passes through air without noticeably affecting its temperature. Radiant heat is not apparent until it strikes an opaque surface where it is absorbed and then produces a rise in the temperature of the substance which it absorbed into. The circulation of air will not control or reduce a concentration of radiant heat. The presence of radiant heat is measured in what is called mean radiant temperature or MRT. With a presence of radiant heat at a MRT of about 85.5 degrees F, a person will start feeling slightly warm at an air or ambient temperature of about 79.6 degrees F. With an increased MRT of 98 degrees F, a person will start feeling slightly warm at a lower air temperature of about 72.5 degrees F. If you increase the MRT, people will start feeling warm at a lower air temperature and if you decrease the MRT, people will start feeling warm at a higher air temperature.

The heat given off by people at room temperature is about 40 percent radiant. The radiant heat given off by appliances varies depending on the surface temperature of the equipment. The higher the surface temperature the greater the radiant heat loss. The heat gain from appliances will be from 20 to 80 percent radiant heat. The heat given off by incandescent lights is 80 percent radiant heat. The heat given off by fluorescent lights is 50 percent radiant heat.

Air Velocity

The last factor which causes the body to have a varied reaction to same air temperature is air velocity. Air velocity is the movement of air measured in feet per minute or FPM. At an air velocity of 10 FPM, a person will feel slightly warm at an air temperature of about 72.7 degrees F. At a higher air velocity of 50 FPM, a person won't feel slightly warm until a higher air temperature of about 75.7 degrees F.

Air circulation, the motion and distribution of air, removes the moisture which is evaporating from the body's skin pores. This moisture is constantly generated through a body process called insensible perspiration. Insensible perspiration is an attempt by the body to maintain a constant humidity level on the skin's surface. If the air circulation is insufficient, the shelter occupant will feel uncomfortable even though the relative humidity of the air inside the shelter is comparatively low.

When air temperatures rise above the comfort zone, the body loses increasingly more of its heat mainly through evaporation, and when the air temperature dips below the comfort zone, the body loses increasingly more of its heat mainly through radiation.

Air Conditioning

In most shelters cooling, not heating, is the major challenge and energy requirement. Lights, people, equipment and appliances give off heat. This heat has to be removed when its net accumulation rises above 77 degrees F.

If you plan on having mechanical air conditioning in your underground shelter, its main advantage is that less fresh air needs to be induced into the shelter because you are relying on incoming fresh air only for chemical requirements. Usually the primary concern of shelter temperature control will be cooling and the primary concern of humidity control will be dehumidification.

Mechanical air conditioning involves an expenditure of electrical power. Evaporative coolers, which use a minimum of electrical power, work in areas where the relative humidity of the outside air is fairly low. In areas of higher humidity a conventional freon-compressor type air conditioner will probably have to be used. Another option is a desiccant type H.V.A.C., unit which utilizes co-generation, and a desiccant wheel to dehumidify incoming air before it goes into the evaporative cooler portion of the unit. This type unit, by the virtue of its use of co-generation, uses considerably less electrical power than a conventional freon-compressor type H.V.A.C. unit.

Air Conditioning Concepts

There are two primary types of heat which are significant factors in psychrometry and the conditioning of air. These two types of heat are sensible heat and latent heat. Sensible heat is the heat most of us are familiar with. Sensible heat is what you detect with a conventional thermometer. Sensible heat is heat which affects the temperature of air, and its presence or absence can be easily sensed due to the change of air temperature.

Latent heat is hidden heat or wet heat. You don't necessarily feel it and it does not register on a thermometer. When a liquid such as water or freon changes its state from a liquid to a gas, heat is absorbed. The heat energy which causes the physical state of a

Desiccant Type Conditioner

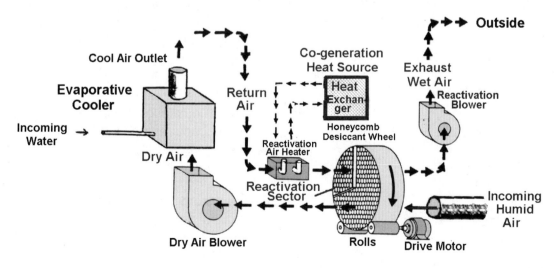

143

substance to change from a solid to a liquid, liquid to a vapor, or vice versa is latent heat. When a liquid is evaporated to a gas, heat is used or absorbed to affect this change in the substance's physical state. Evaporation produces a cooling effect because, in the case of water, it absorbs heat from the surrounding air as it changes its molecular structure from a liquid to a gas. When this same vapor or gas is then changed back into its original liquid state, through the application of pressure, the latent heat which was absorbed during evaporation is now released during condensation.

Some other thermal terminology we need to understand is the concept of "wet bulb" and "dry bulb" temperature. Both wet and dry bulb temperatures are used in determining the sensible and latent content of air. Dry bulb temperature is the common air temperature reading you take with a thermometer. Dry bulb temperature is reflective of sensible heat content in the air.

Wet bulb temperature is measured with a sling psychrometer. The sling psychrometer is a frame with two thermometers fastened to it which has a chain attached. The chain is used to rapidly whirl the psychrometer in a circle through the air. The bulb at the base of one of the thermometers is covered with a small piece of cloth which is dampened with water. When the psychrometer is whirled around, this thermometer comes in contact with the maximum amount of air. As a result the moisture on the cloth evaporates, at a rate directly dependent on the amount of moisture in the surrounding air. As the evaporation of the moisture occurs it absorbs heat, latent heat, as the water changes from a liquid to a gas. This absorption of latent heat is thus registered on the "wet bulb" thermometer as a temperature reading which is less than the reading on the other "dry bulb" thermometer which indicates air temperature.

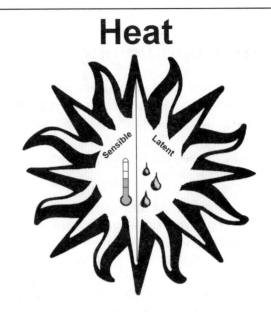

Heat

Conventional Freon Air Conditioners

Changing a liquid into a gas, (evaporation), removes or absorbs latent heat into the gas from the surrounding air. Making a gas more compact or dense, (condensation), releases latent heat which the gas was carrying into the surrounding air. Evaporation cools the surrounding air through the absorption of latent heat and condensation heats the surrounding air through the release of latent heat.

Hot air enters the air conditioner's evaporator, which is a radiator type component filled with liquid freon. The heat energy in the air is absorbed by the liquid freon and the cooled air is blown into the living space. As a result of absorbing latent heat energy, the freon changes from a liquid into a gas. Then the freon gas is compressed and forced through another radiator type component called a condenser. A fan moves air through the condenser and the air picks up the latent heat energy from the freon gas and exhausts it outside. As a result of this removal of latent heat energy the freon changes its state from a gas back into a liquid and the cycle continues to repeat itself.

Evaporative Cooling

If your geographical area has a generally low level of relative humidity, then you can utilize an evaporative cooler to cool the incoming air. If your geographical area has generally high levels of relative humidity then you will not be able to use an evaporative cooler. If the humid incoming air is already saturated or loaded down with humidity and therefore latent heat, then this incoming air won't be able to absorb water vapor from the liquid water on the sponge fabric in the evaporative cooler. Remember evaporation absorbs or uses heat from the surrounding air. So if you can't evaporate water because of high humidity then you can't get a cooling effect. A cooling effect equals the absorption of heat or Btus of energy.

Measuring the Effects of Cooling

A shelter builder should understand the basic terminology involved in measuring heat and the effects of cooling. When you are evaluating or shopping for heating and cooling equipment you will notice that such equipment is rated for its capacity to either produce heat or remove heat.

Heat is energy. The standard unit for measuring quantities of energy in the form of heat is the British thermal unit (Btu.) One British thermal unit equals the amount of heat energy required to raise the temperature of one pound of pure water, one degree Fahrenheit.

The standard unit for measuring cooling effect or capacity is referred to as a ton of refrigeration. One ton of refrigeration capacity equals the cooling effect, (the absorption of heat energy), of melting one ton (2,000 lb.) of ice at 32 degrees F into 32 degree F water in 24 hours. The cooling effect of one ton of refrigeration equals the absorption or assimilation of 288,000 Btu. of latent heat per 24 hours, or 12,000 Btu. per hour. Air conditioning equipment is rated in tons per hour capacity.

Another piece of information you will need when shopping for air conditioning equipment is the Sensible Heat Ratio. The sensible heat ratio is the ratio of the sensible heat loads in relation to the grand total of both latent sensible and latent heat loads.

Sizing Air Conditioning Equipment

When determining the tons of refrigeration required to cool a shelter we must determine the total Btu. per hour of heat gain from people, appliances, light fixtures and equipment. We also need to include in this total Btu. figure the amount of heat that will have to be removed from the incoming outside fresh air. This total heat load in Btu. per hour divided by 12,000 (the cooling effect measured in Btu. of one ton of refrigeration for one hour), will produce the total refrigeration, in tons, required to maintain control of the temperature in the shelter and prevent accumulated buildup of heat.

When sizing air conditioning, (cooling equipment), you need to base your

$$\text{Sensible Heat Ratio} = \frac{\text{Total Sensible Heat Btu}}{\text{Grand Total Heat Load Btu}}$$

(total sensible + total Latent)

calculations on the peak cooling load. The air conditioning equipment needs to be able to cool air coming from the outside into the shelter down to a bearable comfort level under the worst or most extreme conditions of heat in your geographical area. You will also need to determine the highest level of relative humidity that could be generally expected at this high air temperature. The presence of humidity in the air indicates a proportionate presence of latent heat in the air.

Not only does the air conditioning equipment have to neutralize or absorb the quantity of heat in the air coming in from the outside which exceeds the desired room temperature, but it must compensate for the heat generated by people, lighting and appliances inside the shelter. In other words, the air leaving the cooling equipment must be lower in temperature and humidity than the desired interior room temperature and relative humidity. The difference in temperature and humidity between the air leaving the cooling equipment and the desired inside temperature and relative humidity should be equal to the interior heat and humidity being generated.

When doing the initial cooling load survey of the shelter, the following factors should be considered:

The cubic feet of interior air space which is a product of multiplying the height X width X length of the structure.

The number of people who will occupy the shelter and their activity levels.

How many cubic feet per person in the shelter.

Lighting considerations include determining peak total wattage, type of lighting (either incandescent or fluorescent), and a calculation of watts per square feet of floor space.

Electrical motors need to be evaluated. The name plates on the motors will give the brake horsepower. The amount of time the motor will be operating must be determined.

Appliances need to be identified and their wattage needs to be determined along with when and how long they will be used during a day. Don't pyramid or total heat gains from appliances. Not all appliances will operate at the same time. The overall pattern of total appliance use needs to be evaluated and an average heat gain determined.

Ventilation considerations need to determine CFM per person, CFM per square foot, and air changes per hour, (shelter's cubic feet divided by cubic feet of air per hour). The need for exhaust fans in bathrooms and in the kitchen needs to be determined. Generally bathrooms should have up to 2 CFM per square foot of floor space and institu-

Psychrometric Chart

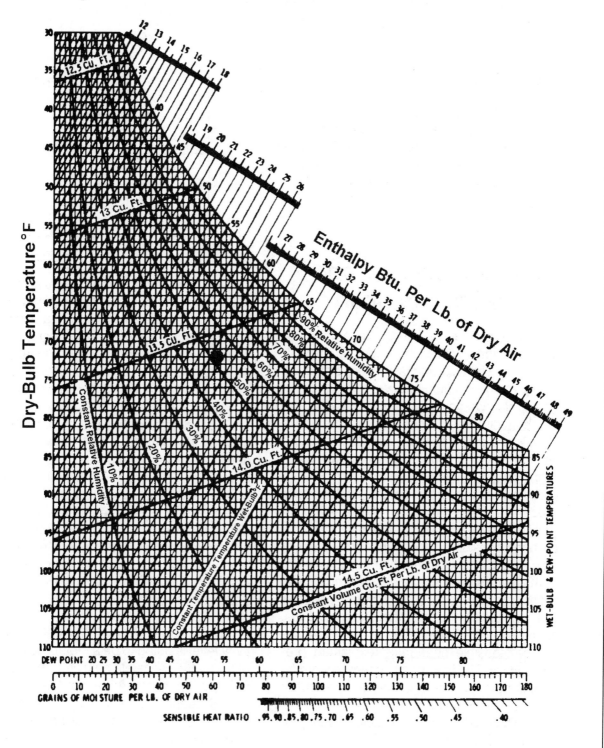

tional kitchens should have up to 4 CFM per square foot.

Outdoor Loads, as we have discussed, will include the humidity and temperature of the air entering the shelter from the outside.

Psychrometry

Psychrometric charts are used for making air conditioning calculations and sizing cooling equipment. Psychrometric charts graphically display the fundamental mathematical relationships of all the various factors involved in the thermodynamic properties of moist air.

Factors Displayed on Psychrometric Chart

This chart can be used to determine unknown factors involved in air conditioning calculations. The following factors are displayed on the chart: dry bulb and wet bulb temperature, which we have already explained; relative humidity (the percentage of moisture in the air compared to the amount of moisture which the air can hold at given air temperature); grains of moisture per pound of air (a measurement of the weight of water vapor present in a given cubic foot of air); dew point (the temperature at which water vapor in the air begins to form or condense into liquid or dew — this temperature point varies with the relative humidity and changes in atmospheric pressure); The total heat per pound of dry air (an indication of the total heat energy, in Btus, which the air is holding in a given circumstance); sensible heat ratio (explained previously); and volume of air (varies according to its moisture content and temperature). Several known factors can be plotted on this chart which will result in revealing figures for the other unknown factors.

Cooling Load Per Hour Estimating Form

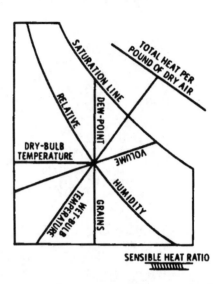

On the following page is an example of the Cooling Load per Hour Estimating Form. It has been filled out with a hypothetical calculation of the necessary tonnage required to cool a shelter located near Minneapolis, Minnesota. If we reference Appendix B, the geographical listing of high temperatures and humidity, we would see that the outdoor dry bulb air temperature is 95 degrees F. and the wet bulb temperature is 75 degrees F. The grains of moisture per pound of air is listed as 103 gr./lb. Furthermore, let us assume that we want to maintain an inside shelter air temperature, (dry bulb), of 77 degrees, a relative humidity of 40 percent and the shelter air system is providing an air flow rate after the filters of 70 CFM.

We will specify that the people's activity level is going to be a metabolic rate of 550, with a sensible heat output of 223 Btus per hour (see Appendix A). There are 10 people in the shelter. We have two electric motors, one 1/2 horse and one 3 horse (see Appendix C for the total

Outdoor Conditions

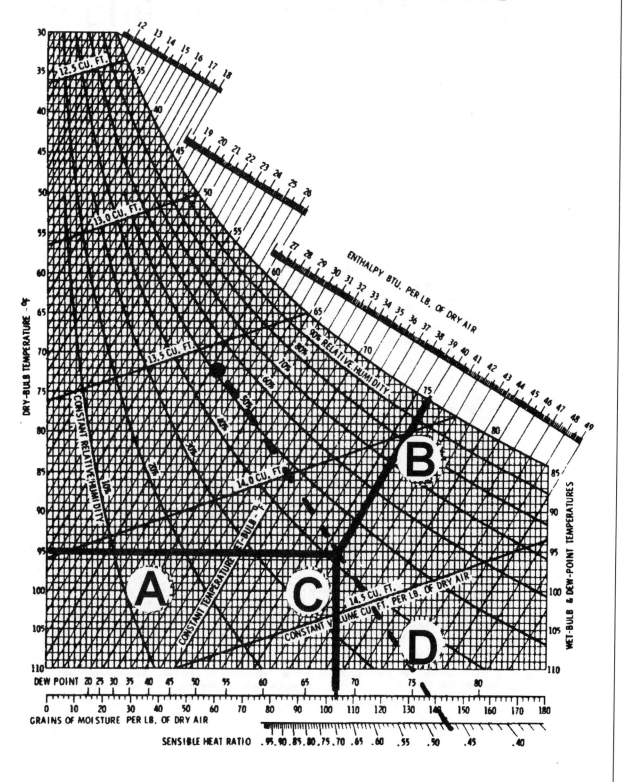

Desired Indoor Conditions

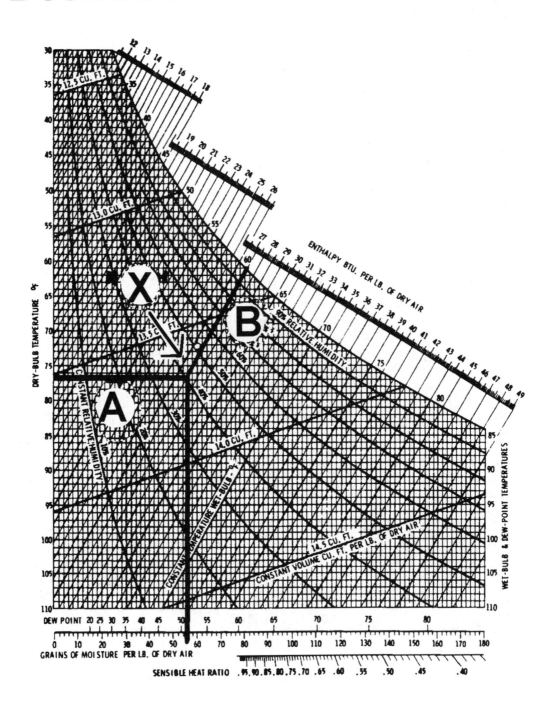

Btu output of these two motors). We will also specify that the lighting is 280 watts of incandescent and 625 watts of fluorescent (see Appendix C). The cooking sensible heat load is from one kerosene burner for 3 hours a day. The appliance sensible heat load is a coffee pot brewing for 2 hours and warming for one and some other minor appliances producing an additional 595 Btus. (See Appendix D and use figures for sensible heat only).

The next step is to calculate the latent loads. At the given metabolic rate of 550, a person will generate 237 Btus of latent heat. The cooking latent heat load is from one kerosene burner for 3 hours a day. The appliance latent heat load is a coffee pot brewing for 2 hours and warming for one. Some other minor activities produce an additional 600 Btus. (See Appendix D and use figures for latent heat only).

Our objective is to determine how much cooling effect is needed to neutralize the total heat, both latent and sensible, from the outside and total heat from the inside so that an average inside air temperature of 77 degrees F. is maintained with a relative humidity of 40 percent.

Outdoor Conditions Psychrometry Chart

If you look at line "A" on the Outdoor Conditions Chart, it represents a dry bulb temperature of 95 degrees F. Line "B" represents a wet bulb temperature of 75 degrees F. Line "C" is extended upward from the 103 point on the grains of moisture per pound of dry air scale. These are three known factors plotted on the chart. You will notice that when the factors are charted all the resulting meet at the same point. If you look closely at where these lines intersect and compare the point to the lines sloping upward in a curve to the left, you will be able to determine that the relative

humidity for air at 95 degree dry bulb and 75 degree wet bulb is 42 percent. Now examine dashed line "D" moving from the center point of the chart through the intersection of lines "A", "B", and "C" and extending downward until it crosses the sensible heat ratio scale. Reading where line "D" crosses the scale tells us that air at a dry bulb temperature of 95 degrees and a wet bulb temperature of 75 degrees has a heat energy content with a sensible heat ratio of .47.

Desired Indoor Conditions Psychrometry Chart

Given our known factors, line "A" of the desired indoor conditions chart represents a dry bulb temperature of 77 degrees and this lines stops where it crosses the 40 percent humidity line, point "X", our other known factor. Extending line "B" from point "X" we see that the wet bulb temperature for air with a dry bulb temperature of 77 degrees and a relative humidity of 40 percent is 60 degrees. If we extend a line straight down from point "X" to where it crosses the grains of moisture per pound scale we can see that air with a dry bulb temperature of 77 degrees and a 40 percent relative humidity has a moisture content of 56 gr./lb.

151

Cooling Load per Hour Estimating Form

Average persons metabolic rate **550**

Desired inside shelter dry bulb temperature **77** F.

Desired inside shelter relative humidity **40** %

Number of people **10** C.F.M. per person **7** Total C.F.M. **70**

INTERNAL SENSIBLE HEAT LOAD

Btu per Hour

People   # **10** People X **223**   = **2230**

Electrical Motors   = **2340**

Lights   Total _____ Watts X (3.414)   = **3090**

Appliances, Etc.   = **3695**

Cooking   + = **12,600**

Total Internal Sensible Heat Load   = **23,955** Btu

INTERNAL LATENT HEAT LOAD

People   # **10** People X **237**   = **2370**

Cooking   = **12,600**

Other   + = **1400**

Total Internal Latent Heat Load   = **16,370** Btu

OUTDOOR AIR SENSIBLE HEAT LOAD

Sensible   **70** CFM X **95** F. = **6650** Btu

OUTDOOR AIR LATENT HEAT LOAD

Latent   **70** CFM X **103** Gr/Lb = **7210** Btu

INDOOR AIR SENSIBLE HEAT LOAD

Sensible   **70** CFM X **77** F. = **5390** Btu

INDOOR AIR LATENT HEAT LOAD

Latent   **70** CFM X **54** Gr/Lb = **4158** Btu

Cooling Load per Hour Estimating Form

Average persons metabolic rate_____

Desired inside shelter dry bulb temperature_____F.

Desired inside shelter relative humidity_____%

Number of people_____C.F.M. per person_____ Total C.F.M._____

INTERNAL SENSIBLE HEAT LOAD

Btu per Hour

People   #_____ People X _____ = _____

Electrical Motors   = _____

Lights   Total _____ Watts X (3.414)   = _____

Appliances, Etc.   = _____

Cooking   + = _____

Total Internal Sensible Heat Load   = _____Btu

INTERNAL LATENT HEAT LOAD

People   #_____ People X _____ = _____

Cooking   = _____

Other   + = _____

Total Internal Latent Heat Load   = _____Btu

OUTDOOR AIR SENSIBLE HEAT LOAD

Sensible   _____CFM X _____ F. = _____Btu

OUTDOOR AIR LATENT HEAT LOAD

Latent   _____CFM X _____Gr/Lb = _____Btu

INDOOR AIR SENSIBLE HEAT LOAD

Sensible   _____ CFM X_____ F. = _____Btu

INDOOR AIR LATENT HEAT LOAD

Latent   _____ CFM X_____Gr/Lb. = _____Btu

Cooling Load Heat Totals

	Sensible	Latent
	Btu	Btu
Outside Heat	6650	7210
+Internal Heat	23,955	16,370
Sub Total	30,605	23,580
-Desired Inside	5390	4158

Total Heat loads = **25,215** Btu(sensible) **19,422** Btu(latent)

Grand Total Heat Loads = (total sensible + total latent) = **44,637** Btu

$$\text{Tons of Refrigeration} = \frac{\text{Grand Total Heat Loads}}{12,000}$$

$$\text{Tons of Refrigeration} = \frac{44,637}{12,000}$$

Tons of Refrigeration = **3.71** Tons

$$\text{Sensible Heat Ratio} = \frac{\text{Total Sensible Heat Btu}}{\text{Grand Total Heat Load Btu}}$$

$$\text{Sensible Heat Ratio} = \frac{25,215}{44,637}$$

Sensible Heat Ratio = **.564**

Cooling Load Heat Totals

	Sensible Btu	Latent Btu
Outside Heat	_____	_____
+Internal Heat	_____	_____
Sub Total	_____	_____
-Desired Inside	_____	_____

Total Heat loads = _____Btu(sensible)   _____Btu(latent)

Grand Total Heat Loads = (total sensible + total latent) =_____Btu

$$\text{Tons of Refrigeration} = \frac{\text{Grand Total Heat Loads}}{12,000}$$

$$\text{Tons of Refrigeration} = \frac{\rule{6cm}{0.4pt}}{12,000}$$

Tons of Refrigeration =_____**Tons**

$$\text{Sensible Heat Ratio} = \frac{\text{Total Sensible Heat Btu}}{\text{Grand Total Heat Load Btu}}$$

$$\text{Sensible Heat Ratio} = \frac{\rule{6cm}{0.4pt}}{\rule{6cm}{0.4pt}}$$

Sensible Heat Ratio = _____

Review of Calculations on Cooling Load per Hour Form

Starting with internal sensible heat load, we have ten people multiplied by 223 sensible heat Btus. We obtained this figure from Appendix A in the back of the book, and using the given metabolic rate of 550 and the inside temperature of 77 F. we located the listing for latent heat. Calculating electrical motors, we turn to Appendix C and looked up the Btus given off by a 1/2 horse and a 3 horse motor. Turning to Appendix D, we located the sensible heat given off by appliances and cooking.

Moving on to internal latent heat load, again we reference Appendix A and locate the latent heat listing for 77 F. at a metabolic rate of 550 which is 237. Turning to Appendix D we locate the latent heat given off by cooking and appliances.

Calculating the outside air sensible heat load we multiplied the 70 CFM times the outside 95 F. dry bulb temperature. Calculating the outside air latent heat load we multiplied 70 CFM times 103 gr./lb. In order to come up with this figure of 103 gr./lb., we had to go to Appendix B and locate Minneapolis, Minnesota.

Calculating the indoor air sensible heat load we again multiply the 70 CFM by the desired inside dry bulb temperature of 77 F. As for the indoor air latent heat load we multiplied the CFM times 54 gr./lb. This was obtained by locating 77 dry bulb at 40 percent humidity on the desired indoor conditions psychrometry chart and extending a line straight down to the grains of moisture per lb. of dry air scale. You will notice it crosses the scale at 54.

Air Intake and Exhaust Ducts

Another aspect of air supply to consider is the placement of the incoming air duct in relation to the generator exhaust duct. If the generator exhaust pipe is upwind of the incoming air duct, you will be in danger of contaminating your incoming fresh air with carbon monoxide. Try to place the generator exhaust pipe as far away from the incoming air duct as possible and in a positional relationship which will be least prone to cross contamination due to wind patterns. Also, if your engine room has a separate air system specifically for cooling the diesel generators, make sure that the resulting heated exhaust air is far enough from any air intake ducts that you are not recycling this heat back into the shelter.

Special attention should be given to the configuration of the surface portion of your incoming and exhaust air ducts. The two major concerns are vulnerability to blast effects and vulnerability to sabotage and tampering. Also, considerations should be made for the height that the incoming air duct extends above the ground. You definitely don't want the air supply cut off by drifting snow, volcanic ash, etc. If your air is cut off you can only stay in the shelter for as long as the CO_2 level inside remains safe. At a certain point, when CO_2 levels get too high, you will have to open the door and face the hostile environment outside or perish.

The normal amount of air drawn into the shelter for meeting chemical requirements can be temporarily eliminated or reduced if the shelter has either a capacity to mechanically scrub CO_2 out of the

Configuration of Air Supply Vent

Surface Vent → **Contaminated Air**

Steel Pipe

Gravel Sump Drain

Blower

Air Filter

Blast Valve

Clean Air

Buried Tank

Air Supply Vent

Exhaust Vents

atmosphere or if the shelter has some compressed air reserves.

Only a small portion of the air we breathe is oxygen. Air is basically a mixture of two gases, being approximately 23 parts oxygen and 77 parts nitrogen.

Sealed Shelter Atmosphere

Circumstances could arise where the occupants of a shelter may be forced to completely close up a shelter and discontinue bringing in fresh air from the outside. These circumstances might include sabotage, surface fires, blast effects, concentrated contaminates, etc. This being the case, the shelter occupants will either have to make do with the volume of air contained in the shelter for as long as possible or have some way of removing carbon dioxide and/or supplementing the oxygen supply. In a nonactive mode, a person produces .67 cu. ft. of CO_2 per hr. and a 3% concentration of CO_2 is the ceiling of the safe limit.

Below in the chart is an equation to determine how long you can safely maintain a sealed atmosphere in your shelter.

Example: The shelter's inside open volume equals 3,000 cu. ft. This open area does not include any area occupied by storage, cabinets and closets. There are 10 people in the shelter. How many hours can the shelter maintain a sealed atmosphere, (Z), before the CO_2 level reaches 3% of the shelter's interior air volume?

Air Scrubbers

An air scrubber is a filtering system which removes carbon dioxide from the air. Air scrubbers utilize either sodium hydroxide or lithium hydroxide as the filtering agent. Sodium hydroxide seems to be preferable for shelter operations because it is less caustic. A shelter can extend the amount of time that it can remained sealed if it has the ability to remove the CO_2 from the air. In a sealed shelter situation, CO_2 reaches dangerous levels long before the oxygen is depleted. For example, if a shelter had enough interior cubic air volume to maintain a sealed environment for three hours before CO_2 reached dangerous levels, then they would have enough remaining oxygen to continue maintaining a sealed environment for about nine hours as long as they had the ability to scrub the CO_2 out of the air.

Sealed Atmosphere Formula

Z = # of hours a shelter can safely maintain a sealed atmospheretain a sealed atmosphere

$$Z = \frac{\text{(volume of open air inside the shelter in cubic feet)} \times (3\% \text{ or } .03)}{\text{(\# of people in the shelter)} \times \text{(one person's hourly production of } CO_2)}$$

$$Z = \frac{3{,}000 \text{ cu. ft.} \times .03}{10 \text{ People} \times .67} \qquad Z = \frac{90}{6.7} \qquad Z = 13.43 \text{ Hrs.}$$

Oxygen Replacement

Having a stockpile of compressed air would enable a shelter to maintain a sealed atmosphere for an even longer duration. It is important to remember that additional oxygen is of little use if the accumulated CO2 is not removed by some means. One economical way to store air is to purchase a new propane tank which has never had propane in it and fill it with pressurized air. You must use a breathing grade compressor, not a regular compressor for air tools. Pure oxygen is dangerous when released inside a shelter. Concentrations of pure oxygen can stimulate combustion. Pure oxygen in contact with certain oils will cause spontaneous combustion.

Carbon Monoxide (CO) Poisoning

Carbon monoxide, (CO), which is produced by the incomplete combustion of carbon, is fatal to humans when the air contains a concentration in excess of 1/10 of 1%. It is unseen, odorless, and tasteless. If there are structures burning outside the shelter or if there is a leak in the exhaust from the generator, CO could contaminate the air inside the shelter.

Carbon Dioxide (CO2) Monitor

A carbon dioxide (CO2) monitor is a valuable consideration for any shelter. CO2 is another unseen, odorless and tasteless killer. Carbon dioxide is safe and natural in the atmosphere up to .04% by volume. In situations where the shelter may have been populated beyond the capacity of the air supply system, CO2 levels will begin to rise. Also, if outside conditions force the closure of the

Gas Tech Model 3252
CO2 Monitor

incoming air ducts, the result will be that CO2 levels will eventually rise to unsafe levels. When CO2 levels rise to 2%, breathing becomes deeper and the volume of air an individual inhales increases by 30%. When CO2 levels rise to 4%, breathing becomes labored and an individual experiences great discomfort. When CO2 levels reach 5%, individuals may experience nausea. Humans can tolerate up to about 10% CO2 and then unconsciousness results within 10 minutes. Within several hours, death will occur.

Like a Survey Meter, the carbon dioxide monitor can indicate when you have a dangerous environment and give you the opportunity to rectify the situation before it is too late. Gas Tech makes a portable infrared O2 - CO2 monitor Model 3252OX. This unit detects 0-25% Oxygen and 0 - 5% Carbon Dioxide. The cost is $2,500. A less expensive unit is the Bendix Gastec model SG-4010 which will indicate the level of oxygen or CO2. This unit uses vapor detector tubes to detect the CO2. The unit costs $200. and can be ordered from Linde National Specialty Gases Office, 100 Davison Ave., Sommerset, N.J. 08873, phone # (201) 356-8000.

Air Filtration

Air filtration is probably the third most important element in any shelter following radiation shielding and air supply itself. During a nuclear disaster, the bulk of the airborne contaminates is made up of radioactive fallout which is carried through the air by dust particles. At the minimum, you should be able to filter airborne dust contaminates out of the incoming air supply. This can be done with a HEPA filter. If you are planning a shelter in close proximity (within 10 miles), to a known target area, you may have to deal with more serious contaminates in the air. Other serious airborne contaminates—such as chemical war gases, biological agents and radioactive iodine gas—can be filtered out of the air by more sophisticated carbon air filtration equipment. Carbon filtration involves a series of filters. My advice is that before investing in sophisticated carbon air filtration equipment, consider distancing yourself and family from any potential target.

The air filtration process can be implemented to its ideal, if resources are available, or implemented to its minimum, depending on the budget available and the level of protection desired. Nuclear airborne contamination threats fall into two categories; radioactive fallout and radioactive iodine gas. Radioactive fallout particles are carried by airborne dust particles. If you can filter the dust out of your incoming air supply, you have taken care of the radioactive fallout problem. The removal of airborne dust particles from your incoming air supply is the easiest and least expensive form of filtration to implement.

The removal of radioactive iodine gas, on the other hand, is more complex and more expensive. The potential of guaranteeing complete removal under all circumstances is less reliable. Radioactive iodine gas is a by-product of the "blast" and its presence would only be experienced within ten miles of the detonation of a nuclear device. Its presence, under the worst circumstances, could last up to three days. Fallout shelters located more than ten miles from a known target area shouldn't be concerned about radioactive iodine gas. Complete nuclear air filtration may or may not remove all the radioactive iodine from the air. The effectiveness of radioactive iodine filters depends on how concentrated the iodine gas is in the incoming air and how quickly the filtering agent becomes saturated with radioactive iodine. Once the filter medium becomes saturated, it starts to diminish its absorption and removal of the contaminate from the incoming air.

The best solution is not to bring any outside air into the shelter for twenty-four hours or so. The problem with not taking in air from the outside is that CO_2 levels start building to life threatening levels at a variable rate which is dependent on the number of occupants that are in the shelter, the cubic volume of air contained inside the shelter and the occupants activity level. So the first requirement for safely sealing up your shelter for any length of time is having a CO_2 monitor to determine when CO_2 levels have become dangerous (See section on CO_2 monitor). At this point either outside air would have to be induced into the shelter or an available reserve of compressed air would have to be tapped and released to displace the CO_2 concentrations which have built up in the shelter. The other option as we have previously discussed is to scrub the CO_2 out of the air with the use of either sodium hydroxide or lithium hydroxide. (See section on Air Scrubbers.)

Contaminate Particle Size

One thing that we need to understand when we are talking about filtering systems is the relative size of the material we are trying to filter out of the air.

In general, contaminates are measured in microns. One micron equals 1/25,000th of an inch. To give you a comparative idea, 50 microns is equal to the size of a sharp pencil dot.

Efficiency

Another thing we should consider in relation to air filtration is the efficiency of the filter. A filter's efficiency relates directly to the level of filtration or how clean the air will end up being. Higher efficiency filters cost more and produce more air resistance.

Resistance

When designing a filter system be aware that the determined air requirement for the shelter must be available at the end of the filtering process. This

Contaminate Size Chart

Contaminate	Approximate Particle Size Range in Microns
Dust (inside)	0.0001 - 20
Fumes	0.001 - 1
Viruses	0.003 - 0.05
Tobacco Smoke	0.01 - 1
Oil Smoke	0.03 - 1
Bacteria	0.3 - 30
Ground Talc	0.5 - 50
Fly Ash	1 - 100
Lint	10 - 100
Plant Spores	10 - 30
Pollen	10 - 100
Human Hair	30 - 100

Contaminate / Filter
Comparison Chart

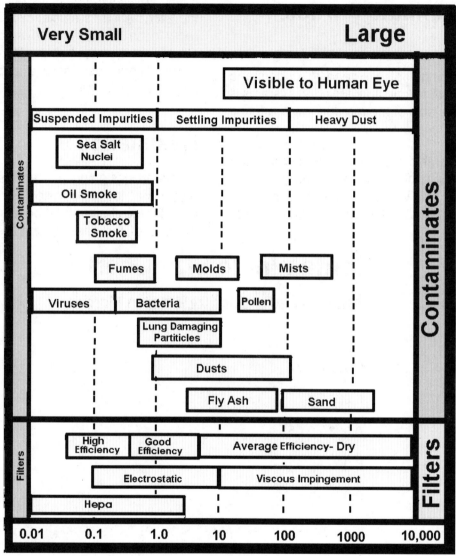

Size in Microns

means that the resistance of the filters to the incoming air needs to be taken into consideration. This is because the air passing through the filter system has to overcome the resistance that the various filters impose on it. The resistance to air flow is a measure of the fan or blower horsepower required to force the air through the filter. The resistance values supplied by manufacturers for various filters is based on the performance of a clean filter at the rated air volume in CFM. Five hundred CFM, (cubic feet per minute), of air coming into a filter system is going to be less than 500 CFM after it passes through or comes out of the filter system.

Such a resistance to air moving through a filter is referred to as static pressure drop and is typically measured in units called inches of water column. When you are shopping for filters, the manufacturer normally specifies the CFM rating for the filter and the static pressure drop which the filter creates.

Residence Time

The final consideration when putting together a filter system is what is called residence time. Residence time is the time the entering contaminated air must be present in or "reside" in the filter medium or, in the case of an electrostatic air cleaner, be in proximity to the grounded collector plates, in order for the medium to absorb, neutralize or particulate the contaminate out of the air. The velocity (speed of the incoming air) must be matched with the residence time requirement of the filter, or vice versa. In order to remove the contaminates from the air, carbon filters require that the air passing through the filter have a residence time of 0.35 sec.

Filters

Generally speaking, air filtration is not accomplished by one filter but by a combination of a series of filters.

Gross / Pre-Filter

The first filter in any filtering system is a gross or a pre-filter. The most common gross filter used is a 40 to 60 percent efficient conventional furnace type filter. This filter will remove the larger dust particles which are the bulk of the contaminates. In the most basic air filtration system, this would be the only filter used.

Electrostatic Filter

Gross Filters

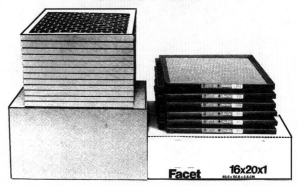

If your shelter has a power generation system you may want to consider an electrostatic filter as the second filter after the pre-filter. This type system must be sized for the specific CFM rate because it needs a specific residence time to function. The electrostatic filter charges smoke and dust particles and then separates them out of the air through electrostatic precipitation. Contaminated air is drawn into the air cleaner and first passes through a high intensity electric field, created by imposing a positive charge of approximately 12,000 volts DC through ionizer wires. Dust and fume particles are positively charged in this intensive field and then enter a sec-

Emerson Model #3C567 Electrostatic Filter

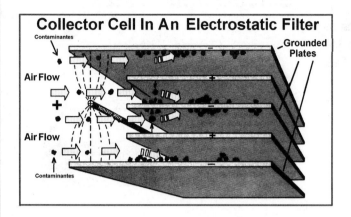

Collector Cell In An Electrostatic Filter

ond field of approximately 5,800 volts where the contaminates are collected on grounded plates. The cleaned air is then pulled out of the filter unit by the suction fan and distributed.

There are a number of companies which make electrostatic/electronic air cleaners including American Air Filter, Ultra Air and Emerson, to name a few. The Emerson filter model # 3C567 is pictured on the previous page and can be ordered from Grangers. It removes contaminates down to 0.01 micron in size which includes 98 percent of all dust, dirt, smoke and bacteria.

Electronic air cleaners need 110 volts A.C. power and the self cleaning models require water at domestic pressure. Different models have varying methods for cleaning the accumulated contaminates off the collector plates. Be aware that this type of electronic equipment must be protected against the effects of electromagnetic pulse.

HEPA Filter

The next step up in a basic air filtration system would be a HEPA filter used in combination with a pre-filter. These two filters in series, gross filter and HEPA filter, will elimi-

nate any airborne nuclear fallout dust particles but will not remove radioactive iodine gas which could be present in or near blast areas. In a more complex system, the HEPA filter would be the third filter in series after the electrostatic filter. HEPA filters are 95 to 99 percent efficient and remove particles down to about 0.1 or 0.3 micron size which a gross filter will not remove.

The HEPA filter will also filter out some smoke and biological contaminates. Even though biological viruses are smaller than 0.1 microns, an aerosol is used to disperse biological agents. The aerosol is large enough in size that it can be strained out by the HEPA filter.

When shopping for HEPA filters specify that you want a filter which meets military and nuclear specifications, (MIL-F-51068), which require special cell side material, radiation resistant media, rabbeted joints and special testing.

Hepa Filter Cut Away Drawing

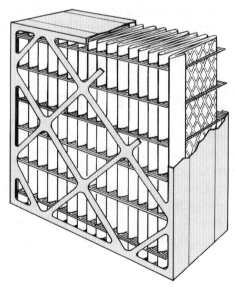

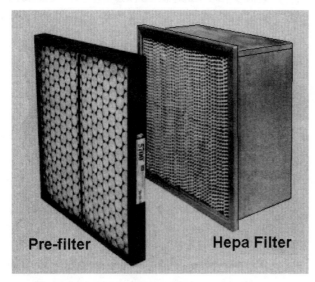

Pre-filter   Hepa Filter

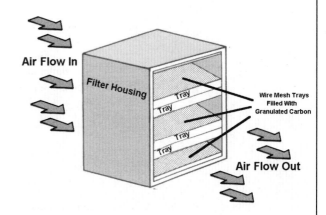

Air Flow Through Carbon Filter

Air Flow In

Filter Housing
Tray
Tray
Tray
Tray

Wire Mesh Trays
Filled With
Granulated Carbon

Air Flow Out

Carbon Filters

The fourth, fifth and sixth filter considerations are carbon filters. The fourth filter is an activated carbon filter impregnated with chemicals which have an affinity to remove radioactive iodine gas. Make sure when you are shopping for an activated carbon filter that you don't purchase a residential type activated carbon filter. There are a number of different types of activated carbons. Use only those which have been treated or "activated" with the specific chemicals necessary for neutralizing radioactive iodine gas.

The fifth potential filter in a filter series would be an activated carbon filter impregnated with chemicals which have the ability to remove acid gases. These are gases produced by burning forests, plastics and industrial chemicals. These gases include: carbon

Bed /Tray System for Carbon Air Filtration

dioxide, nitrogen oxide, benzene gas, toluene gas, nitric acid gas, chlorine gas, chlorinated dioxide gas, sulfuric acid gas, hydrochloric acid gas and acrolein gas.

The sixth potential filter in a filter series is the Whetlerite carbon filter. This carbon has been impregnated with copper, chromium, and silver to remove chemical warfare agents. These gases include a wide variety of chemical warfare agents.

Filter Systems

Luwa of Switzerland makes nuclear/biological/chemical (N.B.C.) filter systems. Their smallest filter systems is a 100 CFM unit which is suitable for a small shelter. This system includes all the filters and the 100 CFM unit can be hand cranked if necessary. The units cost a minimum of about $5,500 and have to be air freighted from Europe.

Various American manufacturers make filter systems for commercial applications. These units can be used as

165

Barney-Cheney Carbon Filter System

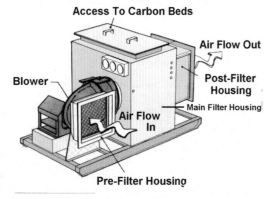

Access To Carbon Beds

Blower

Air Flow Out

Post-Filter Housing

Main Filter Housing

Air Flow In

Pre-Filter Housing

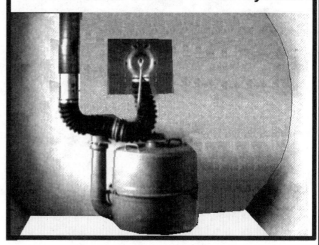

a basis for building a N.B.C. filter system.

There are manufacturers here in the United States which make N.B.C. filter equipment for the U.S. military, like the M93, 100 CFM filters for the Abrams tank. Unfortunately, they are <u>very expensive</u>. At this time, until civilian demand reaches a reasonable volume, no American manufacturer makes an off the shelf military grade N.B.C. filter system which can compete in price with the Luwa. However, a person could use commercially available components to put together their own system in the 600 to 1,200 CFM range and still be cheaper than Luwa.

All Ban M93 100 cfm Filter Unit

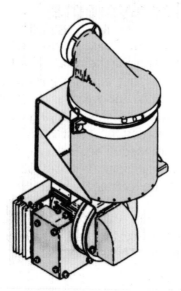

Remember that the bottom line in the air filtration issue is that most people do not need expensive carbon filtration systems and will do just fine with a combination of a pre-filter and a HEPA filter.

The newest innovation in air filtration technology is a system called TSA, or Temperature Swing Absorption. This involves the use of two parallel absorption beds. While one bed is being used to filter the incoming air the other bed is being regenerated with heat which vaporizes contaminates. As far as I know, this technology has not yet migrated down to the level of civilian access, and thus only the military currently has this type of equipment.

Post Filter

A post filter is necessary when carbon filtration is being used. The post filter captures the carbon fines which end up being picked up by the air when it passes through the carbon filter beds. These "carbon fines" can contain contaminates which the carbon has absorbed or scavenged from incoming contaminated air. Since they are so fine and move with the air they tend to get into everything and can be easily inhaled. The post filter is usually another HEPA filter.

Chemical Warfare Agent Monitors

BxICAD Chemical Agent Detector

Chemical warfare agent monitoring instruments are a consideration. A company called E.T.G. makes the BxICAD, individual chemical agent detector. This unit has an actual alarm which provides instantaneous detection. Some sort of detection alarm would be important since you generally won't be able to smell or see chemical warfare agents. Without a detection system, the first indication of the presence of chemical agents may be death. This unit is currently being used by NATO forces. The BxICAD costs about $3,600.

Another more economical option is the M256A1 Chemical Agent Detector Kit. It sells for $150. and is made by Chemical Compounding Corporation, in Riverhead, N.Y. The M256A1 needs 15 minutes to take a reading.

Air filters should be located where they can be changed or serviced without having to go outside. Filters could become clogged or saturated. For this reason spare filters should be stocked. Air filters should also be located away from the inhabited portions of the shelter. Since the filters are filtering out contaminates, they are concentrating contaminates. In the event of nuclear activity which requires the use of the shelter's air filtration equipment, the area around a filter box or canister will tend to be radioactively hot. It is not something you want in or near your shelter living area. Also when placing your filtration equipment, consider that you may have to decontaminate the person who has to change the filter. Consider that the spent filter is very hot and needs to be disposed of in a place where it will not be endangering the inhabitants of the shelter.

If surface conditions, due to blast, fires or sabotage, produce high air temperatures around the incoming air duct, outside air must not be brought into the shelter or into the air filtering system. Besides the obvious damage to the occupants, the high temperatures would destroy the filters.

Blast Valves & Gate Valves

Blast valves are used to protect the inhabitants of the shelter from sudden increases in overpressure which accom-

Blast Valves

Swung

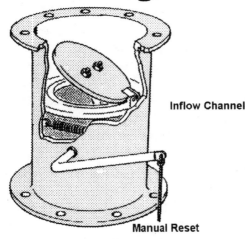

Manual Reset

Swedish

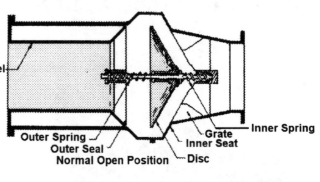

Inflow Channel

Outer Spring
Outer Seal
Normal Open Position

Grate
Inner Seat
Disc

Inner Spring

Stephenson

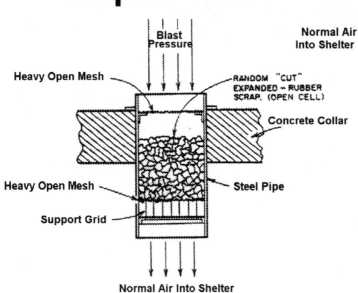

Blast Pressure

Heavy Open Mesh

RANDOM "CUT" EXPANDED ~ RUBBER SCRAP. (OPEN CELL)

Concrete Collar

Heavy Open Mesh

Steel Pipe

Support Grid

Normal Air Into Shelter

Breckenridge

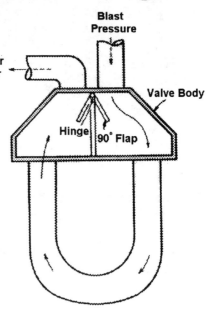

Blast Pressure

Normal Air Into Shelter

Valve Body

Hinge

90° Flap

pany and are the result of the detonation of a nuclear device in close proximity to the shelter. These valves not only protect against positive pressure but also negative pressure which follows immediately after the initial overpressure. These valves typically close off the incoming air and exhaust ducts at 3 lbs. of overpressure.

There may be situations other than overpressure in which the inhabitants of a shelter might need to close off or seal the incoming and exhaust air ducts. This closing up of all connection to the outside is generally referred to as buttoning up. One reason to button up is an extremely high level of contaminates outside. Another reason would be people on the surface intentionally trying to tamper with or pollute the shelter's incoming air duct.

There are blast valves which act both as a blast valve and a buttoning-up gate valve. It is advisable that the blast valve is located where it can't be tampered with by disgruntled people on the surface and where the inhabitants can service it if it jams or malfunctions. If you are not building a blast shelter you don't need a blast valve, but you should have some sort of valve to isolate the shelter from the outside environment if necessary.

Dobeck Blast Valve

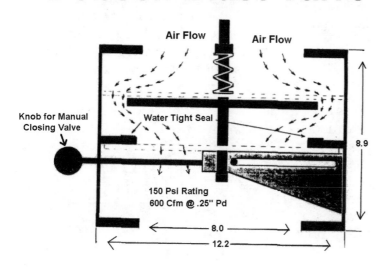

Dobeck Blast Valve

Chapter 13
Power Systems

Power is the one most vital element which supports our present standard of living and level of technology. Unfortunately, the national power distribution grid is vulnerable to the effects of war and major natural disaster. We tend to taken for granted the power which this system provides. The absence of power would radically change our lives.

If the power goes off, most people won't have running water for any use, let alone getting a drink or flushing the toilet. Unless a person has a wood stove, he won't have any heat once the power grid goes down. Even natural gas and propane furnaces require electricity to operate. Communications terminates without power and most of our present day advanced medical services are supported by power.

Power is the heart of every shelter system, because most of a shelter's critical life supporting systems are dependent on power in some form. Most shelter systems which are set up to operate for any length of time require the support of some sort of active power generation system.

Alternative energy sources, such as solar photovoltaics, and wind are a nice idea and may be useful as a power source after a disaster has occurred, but are not necessarily dependable primary power sources for an underground shelter. Solar panels on the surface can be easily damaged by weapons' effects, abnormally high surface winds or sabotage.

Charged batteries can supply short-term power. Smaller systems can operate for significant periods of time with battery stored D.C. current, but ultimately the batteries need to be recharged once the stored electrical energy is expended. This recharging can only practically be accomplished by a gas or diesel generator.

Electrical Terms

Certain terms are used to measure the quantity and force of electrical power. **Amperage** represents volume and **voltage** represents the strength or strength of the electrical current The similarities in the flow of water and electricity can help us understand amperes and volts. **Amperes**, or **amps**, represent the rate of flow like the volume or gallons per minute of flowing water. **Volts** represent the force of electricity similar to the pressure of water in a pipe or the pounds per square inch.

Another electrical term that needs to be understood is **watt**. Light fixtures and electrical appliances are rated in terms of watts. A **watt** is a measurement of electrical power which takes into consideration both the volts and the amps used by an appliance. If you multiply the volts an appliance uses by the amps it draws, the result will equal the total wattage.

Volts X Amps = Watts.

A.C. or D.C.

Electrical power comes in two basic forms. The first is **A.C., alternating current**, the power utilized when someone plugs into an outlet in their home or business. With A.C., the electrical current flow changes direction or "alternates" 50 to 60 times per second. The other form is **D.C., direct current**, which powers automobiles and many of the appliances in recreational vehicles. With D.C., the electrical current flows "directly" (in one direction). A.C. power is usually available in 110 and 220 volts. D.C. power systems usually operate in the 12, 36 or 48 volt range.

One of the initial questions that needs to be addressed is whether your shelter is going to operate off a D.C., direct current system, or A.C., alternating current or both. One of the advantages of an A.C. system is that it will power conventional appliances which are relatively inexpensive. The other advantage to A.C. is that it is not subject to extreme line voltage loss.

D.C. power, between 12 and 36 volts, is a little safer than 110 to 220 volt A.C., in regards to electrical shock. Also, D.C. power systems are generally made up of lower technology components making them less susceptible to the damaging effects of EMP. Electrical wir-

ing for D.C. systems are easier to install and less expensive than an A.C. system, as long as the system does not involve any long power line runs.

D.C. Voltage Line Loss

Long power line runs for D.C. systems are not efficient. There is a phenomena referred to as voltage line loss which effects the transmission of D.C. power through wires. This becomes an acute problem on long runs of wire. The only solution to this line loss problem is to greatly increase the diameter of the power cable. This increase in power cable size greatly increases the cost of the system.

A "Wire Loss Table" #13-1 found at the end of the book in Appendix "D" shows the voltage loss experienced at various sized cables and at 120, 12 and 24 volts. The tables shown in #13-1 are five percent tables, which means that at the listed amperage ratings and at the listed distances, 5 percent of the power would be lost to friction. A five percent voltage loss is normally acceptable in a low voltage system. If you desire a reduced voltage loss of 2 percent, divide the given distance by 2.5. For a 10 percent loss, multiply the distance by 2. For a 48 volt system double the 24 volt distances in order to obtain a 5 percent loss figure.

Another disadvantage to a purely D.C. power system is that conventional A.C. household appliances cannot be used. This is offset by the availability of some appliances developed to fill the needs of the RV (recreational vehicle) community and the marine industry, but nonetheless D.C. appliances are much more expensive than A.C. appliances. In most cases D.C. appliances use less energy than A.C. appliances. In general, low voltage systems by their nature are not conducive to high-energy-use appliances and equipment. Blower fans and deep well pumps are two appliances that only work efficiently with A.C. Energy conservation seems to be an essential part of the nature of low voltage systems.

Inverters

In a limited sense, with an inverter a person can have their cake and eat it too. Inverters bridge the gap between D.C. and A.C. systems. An inverter can be added to a D.C. power system and will change direct current into alternating current. This allows a shelter to have A.C. power for equipment and appliances which cannot be obtained, or are too expensive in a low voltage version or function more efficiently in an A.C. version.

An inverter is a piece of high tech electronic equipment and is thus susceptible to

being disabled or destroyed by electromagnetic pulse. Inverters need to be shielded from E.M.P and have M.O.V.s installed on all of the lines coming into and leaving the inverter. This is discussed in the chapter on E.M.P. protection.

Be aware that an inverter is not 100 percent efficient. Depending on the make and model of the inverter, it is going to have an inefficiency of between 5 and 20 percent. This represents electrical power which is lost as heat during the inversion process of changing D.C. to A.C. Also, inverters are not efficient when they are operating at a demand which is less than 10 percent of their rated capacity. Inverters have optimum efficiency operating in a range between 20 and 50 percent of their rated capacity. Large, late generation, inverters draw about 0.3 watts when they are in the idle mode. This same inverter will draw 10 to 15 watts when it has to start up and run a 2 watt load.

Trace Invertor

Power Waveform Problems

Inverters imitate the A.C. sine wave form. This is not a perfect imitation. Most of the newer versions on the market today produce what is called a "modified sine wave" form of A.C. power. The first generation of inverters produced in the 1980s and a few cheaper ones on the market today produce what is called a "square wave" form of A.C. power. Sophisticated electronic equipment doesn't like the power produced by square wave inverters. In some cases it will destroy the equipment. Even with a modified sine wave inverter, electronic equipment with silicon controlled rectifiers or triacs as filters on the incoming power can be damaged by inverter power. This modified sine wave can produce a buzz in radios and amplifiers. Inverters also broadcast radio noise, mainly on the AM band, which can interfere with reception if the radio is too close to the inverter.

Limitations

Most appliances will operate off inverter power as long as the inverter and the battery bank have sufficient capacity. But some equipment such as refrigerators, large motors, air conditioners, electric heaters and power tools are such exorbitant power consumers that they will in short order deplete the energy stored in the battery bank. Large inductive or split-phase type A.C. motors will draw up to six times their rated operating wattage when starting up. A typical 1/2 horsepower motor can draw 6,700 watts while starting up. In most cases, it would be best to have an A.C. generator that can be fired up when the need arises to directly power any of this type of equipment.

Inverter-Charger

The last consideration in inverters are battery chargers. A battery charger "converts" A.C. power into D.C. power. If your main generating power source is an A.C. generator, you will have to obtain a battery charger. A company named Trace makes an inverter that is also a converter. In essence, the power can run either way through the Trace unit. Power running from the A.C. generator through the Trace into the batteries is converted to D.C. D.C. power running from the batteries through the Trace to the shelter's electrical distribution system is inverted into A.C.

Generator Hybrid System

One practical solution to stabilizing power loads is a generator hybrid system. This involves coupling a generator to a battery charger, a battery bank and an inverter. The battery charger operates whenever the generator is running and the inverter powers the A.C. loads whenever the generator is not running. As we previously discussed, there is a certain amount of efficiency loss involved in changing or inverting battery stored D.C. power to A.C. power with an inverter, but it is much more efficient than running a generator for extended periods of time with only minimal loads.

Ultimately, the question of continual power generation versus operating off batteries is a question of system size and the system's demand for electricity.

Batteries

Batteries are sensitive and expensive. Experience has shown that batteries are the weak link in an independent power system which requires close attention and maintenance. I have seen thousands of dollars of battery investment in shelter power systems go down the drain due to poor maintenance, self discharge and freezing.

When shopping for batteries, the important factors to compare are how many amps will the battery deliver over a long slow discharge and how many times the battery can be cycled before it wears out. The battery's **amp-hour rating** and its **cycle life** are the primary factors which should be evaluated in choosing batteries for a shelter power system. Battery manufacturers rate their batteries at a particular **hour rating**. A 20-hour rating is the minimum standard for a battery to be used as part of an independent power system. A battery's energy storage capacity is also rated in **amp-hours**. One amp of power supplied for one hour = one amp-hour. Batteries with a higher amp-hour ratings and greater expected cycle lives, are going to produce more power and last longer.

Another term used to indicate the potential life span of a battery is the total charge-discharge cycles a battery is designed to endure at a given **DOD (depth of discharge)**. The DOD is the optimum amount that the battery can be discharged to without shortening the battery's expected lifespan in terms of charge-discharge cycles. If a battery is discharged beyond the optimum point, its lifespan will be shortened. If it is discharged completely and left, it will be ruined.

Cold Cranking Amps is a term used to classify "car" batteries. Car batteries have no application in alternative battery

Generator Hybrid Power System

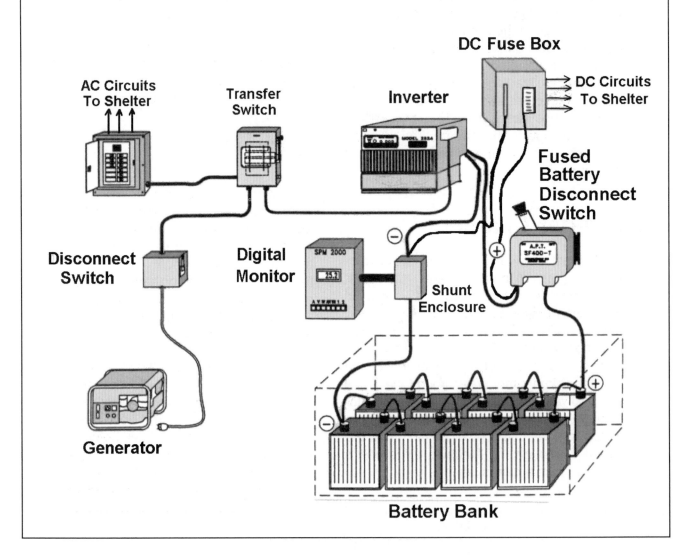

Generator Hybrid System
With Solar and or Wind

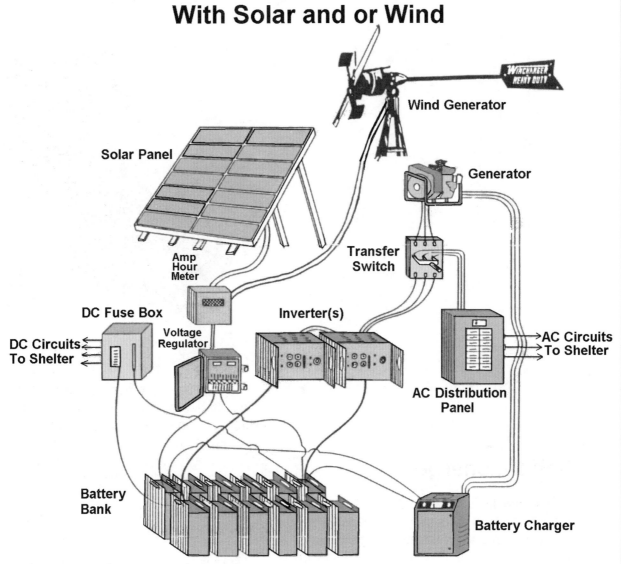

Wind Generator

Solar Panel

Generator

Amp
Hour
Meter

Transfer
Switch

DC Fuse Box

Inverter(s)

DC Circuits
To Shelter

Voltage
Regulator

AC Circuits
To Shelter

AC Distribution
Panel

Battery
Bank

Battery Charger

storage systems because they are not capable of "deep" cycling, which is a primary requirement. A Cold Cranking Amps rating indicates a "shallow" cycling battery.

A shelter power system utilizing battery storage, requires rechargeable "deep cycle" batteries. A "deep cycle" battery is one that can stand repeated cycles of deep discharge and still function. A cycle is a repeated sequence of charging and discharging the battery. Automobile batteries are designed to give a vehicle's starter a strong short surge of electricity to get the engine started. At this point the alternator takes over and recharges the battery immediately. Automobile batteries degenerate when they are left for extended periods of time partially charged. Deep cycling also degenerates automobile type batteries very rapidly.

Industrial Battery

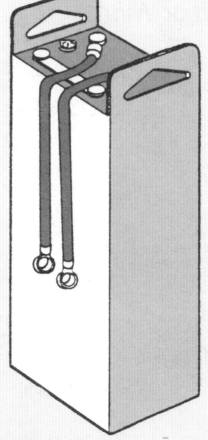

Self Discharge

All lead acid batteries that are charged and left to stand will slowly lose this charge. This is referred to as self discharge. This is an important thing to remember in shelter applications where the system is on standby and generally not being used. If the batteries are not given scheduled maintenance and recharging they will eventually go flat and be ruined.

A company named Battery Doc makes an electronic device which is purported to be able to restore dead batteries by removing lead sulfate coatings which accumulate on the interior plates when batteries go dead. For information call Battery Doc at 1-(800) 866-3750.

One thing to remember with lead acid batteries is that during the charging cycle they generate potentially explosive hydrogen gas. This needs to be vented off in a safe manner. One option to venting is a product which is called a Hydro Cap. Hydro Caps convert the hydrogen back into water and put the water back in the battery.

Temperature affects battery performance. Cool temperatures slow the chemical reaction which takes place in a battery. Ideally, batteries should be kept at a temperature of approximately 60 degrees F. in order to maintain maximum performance efficiency. Underground shelters, in the Northern United States, maintain a constant 40 to 50 degree F. temperature year round.

When sizing your battery bank remember that bigger is better. A larger battery bank will last longer than a

Battery Amp-Hour Capacity & Deep Cycle Comparison

Battery Type	Amp. Hr. Capacity	Tolerated Deep Cycles
Auto - 12 Volt	60 - 70	20
R.V. - 12 Volt	100	200
Golf Cart - 6 Volt	20	250
Mining Vehicle - 6 Volt	350	750

Hydro Cap

smaller one with similar loads because the demand is less on each individual battery and you will have the ability to expand your system to handle unforeseen loads.

Battery Types

There are three types of lead acid batteries but the only one which is applicable for shelter power systems is the lead-antimony.

Lead-Antimony

Lead-Antimony batteries have high discharge rates, good charge/discharge efficiencies and high reliability in fast charging and deep cycling use. These type of batteries are especially appropriate for systems with generators and battery chargers.

Lead acid batteries can be obtained in the conventional "flooded" type or in the "sealed-gelled" type. A flooded type uses a liquid sulfuric acid electrolyte and the "sealed-gelled" type battery uses a fully gelled electrolyte. The advantage of the sealed-gelled battery is low maintenance, but they cost about 35 percent more than the conventional flooded type.

Nickel-Cadmium

Another shelter battery option is the Nickel-Cadmium battery which has many advantages over the lead acid battery with the exception of cost. Nickel-Cadmium batteries cost 6 to 10 times more than a comparable lead acid battery. 100 percent of the stored power in a Ni-Cad can be used without seriously harming the battery and it can be left for years at a partial or depleted state of charge without causing severe damage. Ni-Cads achieve the longest life when they are cycled no lower than 20 percent below their total capacity. Also, low temperature does not affect the performance of the Nickel-Cadmium battery. These batteries can last 20 or more years.

There is also a new variation of the Ni-Cad called a fiber-nickel-cadmium which has an internal fiber system in-

177

Battery Cranking Power Available At Different Temperatures

80° F.	100%
30° F.	65%
0° F.	40%

stead of conventional plates. The rated lifespan for this new Ni-Cad is 15,000 cycles at a 25 percent depth of discharge.

Surplus Lead-Calcium

From time to time the telephone companies end up retiring Lead-Calcium batteries which they use for backing up their systems. These batteries are usually relatively inexpensive and normally have a lot of life left in them but they have certain drawbacks. Lead-Calcium batteries don't like to be cycled below 15 percent of their capacity. Also the cells are 4 to 5 times larger than the cells of a comparable lead-acid battery. This means they take up more room and they are harder to move.

It is not a good idea to mix batteries of differing type and age. This tends to shorten the life of the battery bank.

One final consideration regarding batteries is the dry storage of spare batteries for future use; keeping the liquid electrolyte separate. This is a very good idea but you have to be careful. Not all batteries are manufactured with a truly dry method. Even though you can purchase most batteries special order without the electrolyte in them, the lead plates will still sulfate and ruin during dry storage if the battery was not manufactured with a truly dry process.

Batteries In Parallel vs. Series

Batteries may be wired in either series or parallel configuration. When a battery is wired in series, the positive terminal is wired to the next battery's negative terminal. This increases the voltage while maintaining the amperage of the two batteries. With parallel wiring, the positive terminal is wired to the next battery's positive terminal, and the negative terminal is wired to the next battery's negative terminal. This arrangement increases amperage while maintaining voltage. A battery bank may combine both series and parallel wiring configurations. Batteries are usually wired in series in order to achieve a specific voltage. A number of batteries wired in series can be connected together in parallel to increase the amp hour capacity of a given battery bank.

Equipment and technical advice for remote power systems can be obtained from Sunelco Inc., 100 Skeels St., Hamilton, MT 59840-1499, (406) 363-6924, order # (800) 338-6844.

Generators

A generator is the ultimate source of power in a shelter which will facilitate the extended support of life in an isolated environment.

One possible power generation option for those building small to medium-sized shelters, is a small Balmar air or water cooled diesel powered generator unit. Balmar makes both A.C. generators and D.C. battery chargers. Balmar power plants are made for the marine industry and have shown themselves over time to be very dependable. The requirements of a marine environment are very similar to that of underground shelters

Balmar AC generators range in capacity from 120 volt (3.5 KW) units to 120/240 volt (10 KW) units. Be advised that these A.C. units are comparatively expensive. Balmar DC power plants can be obtained in both 12 and 24 volt configurations and models vary in capacity from 75 amps to 200 amps. The Balmar PC-75 battery charger is powered by a water cooled 4 HP Kubota diesel which will produce 70 amps of 12 volt DC power continuously for 24 hours using only three gallons of #2 diesel fuel. Depending on the size, most Balmar units are diesel powered with either a Yanmar or a Kubota engine. Balmar units with water-cooled engines can be ordered with marine heat exchangers.

A number of companies make smaller diesel powered engines. If a person has good mechanical abilities, they can build their own power plant by putting a diesel engine together with a high output D.C. alternator or an A.C. generator. Balmar makes marine/military spec, high output D.C. alternators in a number of models, which range in capacity from 75 to 200 Amps.

Propane powered generators are the next consideration. Many people in rural settings already have a propane tank and use it for heat, cooking or hot water. Propane powered generators like the Generac pre-packaged emergency power systems are user friendly, self contained, and uncomplicated to install. They are built specifically to provide back up power in the event that your local power supply fails. Propane generators cost less than their comparable diesel counterparts, but their expected life span in terms of running hours is less. As we will discuss later in this chapter, **CAUTION: Propane is heavier than air and very explosive.**

Generac propane generators, Balmar diesel generators, high output alternators and inverters can be ordered from Yellowstone River Trading, P.O. Box 206, Emigrant, MT 59027, (406) 333-4707 or (800) 327-7656, e-mail <yrpub@alpinet.com>.

For the larger systems there are a good number of water-cooled diesel generators available on the market. China diesel generators are fairly noisy and they don't have an industrial performance track record, but they are relatively inexpensive.

It is advisable that you have a back-up generator in the event that the primary unit fails. When you have a larger shelter, not only do you have the potential for better facilities, but also you have accountability for more human lives. This being the case, larger shelters have a greater responsibility for system redundancy. Without redundancy, during a real emergency a mechanical breakdown or system failure could produce tragic consequences. The other related essentials are: spare parts, know-how and the right tools to fix the problem.

A scheduled routine shelter maintenance program is an absolute necessity. Systems must be maintained and ready for use prior to an emergency. Maintenance will also be very important during the actual emergency operation of a shelter. The local hardware store might not be open during a real crises. It is too

Electrical Demand Planning Worksheet

AC/DC	Appliance	Qty	Watts Per Hour (Volts X Amps) Mult by 1.1 for AC			Hours In Use Per Day		Total Daily Wattage
			x			x		
			x			x		
			x			x		
			x			x		
			x			x		
			x			x		
			x			x		
			x			x		
			x			x		
			x			x		
			x			x		
			x			x		
			x			x		
			x			x		
			x			x		
			x			x		
			x			x		
			x			x		
			x			x		

Maximum AC Wattage At One Time		Total Wattage (Watt Hours) Per Day

Total Wattage Per Day		Battery Inefficiency Factor		Total Adjusted Daily Wattage
	x	1.25	=	

AC Loads Which Will Run Directly Off the Generator

Appliance	Qty	Wattage Per Hour	Hours Per Day	Watts Per Day

Battery Sizing Worksheet	
Total Adjusted Daily Wattage (from Demand Worksheet)	
System Voltage? (12 or 24)	÷
Total Amps Per Day	=
Desired Capacity of Battery Power in Days	X
Required Amp Capacity for Battery Bank	=
Depth of Battery Discharge - DOD	÷ %
Total Amp Hour Battery Capacity	=

easy to let valuable equipment and systems go into decay and neglect.

Generator Capacity

A generator's capacity to produce electrical energy is rated in terms of **kilowatts (KW)**. A kilowatt equals 1,000 watts. Most generators are rated based on their capacity to meet a momentary surge and their continuous output capacity is generally less. Also, a generator's rating is based on performance at sea level. The real rating of a generator is 3.5 percent less for every additional 1,000 feet of altitude.

The generator size needed cannot be determined until the total electrical demand for the shelter has been calculated. For example, if a person had a trailer that needed to be towed somewhere, but they didn't have a vehicle available to do the towing, the first logical step would be to determine how heavy the load in the trailer was. A person dosen't want to be pulling a 20 ton trailer with a

Volkswagon and on the other hand they don't want to be pulling a 200 lb. load with a diesel tractor. So don't go out and blindly buy a generator without first determining the electrical load.

In order to achieve optimum fuel consumption efficiency and reduce premature engine wear, a generator should operate at 70 percent of its load rating. In shelter applications, where you want to maximize the use of your fuel, this is important. The length of time you can stay in the shelter and operate the life support systems is dependent on how long your fuel lasts.

Load Scheduling

In a shelter during the course of a day there will be periods of peak demand for electrical power and other times there will be minimal demand for electricity. As long as the generator is run close to its rated capacity, efficiencies are maximized. This, in reality, usually does not happen. The trouble is that the majority of a generator's run time is spent powering only a few small loads. Typically in a home application, a generator will run for 16 hours a day waiting to power a large load for only an hour or two. Not every piece of electrically powered equipment is going to be running at the same time.

In many cases, loads can be scheduled in order to try to maintain a somewhat constant power demand. For instance, if you are running a deep well pump, requiring a substantial power demand, which is supplying a water holding tank, the use of this pump could possibly be scheduled during a time in the day when loads are typically low. Also, certain appliances can be scheduled so that they are not running at the same time. Certain loads will be fairly constant most of the day, mainly the lighting and air supply blowers.

Peak Loads and System Sizing

The first step in designing your power systems is to make up an electrical load schedule map for a typical 24 hour period. This will help you to see what your peak loads are, when they occur and how, if possible, they can be rearranged to balance the generator load.

The second step is to fill out the Electrical Demand Planning Worksheet and the Battery Sizing Worksheet on the next several pages. The electrical demand of all the systems needs to be calculated and totaled. This includes all appliances, pumps, blowers, water heaters, cooling equipment, lights, radios, etc.. See Appendix "D" table 13-2 for a list of the typical wattage requirements of common appliances.

When it comes to generators, bigger is not necessarily better. Diesel generators obtain maximum fuel efficiency when operating at about 70 percent of their rated capacity. When it comes to battery systems, bigger is better. Another battery consideration is the fact that batteries are only about 80 percent efficient. This means you have to put more power into them than you will ever be able to take out. It takes 1.2 amp hours of battery charging to produce 1 amp-hour of retrievable battery power. The battery's internal losses consume 20 percent of the power.

As we previously discussed the total energy = watts, and watts = amps X volts. Thus the total energy storage capacity of a battery is referred to as **watt-hours. Watt-hours = amp-hours X volts.**

Power Generation Fuel Options

There are three fuel options for powering shelter generators, propane, diesel and gasoline. Each fuel option has its pros and cons.

Propane

Propane stores well, but it is extremely explosive and some people feel it should not be used in below-ground confined areas. Many internal combustion engines can be adapted to operate on liquid propane gas, LPG. Also, other appliances needed elsewhere in the shelter, such as

hot water heaters, cooking stoves, furnaces, refrigerators and clothes dryers, are readily available to operate on propane. If using propane make sure you have a drain to daylight under every LPG appliance and use LPG sensors. Propane is heavier than air. A leak, resulting in an explosion in a confined area would be disastrous and quite possibly fatal if the shelter were occupied at the time.

Gasoline

Gasoline powered generators are inexpensive. The problem with gasoline is that gasoline tends to separate and lose octane over time, if a preservative is not added. Gasoline vapors are very explosive. Avoid gas generators if you can afford a diesel generator.

Diesel

Diesel is the fuel of choice. It is not explosive, and if it is stored below ground with the proper anti-fungal treatment, will last 20 years or more without losing significant octane. Diesel generators tend to last longer than gas powered generators, but they cost considerably more.

Algae will start growing in diesel fuel which is stored above ground and exposed to warm temperatures. This degenerates the quality of the fuel and the algae tends to clog fuel filters. It is best to bury your diesel fuel tank.

Underground Storage

One draw back to gasoline and diesel is that they are environmentally hazardous liquids. Consequently

Balmar Power Charger
Model A-P 5
Kubota OC-60 Oil Cooled Diesel
12 Volts D.C. 140 Amps

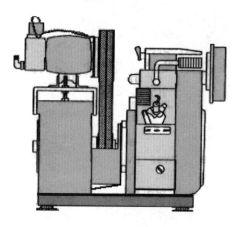

underground storage tanks containing gasoline and diesel are closely controlled by state governments. Permits must be obtained and generally the installation of a tank has to be done by a licensed tank installer. Underground gasoline and diesel tanks must also have vapor monitoring wells. Vapor monitoring has to be done monthly by someone who has a vapor monitoring device. These devices are very expensive. However, there are companies who have this equipment and specialize in doing monthly monitoring.

Compliance with all regulations is definitely in the shelter owner's interest. The cost of a fuel spill cleanup is beyond most people's means. Besides the cost of a leak, the shelter occupants do not want to accidentally lose their fuel during an emergency when their lives may depend on it. The best way to lessen the chances of an accidental fuel loss is to stringently adhere to the standards which experts in the industry formulated. Taking shortcuts or having non-certified people do the tank and piping installations will greatly increase the chances of a system failure.

In some states there is not much regulation of above ground fuel tanks under a certain size. It may be feasible to have a hardened above ground fuel storage tank.

The best fuel storage option in my opinion is a fuel tank inside of a tank, culvert or concrete crypt below ground. The fuel tank needs to have enough space around it so it can be inspected. The tank, culvert or concrete crypt which houses and contains the fuel tank needs to be water or fuel tight so that it will contain any leak from the fuel tank. It should be connected to the shelter in some way but have a tight fireproof door to isolate the area. Such a fuel storage system is not considered an "underground" buried fuel tank and should not come under regulations for underground fuel tanks.

Don't fool yourself into thinking that you can covertly install an underground fuel tank and have it filled. Fuel suppliers in many states must file reports regarding fuel they have delivered to any underground tanks. The current laws controlling underground fuel tanks were implemented for good reasons. Underground storage tank regulations, if fol-

Racor Fuel Filter

Perkins Diesel w/ Marine Heat Exchanger

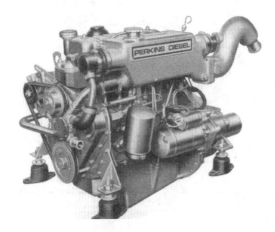

lowed properly, will minimize the potential of a fuel leak.

Diesel fuel, in various quantities, can be stored inside a structure. In the case of a structure like an underground fallout shelter, fuel could be stored in barrels and tanks as long as they could be inspected for leakage. Obviously, the room where such storage occurs should have a floor which would contain a spill without leaking it into the ground. Also, this room should be isolated from the rest of the shelter with a fireproof/vaporproof door. Diesel fuel can be purchased for off road use from the local bulk plant. If privacy is a concern, the bulk plant will fill any tank or barrel which you bring to them in the back of a truck.

Consult the local fire marshal regarding the inside storage of diesel fuel.

Long-Term Fuel Storage

Gasoline usually contains certain undesirable constituents which are subject to oxidational changes. This results in the formation of gums and a reduction in octane.

B.H.T. is an effective treatment used by the U.S. Government in gasoline. The addition of B.H.T. inhibits oxidation and attendant gum formation but does not form deposits in carburetor fuel induction systems. B.H.T. can be obtained from Tri-Ess Inc., 622 W. Colorado St., Glendale, CA 91204.

The military uses 1 lb. of B.H.T. per 1100 gallons of gasoline. The supplier suggests that more could be added for longer protection. In tests where the treated gasoline was exposed for two years to hot desert temperatures the fuel performed adequately. If the fuel were stored at a more constant, cool temperature it could maintain quality for possibly 5 - 10 years. For such long-term storage, a chelating agent such as disodium EDTA, in the amount of 100 mg. per gallon will provide further protection by catching free radical metallic ions.

Diesel fuel over time tends to accumulate water. The water condenses out of the air and collects in the bottom of the tank. The interface between the layers of fuel and water provides all the nutrients that micro-organisms need to grow and multiply. These micro-organisms produce acids that corrode metal components and waste products that block fuel lines, clog filters, and can generally foul entire fuel systems with sludge. Diesel fuel stored undisturbed for long periods of time is especially vulnerable to this microbial infestation.

Raycor makes a Biocide treatment for diesel fuel called RX-300. Most fuel suppliers can supply you with a Biocide fuel treatment. The RX-300 treatment is 10 oz. per 50 gallons of diesel fuel. RX-300 can also be obtained from most truck parts or marine supply stores.

Any liquid hydrocarbon fuel which has been stored for a long time should be filtered through a fine filter before use. This is especially true with diesel fuel.

With diesel not only do you need to be concerned with filtering out sediment, but water needs to be removed from the fuel. Racor makes a full line of filter systems and water separator filters.

Co-generation

Co-generation is using waste heat for practical purposes. Power plants, in the process of converting fuel into mechanical motion, not only produce electricity, but also produce a by-product called heat. There are basically four sources of heat produced during power generation: radiant heat from the engine block, heat removed from the engine by the liquid cooling system, heat ejected with the exhaust gasses and friction heat from internal engine moving parts. There is about 95,000 BTU in one pound of diesel fuel. During the process of making mechanical motion, approximately 25 to 30 percent is converted to mechanical power, 10 percent is converted into radiant heat which comes off the engine block, 25 to 30 percent is converted into heat which is removed by the radiator and the water cooling system, 30 percent is converted into heat which goes out the exhaust pipe, and 5 to 8 percent is converted into heat through friction in the engine's moving parts. This heat can be both a resource and a problem.

Most of us do not consider it a big problem having to deal with heat removal based on our experiences with

Bowman Marine Heat Exchanger

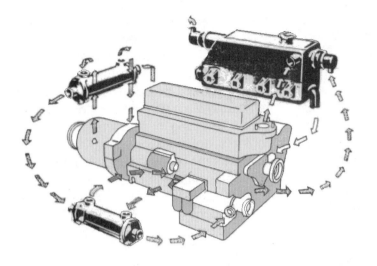

Young Heat Exchanger

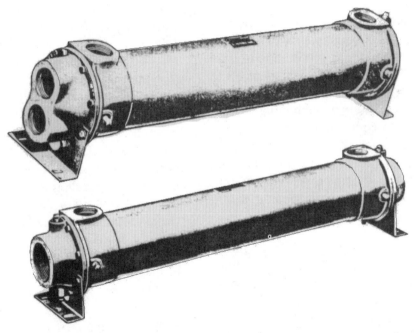

automobiles, but there are distinct differences between a combustion engine in an automobile and a combustion engine below ground in a confined area. The first difference is that the automobile has an unlimited amount of air around it when it is sitting above ground. The other aspect of automobiles is that when they move a large volume of fresh air is constantly passing over the block and through the radiator. The above ground environment facilitates heat removal.

A below ground shelter is a confined area with a limited volume of air. A typical 22 KW water cooled diesel generator in a confined area needs about 1,200 CFM just to remove radiant heat off the block. The operation of the water cooling of the engine, i.e. the radiator which is a water to air heat exchanger, requires another 1,800 CFM. This is a considerable volume of air especially when you are having to contend with a contaminated environment outside and filter all the air you are bringing in. In most cases, in a confined shelter environment, it is not practical to use a radiator, i.e. a liquid to air heat exchanger, as a means of cooling a diesel engine. It requires too much air flow.

In the same sense, large air cooled diesel generators do not work well in the below ground shelter environment because they require the induction of massive volumes of fresh air to accomplish waste heat removal.

Ships have a similarly confining environment. In the marine industry, they use a water to water heat exchanger. This marine heat exchanger uses sea water instead of air to remove the heat that the liquid engine coolant typically brings to the heat exchanger. When using a liquid to liquid heat exchanger, the heat, which is usually discarded, is used for some practical purpose such as domestic hot water or heating incoming air via another liquid to air heat exchanger located in the H.V.A.C. unit.

Another engine cooling option you may have considered is having a remote radiator on the surface. This has certain problems associated with it. First of all, it has to be protected from over pressures if the shelter is built for blast protection. Also, it is vulnerable to tampering. Besides this, there is the logistical problem of having an electrical fan to induce air movement through the radiator and the long, and many times inaccessible, coolant run from the engine to the radiator on the surface.

It should be noted that co-generation cannot be done directly off the engine's coolant system but only indirectly through a heat exchanger. You do not want to tamper with the engine's coolant loop. It is engineered to maintain an optimum operating temperature. As long as you are taking excess heat from it indirectly via a heat exchanger you will not be affecting or interfering with the engine's optimum operating temperature. Modine manufactures liquid to liquid heat exchangers. They are located at 1500 DeKoven Ave., Racine, WI 53401, phone # (414) 636-1200. Young Company also makes liquid to liquid heat exchangers. They are located at 2825 Four Mile Road, Rancine, WI 53404, phone # (414) 639-1011.

Heat recovery from engine exhaust gases is another co-generation resource. This is usually accomplished with an air to liquid heat exchanger. One caution is that the process of removing heat from exhaust gases produces condensation. Contingencies should be made to recover or isolate this condensed moisture so that it does not run down the inside of the vertical ascent of the exhaust pipe and back into the engine. If multiple generators are being used, one operating and one on stand-by, each generator set should have its own independent exhaust pipe. This is so that there is no possibility of condensed moisture from one operating unit running back into the nonoperating unit on stand-by. Water accumulating in a nonoperating unit can cause considerable problems. Also, if the exhaust gases are kept above 350 to 400 degrees, there will not be any condensation problem.

For special applications, air to air heat exchangers are available. These heat exhaust heat exchangers are very expensive and the manufacturers do not recommend trying to run multiple gen sets through the same heat exchanger because of the condensation problems previously discussed. Vaporphase, 600 South Holmes, Saint Louis, Missouri 63122, phone (314) 821-7900 is one manufacturer of exhaust heat exchangers. Exothermic Eclipse Inc., phone # (419) 729-9726 is another. Cain Industries, located at P.O. Box 189, Germantown, WI. 53022, phone # (800) 558-8690. makes the HRS Heat Recovery Silencer.

It should be noted that in some cases oil coolers may need to be installed when engines are installed in a stationary, confined area. Oil temperatures should be monitored to make sure the temperature is not exceeding the manufacturer's recommended operating temperature. These heat exchangers are liquid to liquid and can be used in series with other heat recovery-water heating systems. Bowman is a manufacturer of these oil cooling heat exchangers.

One viable alternative to having both a liquid to liquid heat exchanger, for the engine liquid coolant system, and an additional exhaust gas air to water heat exchanger is to purchase and modify a marine heat exchanger. Marine heat exchangers not only recover heat from the engine liquid coolant loop, but also recover a reasonable amount of heat off the exhaust gases. The modification necessary is replacing the low pressure rubber connections with conventional copper plumbing fittings. Bowman of Birmingham, England is one manufacturer of marine heat exchangers. The American distributor is Trans Cool Ltd, 7700 River Road, North Bergen, N.J. 07047.

Balmar Model FST-3.5
3.5 KW AC Generator
6HP Water Cooled Kubota Diesel

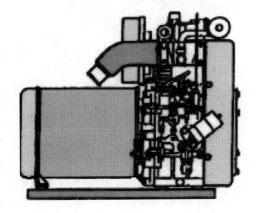 

Chapter 14
Cooking Fuel Options

The three factors which should be considered when choosing a fuel type for cooking in a shelter are Btu performance, safety and how clean does the fuel burn. No one fuel option gets a triple A rating for all three of these factors.

In smaller shelters alcohol, (Ethanol), or candles may be the only safe means of cooking. Many small shelters, like the pre-made fiberglass shelters, are bringing in such a small quantity of fresh air that using natural gas or kerosene would not work. Ethanol burns fairly clean. Its two by-products of combustion are water vapor and CO_2. The CO_2 could create a danger if the shelter is over populated beyond the capacity of its air system or if the shelter is being temporarily sealed from the outside and fresh air is not being brought in.

The only problem with alcohol is that it does not give off much heat in terms of Btus. In other words, it takes three times as long to cook a pot of beans. This may not be a problem in a smaller shelter, because there are only a few people to feed. In a larger shelter where many people need to be fed, it takes too long to cook a large pot of food with alcohol. Natural gas or propane work well and are very stable fuels to store but (as

Alcohol Stove

Perfection Kerosene Stove

Sibir Kerosene Refrigerator

Stove Top Oven

was mentioned in the last chapter) they are extremely explosive. Kerosene on the other hand is non-explosive and burns fairly clean. Perfection is a company which makes kerosene stoves.

Siber makes kerosene refrigerators and freezers. There may not be much need for a refrigerator in a shelter because most of the food is dehydrated, but a freezer may come in handy in holding perishable and frozen foods and medications which are brought in at the last minute. Both kerosene and propane burners should have an exhaust hood and fan over them.

Perfection stoves, Siber refrigerators, freezers and other non-electrically powered

equipment can be purchased from Lehman's Hardware, P.O. Box 41, 4779 Kidron Road, Kidron, Ohio 44636, phone # (216) 957-5441.

Electric stoves are another option. The advantage to electric stoves is they produce hot and clean heat. The disadvantage is that in order to get that electric stove to work, you have to convert diesel fuel into mechanical motion and then into electrical power and then into resistance heat. This involves using a lot of electricity with a lot of wasted efficiency in converting the original liquid fuel to heat by a very circuitous route. In other words, it takes more liquid fuel to cook with electricity than any other fuel. If you have an abundant supply of diesel, an electric stove may be a good choice.

One drawback to kerosene and alcohol stoves is that they don't have big ovens, or depending on the model, they may not have an oven at all. There are stove top ovens which are heated by the burners that can be purchased. This will help expand oven space.

Chapter 15
Lighting

Adequate lighting is important if sanity is to be maintained during an extended shelter occupancy. Lighting can be accomplished by either D.C. or A.C. electrical current. There is a wide selection of lighting systems available. The main considerations when selecting a bulb or fixture are wattage, lumens, lumens per watt, heat given off, rated life and cost. Not all of the wattage used by light fixtures is converted into light or lumens. Depending on the type of bulb, much of the energy ends up being converted into heat. Heat is something you generally do not need in a shelter.

When selecting bulbs scrutinize the manufacturer's listing of how the bulb performs in terms of lumens per watt. This will tell you exactly how efficient the bulb is. Conventional incandescent bulbs produce a lot of heat and don't have a very long life span. Efficient fluorescent fixtures last a long time, produce more lumens per watt and cost much more. See the lighting Comparison in Appendix "D" table 15-1 for a complete comparison of the performance of various fluorescent, incandescent, tungsten halogen quartz and high intensity discharge lamps.

D.C. Lighting

D.C. has both advantages and disadvantages. D.C. lighting systems that run directly off a battery bank are simple and less prone to failure due to E.M.P. effects. (However, E.M.P. can damage the system's battery charging capability.) Even though the D.C.

battery powered lighting is not 100 percent failure proof, it would operate for a period of time after an E.M.P. effect, allowing for lighting to facilitate repair of damage to the generating-charging system. Even if the power system ends up being an A.C. system, battery storage is an important backup in the event of generator failure. The main disadvantage is that long electrical runs with D.C. power in larger shelters can be difficult due to line voltage loss.

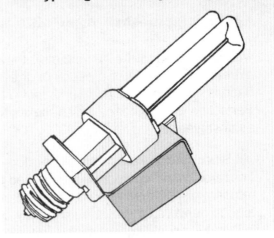

PL Type High Efficiency DC Light Fixture

A.C. Lighting

A.C. powered lighting has certain advantages and disadvantages also. A.C. appliances and lighting fixtures are universally available. In order to utilize a battery power storage system in conjunction with A.C. lighting, an inverter

Efficient AC SL Type Fluorescent Fixtures

  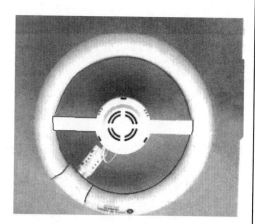

must be used. Inverters convert D.C. power to A.C. power. Inverters are about 90 percent efficient. The lost power, indicated by the lost efficiency, is converted into heat during the inversion process. Inverters have sensitive electronic components, and need to be shielded from E.M.P.

Thus one of the disadvantages to an A.C. powered lighting system is that its ability to operate off reserve battery power is dependent on the inverter. If both the inverter and the generator are damaged by E.M.P., the shelter will be totally without lighting, even with reserve power stored in a battery bank. This means trying to repair sensitive electronic components with flashlights. It seems that some sort of basic emergency D.C. lighting system which can run directly off the batteries would be a good idea. Also, having backups for all sensitive components and the personnel, tools and training to make the repairs is essential.

Calculating Heat Given Off By Light Fixtures

You can calculate the amount of sensible heat a light fixture gives off. Multiply the watts times 3.414, (watts X 3.414 = Sensible Heat). Incandescent and fluorescent have the same watt to heat ratio, but fluorescent bulbs produce more lumens per watt.

The standard incandescent is designed to have a short life (about 1000 hours) and produce seven parts of heat energy to one part of light energy. An efficient bulb such as the compact fluorescent gives off more light per watt and lasts as long as ten incandescent bulbs (10,000 hours). Compact fluorescent bulbs costs more initially, but payback eventually in reduced electricity consumption. A 27 watt compact fluorescent gives off the same amount of lumens as the regular 100 watt incandescent bulb.

Incandescent bulbs operate by passing electricity through a thin filament of tungsten wire. Most of the energy put into the filament is transformed into heat and only about 1/8 of the power is directly converted into light energy.

Compact fluorescent bulbs work on a totally different principle. A high voltage is created in the ballast portion of the light which streams through the fluorescent tube, exciting mercury gas and producing ultraviolet light. This ultraviolet light strikes the phosphorous coating inside the tube, and causes it to glow. The difference in the light quality of these

Energy Conversion Comparison

 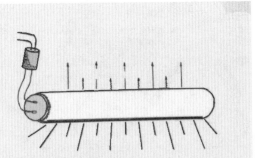

	Incandescent	Fluorescent
Lumens	10%	20%
Radiation	80%	20%
Convection & Conduction	10%	40%
Ballast Heat		20%

lamps and the longer tube fluorescent is in the type of phosphorous coating utilized.

High efficiency lighting is important for shelters because they have to generate their own electricity. Lighting is one of the shelter's bigger electrical loads.

See Appendix "G" in the back of this book for a lighting comparison chart which shows the performance of 27 different light fixtures.

Full Spectrum Lighting

Prolonged stays in shelter environments with poor quality lighting will produce effects ranging from possible depression to chronic illness. Studies have shown that the frequency quality of lighting effects mood, health, thinking processes, organic responses, and the rhythmic functions of the body. The attendant problems associated with the short dark days in Scandinavian countries fostered the development of the full spectrum lamp. The essential element in-

volved is neodymium which absorbs yellow and orange and strengthens other colors to produce a full spectrum light.

One 1973 study by Dr. John Ott found that the installation of full spectrum lighting in a first grade classroom reduced nervous fatigue, hyperactivity and irritability. It also increased attention spans. According to a 1983 report by Dr. David Kripke, when florescent lights in a classroom were replaced with standard incandescent lights, nausea, headaches, nervous fatigue, hyperactivity and irritability were reduced by 32.3 percent. In another instance, a factory documented a 25 percent increase in productivity by changing over to full spectrum lighting.

It would be helpful to install some full-spectrum lighting in shelter common areas, like the dining area, where, in the absence of sunlight everyone can have some daily exposure to and benefit from full-spectrum light.

Flashlight

One more lighting consideration is the flashlight. Every shelter should have flashlights and a supply of spare batteries. Rechargeable batteries are a good idea but be aware that they wear out. Regular lead acid batteries when left in storage will go flat after a while, especially if they are in a warm environment.

There is a special battery called "Code Red" which has a 20 year shelf life. They are activated by twisting a cap. They only come in "D" size and are expensive. These batteries are only available through survival mail order supply houses like Safe-Trek Outfitters and Nitro Pack.

The next hot thing in the battery world is a battery charger that will charge both rechargeable batteries and worn out lead acid batteries. It is called Eco-Charger and can be ordered from Real Goods for $59, 1-800-762-7325.

Specialized flashlight bulbs are another issue to be considered. Delta Light makes a LED bulb for flashlights. It uses 10 percent of the power of a regular bulb and with two "D" cell batteries, it should run for about two weeks straight. The bulbs last virtually forever. LEDs don't come in white, your choice is either yellow or red. Red is good at night because it does not constrict your pupils and thus gives you better vision in the dark. These bulbs do not give off a bright light but have very little electrical drain on the batteries. These bulbs can be ordered for $10 each from Delta Light, P.O. Box 202223, Minneapolis, MN 55420.

Candles

The last lighting option is candles. Candles are not the best option because they consume oxygen and are a potential fire hazard. A number of people make 120-hour candles. In spite of any drawbacks, these candles are an essential backup because they are not effected by E.M.P.

Eco-Charger

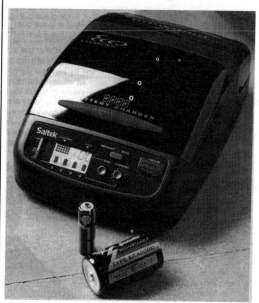

120 Hour Emergency Candle

Chapter 16
Food & Water

Water

The need for an adequate supply of pure water is second only to the need for air in importance. If your shelter is small or expedient and does not have a well or running water you need to consider water storage and rationing. Below is a chart showing the minimum amount of water required per person per day. This varies according to the person's level of activity and the temperature of the environment.

There are several problems associated with water storage. One problem is to keep

Daily Minimal Water Intake

(Person at Rest)

Gallons	Degrees Fahrenheit
.3	60°
.5	70°
.9	80°
1.3	90°

the water from freezing. Having the shelter buried deep enough, (below frost level), solves this problem. The second problem is preventing bacterial growth in the water. Adding one teaspoon of Clorox for every five

gallons of water will prevent the growth of algae and bacteria. The water tank or containers should be emptied and re-filled with fresh re-chlorinated water at least once every six months.

The third problem is storage space for the water. The following gives an estimate of minimal water usage for ten people. At 70 degrees, each person needs .5 gallons of water to drink daily. For cooking and minimal sponge bathing, approximately another .75 gallons per person per day will be needed. The total daily need would be 1.25 gallons of water per person. For ten people the daily need would be 12.5 gallons per day. For a minimum stay of 30 days, this small shelter would need to store 375 gallons of water (approx. 7 55-gallon drums or 75 5-gallon containers). One could solve the space problem by burying a water storage tank adjacent to the shelter. Such a tank should be made of a polyurethane type material or ideally, stainless steel.

The Effect of Radiation on Water Supplies

Ground water would be unaffected by radioactive fallout as long as the well casing was not open at the top allowing contaminates to get into the water source. Eighteen inches of earth will filter all radioactive dust particles out of rain and surface water penetrating into the ground. If stored water is enclosed in sealed containers, any fallout which has settled on the container can be cleaned off before opening. The water in any such container can be safely drunk.

If water is exposed to radioactive fallout, it does not necessarily make the water itself radioactive. Even if the container is open and has been contaminated with radioactive fallout dust particles, the water can be filtered and safely used. The fallout does not contaminate the water chemically, but only through the presence of particulate material which can be removed by filtering. Any material capable of filtering out dust particles such as milk filters, multiple layers of paper towels or layers of cloth from a bed sheet will work as an expedient water filter. A conventional water filter or purification system is obviously the best instrument to use.

If your shelter has a protected well with good water you probably will not need to filter your water. If your shelter does not have a well and is dependent on stored water or some surface source, filtering and purification should be seriously considered.

In war, more civilians die from bad water than from bullets. This is due to the breakdown in sanitation and normal water systems. Civilians resort to drinking contaminated surface water. Even during times of peace municipal water systems can become contaminated. In 1993 in Milwaukee, 4,000 people were treated and 100 died from cryp-tosporidium cyst in the city water system. Chlorine has no effect on crypto-sporidium cyst.

Water Purification and Filtration

There are a lot of ways to purify or condition water. These include reverse osmosis, (RO), activated carbon filters, (ACF), ultraviolet sterilization, (UV), and distillation. No one of these systems is complete in itself and generally needs to be used in conjunction with another system in order to achieve complete purification. Distillation is generally not a viable consideration for shelters due to the power required and the resulting heat and humidity.

Reverse Osmosis

Reverse Osmosis System

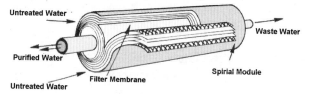

Reverse osmosis is a process that reverses, by the application of pressure, the flow of water in a natural process of osmosis so that the water passes from a more concentrated to a more dilute solution through a semipermeable membrane. This membrane has microscopic openings which are smaller than viruses and bacteria. RO removes viruses, bacteria, parasites, salt, heavy metals, and heavy-molecule chemicals. RO does not sterilize the water. For this reason

RO units should be used in conjunction with a UV system to insure sterilization. An activated carbon filter should also be used in conjunction with a RO unit if there is a concern about chemicals in the water. Most RO units will not work without A.C. current and substantial quantities of water. For this reason, smaller shelters without A.C. power, pressurized water systems, and those dependent upon a limited capacity of stored water, would not be able to utilize RO.

Activated Carbon

Activated carbon filter systems, (ACF), work by passing water through treated carbon. Chemicals, sand and particles in the water stick to the surface of the treated carbon. ACF is the most common type of water treatment. ACF is effective against some chemicals including pesticides, solvents, and chlorine but does not remove heavy metals. In a shelter application, activated carbon filters should not remain wet or filled with water for extended periods of time. When not in constant use filters become incubators for bacteria. Also, once a filter becomes saturated with pollutants it will allow additional

pollutants in the water to pass through the filter.

When selecting an activated carbon filter, a two-stage filter system with 0.1 to 0.3 micron filtering capacity is adequate. The first filter is a pre-filter which removes suspended particles, sand, rust and solids. The second filter removes the bacteria. Multi Pure, P.O. Box 19392, Seattle, WA 98109, (206) 283-4510 or (800) 733-1107, is a supplier for activated carbon filter systems.

Ultraviolet

Ultraviolet sterilization destroys viruses and bacteria. Pura, 1140 So. Aviation Drive, Provo, Utah 84601, (801) 375-3900, makes an ultraviolet water sterilization unit and other water purification units which combine ultraviolet sterilization and activated carbon filters.

If your shelter system operates off a water storage tank, consider the fact that at some point, during or after a disaster, you may have to refill your water tank from above ground water sources which could be contaminated as a result of war.

If you are in a smaller shelter without a pressurized well, you will need a filtering system which does not require excessive pressure to force the water through the filter. Filters can be purchased which have hand pumps for this purpose. Numerous companies supply gravity feed or hand pump Katadyne type water filters. These are used typically by back packers. Such suppliers include Safe-Trek Outfitters, Nitro Pack, the Survival Center and S.I. (See Resource List at end of book).

Ideally, a larger shelter with its own well and a pressurized water system is

Ultraviolet / Carbon Water Filter

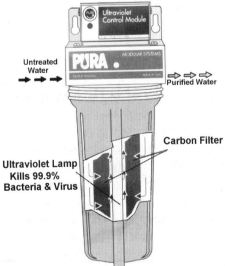

Untreated Water

Purified Water

Carbon Filter

Ultraviolet Lamp Kills 99.9% Bacteria & Virus

the way to go. A larger shelter with an electric well pump should have a replacement electric well pump and a hand pump in the event of an electrical failure. It is desirable to have the well casing coming up into the shelter and to have things configured in such a way so that the well pump can be pulled up for servicing if need be. It is also ideal to have some sort of reserve water cistern as a last resort.

Food

As the first part of this book discusses, the next ten years may be a period of planetary resolution which will outpicture as a syndrome of natural disasters, disrupted weather patterns with attendant crop failures and political and social upheaval. The orderly mechanisms of our civilization are going to be put to the severest tests in recorded history. And though this will bring out the best qualities of self-sacrifice in many persons who rise to the challenge of the times, in general it will be a period of increasing chaos and disintegration.

During such a period, a nuclear attack upon the United States, even a limited one, could destroy the food processing and distri-

bution infrastructure. A total breakdown of the nationwide food supply system could result, plunging local communities into a desperate struggle to feed themselves. It has been determined that far more Americans will starve to death in the year following a nuclear attack than in the attack itself! Food rationing by governmental bodies would be likely wherever it could be organized.

Prepared people can take care of themselves in emergencies and help others. They may be able to retain their freedoms, as well, whereas the helpless would have to comply with governmental restriction and regulation, or go without assistance.

National emergencies have long been used by power brokers to establish greater control over populations. In general, whenever people go to the government for assistance, they pay a price in the loss of some of their previously held freedoms. For example, under an existing FEMA emergency plan, the Soil Conservation Service (a branch of USDA) is to administer the distribution of rationed food county-by-county. Under the same plan, the SCS also administers gun-collection centers. The linkage is obvious, turn in your guns, and you will get your food coupons. In this case, those who planned ahead and stocked up would retain their right to bear arms.

You would not be reading this book if you did not already have an inner conviction that action must be taken to secure the physical well-being of our families and community. We should stockpile food while time and availability remain.

It is recommended that shelters be built and stocked with a minimum of seven months of food and that a total of seven years of food be stored for each person in preparation for an extended period of tribulation.

This section will address practical matters of food storage that have come to our attention in recent years by actual experience in this field. The information will apply both to matters of shelter storage and to individual long-term food storage.

General Comments

Research has shown that the single most important factor affecting a person's physical and mental health during a crisis is the quality of the food available. It is hard to make radical changes overnight. This is especially true for children. In some cases children have been known to starve rather than to eat unfamiliar food. So have a case of jam and some peanut butter to keep the kids from starving. They say that the best food program is to store what you eat and eat what you store, but I personally don't know anyone who does it.

This brings us to the issue of rotation. Some things like dehydrated vegetables, powered milk, cheese and eggs are not that tasty when compared to fresh items. This

makes them harder to rotate into the daily diet.

It would be good to grow a garden every year and get in the habit now of canning fruits and vegetables every year. Stockpile a seven year supply of canning lids and rings. Stockpile at least several years worth of non-hybrid garden seeds. Get together with friends and start buying things by the case wholesale. Start putting away quantities of things you can't can up yourself.

Don't forget sprouts and vitamins. Most of the food in your storage program is dead. It will supply you with bulk, carbohydrates, some protein and sugar but it is devoid of enzymes, and vitamins. Storing seeds and beans for sprouting is highly recommended. The following list gives an idea of what and how much to store. Alfalfa (48 oz.), lentils (24 oz.), alaskan peas (24 oz.), mung beans (48 oz.), soy beans (24 oz.), buckwheat (24 oz.), unhulled sunflower (8 oz.), garbanzo beans (24 oz.) and rye (24 oz.).

Also, digestive supplements are a must! Most dehydrated food is very hard to digest and can create a lot of gas. This can make you fairly unpopular in a crowded shelter environment (it might even get you thrown outside).

The Effects of Radiation on Food

Many people misunderstand the real effects of radioactive fallout on food. If food is exposed to radioactive fallout, it does not necessarily make the food itself radioactive. If the food is enclosed in sealed containers any fallout which has settled on it can be cleaned off before opening the container. The contents of any such container can be safely consumed. Food can be safely grown in soil which has been contaminated by fallout from a nuclear detonation as long as the crops are planted a month after the fallout occurred.

Food such as clams, mussels, and organs of mammals should not be eaten because they act as filters and concentrate contaminates. Livestock can be butchered and eaten if the meat is boned and the fat discarded. Milk from dairy cattle grazing on contaminated ground will be unsafe to drink for 30 days after a nuclear event. After 30 days, the milk can be safely used. Root crops, seeds, fruit and berries grown in contaminated soil should not be eaten because of water-borne contaminates they absorb through their roots and concentrate in the plant. A second crop would be safe to eat.

A good practice, if material and time facilitates, is to cover garden areas with PVC plastic sheeting. This will catch and keep fallout particles from soaking into and contaminating the earth. The PVC can be removed once the fallout is over and radiation levels are low enough to go outside of your shelter. This garden area can be immediately used for planting and any subsequent crops safely eaten.

Access

Unless your food is accessible when needed, it is of no value. While a 7-month supply should be organized and accessible inside the shelter, determination must be made where to place the rest of the stock (if any) during its period of storage. The following considerations will arise: A year supply of basic low-moisture food will require roughly 15 cubic feet of storage space per person. To place additional years of food in a shelter in what could be useful living space, with life-support systems, etc. is probably not justifi-

able. Shelter space is valuable and expensive. Less expensive structures can be constructed outside of shelter space that will adequately store food.

Large amounts of food may be inside the shelter, but they might be so jam-packed that the containers in back are virtually inaccessible. Floor space must be allowed to sort through the mass or careful planning in place to "eat your way" progressively into it.

Food inside the shelter is more easily defended, assuming the shelter is defendable. Food in an obvious container outside of a shelter may be a tempting object of theft. Food buried outside in a non-obvious manner is not likely to be noticed or stolen. But the more hidden the food, the more difficult it will be to retrieve. If excavation is needed to uncover a food storage cache, that activity might expose the cache at a bad time.

When burying food it is wise not to put it directly into the ground (in buckets or cans for example) but rather to put it inside a larger tank or other container first for the following reasons:

1) Buckets are not waterproof or airtight, despite conventional theories to the contrary. Ground moisture will infiltrate the container eventually, thus spoiling the food.

2) All small containers lack the strength to withstand the weight of vehicles on top. They will split and spill the contents.

3) Digging up the cache will be a lot harder than burying it. You will probably dig up most of it just to find one particular container.

Food Storage Life Expectancy

Description	Nutritive Storage Life	Storage Life Before Spoilage
Grains and Beans	Indefinitely	Indefinitely
Rice, Brown	1-2 years	Not good for storage
Rice, White	3-5 years +	Approx. 3 years
Sprouting Seeds	10 years +	Indefinitely
Powdered milk, low-fat	5 years	Indefinitely
Honey	Indefinitely	Indefinitely
Baker's Yeast	1-2 years	5 years w/ refrig.
Powdered Butter	2-3 years	5 years w/ refrig.
Powdered Cheese	1-2 years	Up to 4 years
Garden Seeds	3-5 years	10 years +
Salt	Indefinitely	Indefinitely
Freeze Dried Foods	7-15 years	Up to 15 years
Dehydrated Foods	7 years	Up to 15 years
Herbs Capsulated	3 years +	Indefinitely
Vitamin C	2 years	10 years
Olive Oil	Indefinitely	Indefinitely
Commercially Canned	6 mo-1 year	Up to 3 years

Note: The one factor that has the greatest effect on the longevity of food in storage is a cool stable temperature. Food subjected to the fluxuations of summerheat and winter cold will degenerate quickly. **Nutritive Storage Life** = The length of time a particular food can be stored and still retain most of its original nutrients. **Storage Life Before Spoilage** = The length of time food can be stored before it spoils and thus becomes dangerous to eat.

4) Food that is buried outside can be lost or forgotten. Presence of the food and its exact location must be known to friends and anyone likely to be supervising further work in the nearby proximity. Buried buckets and barrels have been accidentally unearthed by bulldozers (and destroyed in the process). Such a situation does no one any good; its just a waste of time, money and effort.

Optimum Storage Environment

The cooler the storage environment, the longer will be the storage life of the food. Living areas of the shelter will be warmer when in use than buried structures for food storage outside. Food stored in areas exposed to summer heat and temperature extremes will degenerate rapidly.

Identification

Contents of each container should be clearly marked, including the date of packaging. The owner's name should be on each box, can or bucket if more than one party is occupying the storage area.

Organization in Storage and Kitchen

Be organized! Place food into storage according to a workable plan. If the cook can't find the food item she/he wants, what good is it? Whatever the scenario that would require people to take shelter, food can be one of the most comforting and normalizing elements for the shelter occupants. Therefore, the cook is one of the most important people you have. She/he must be given all the assistance possible to prepare decent meals.

This means a kitchen that is intelligently planned out, food that is accessible, an accurate inventory, a pre-planned menu and reasonable privacy for the cook to work in. Post a map in the kitchen of where all the food is stored, so that if the original cook doesn't show up in time, somebody else will know where to start. Practice using the kitchen. Have the ingredients for the first week's menu all laid out. Those who spend some time in their shelter and actually prepare meals there will be miles ahead of those who have never put their system to work. Last of all, don't forget a good can opener and a bucket opener for 5 gallon plastic pails.

Packaging of Food for Storage

Any group shelter must carefully consider the issue of how the stocked food is packaged. Since we have no certain knowledge of when that food will be used, it must remain in a state of readiness - possibly for years. It can only do this if it is packaged for long-term storage. The following are factors to consider on the subject of packaging, and we will discuss each in turn: **a.** consequences of no packaging at all, **b.** shape and size of containers, **c.** the ideal food storage container, and **d.** protection and preservation of contents.

No packaging at all is a formula for disaster! People who have stored food in their shelters in burlap, plastic or paper bags or in cardboard boxes have always regretted their decision. I have seen numerous instances of mice infesting food storage tanks. In one case mice destroyed thousands of pounds of bagged grains. This has happened even though there was not a perceptible source of water for the mice. In this particular instance attempts were made to kill the mice by fumigating the tanks with carbon dioxide. Once the mice were killed, the next job was disposing of the dead mice, collecting the spilled grain, perhaps also separating out the droppings from the kernels, and maybe eventually doing something about the smell.

The other problem has been the presence of ground water seeping into some underground storage areas. Direct contact with water causes paper bags to rupture and spill, then the grains will mold, rot, or ferment. Do not underestimate the amount of moisture in the ground! That moisture, if it finds a way into the storage area, will raise the humidity, soften the paper bags and begin to spoil the unprotected food. Ripped bags and spilled grain are a certainty when damp or weakened bags get moved around.

Shape and Size of Food Containers

Containers need to be manageable in weight. Most people can handle a 5-gallon plastic bucket full of grain, or a case of six metal #10 cans, but find a 6-gallon bucket a bit too heavy. Shape is particularly important in shelters, because it relates to the efficient use of that precious underground space. By far the most space-efficient container available is the 5-gallon rectangular metal tin. Because it has no taper and no rounded edges, it allows nearly 100 percent of available storage space to be utilized. A 5-gallon rounded plastic bucket wastes almost 40 percent of that available space. The tins are however a lot more difficult to handle than buckets because they don't have handles. Cases of #10 cans are nearly as space-efficient as the rectangular tins. There is no perfect food storage container on the market. The perfect container would be airtight, easy to open and close, hard as steel, rustproof, stack nicely in use, and nest within itself when not in use. It would also be inexpensive, have a comfortable carrying handle, have several sizes and finally, be rectangular and without a taper for

space efficiency. Short of perfection, one should choose the most practical container, based on the circumstances at hand. Here is a brief comparison of plastic buckets, metal tins, and metal #10 (1 gallon) cans.

Plastic 5-Gallon Buckets

Pro: Least expensive, don't rust, variety of sizes, easy to carry, commonly available, nest when not in use.

Con: Waste available storage space, allow minor air and water passage through walls, limited stacking strength, not rodent proof, hard to open, stack poorly (unstable column), lids damaged when opened.

Metal 5-gallon Rectangular Tins

Pro: Most space-efficient, airtight, rodent proof, stack well in block, easy to open.

Con: more expensive, harder to carry, less available, one size only, will rust, won't nest.

#10 Metal Cans

Pro: Easiest size to handle (1 gallon), space-efficient, best size for dehydrated fruits and vegetables, most airtight, cases stack beautifully, best size for kitchen use, allows vacuum-packaging in CO_2 or nitrogen.

Con: Most expensive, requires sealing equipment, cans and equipment not readily available.

Protection and Preservation

Most long-term storage foods have a moisture content of 10 percent or less. This is sufficient to prevent the growth of mold. These dry foods will tend to pick up additional moisture from the air, which will decrease their stability (shelf life). A major goal of packaging for storage, then is to keep the dry foods dry. This is accomplished by suitable moisture-resistant containers.

A second challenge is to prevent rodent damage and insect invasion. Since a determined rodent can gnaw through plastic, metal containers are best for extended storage. To control insects, first procure grains and seeds that have been cleaned to USDA standards for No.1 grain. This will greatly reduce but may not eliminate infestation already present. To completely control insects in food requires fumigation by the inert gases of nitrogen or carbon dioxide. Whenever the oxygen in the container is flushed out by one of these gases, the living insects inside will smother from lack of oxygen. Carbon dioxide, unlike nitrogen, is also toxic to insects if present in high enough concentration.

Nitrogen Packing

Nitrogen Packing is effective insect control, as well as a method of preservation. Insects are smothered because oxygen is removed from the container and replaced by nitrogen, an inert gas. Preservation is accomplished because with the removal of oxygen, oxidation ceases. Oxidation is a chemical reaction involving oxygen, which causes food to loose flavor, color, and nutritional value.

True nitrogen packing is a three-stage process: first a vacuum is drawn on the container of food, removing the air; next the container is flushed with nitrogen gas, removing the vacuum; and finally the container is hermetically sealed. Thus the air has been replaced by nitrogen, and no vacuum remains within the food container. Both #10 cans and plastic buckets can be true nitrogen packed. To pack a bucket effectively with nitrogen requires the use of a metalized liner bag (super pails). Specialized equipment exists to draw a vacuum on the metal liner, backfill with nitrogen, and seal. The metalized bag does not admit air passage over time, as does the plastic bucket used alone. This is the only recommended method for bucketing more perishable items, such as dehydrated fruits and vegetables. There are two variations on nitrogen packing:

1) Nitrogen flushing the container without first vacuuming out the air. While called "nitrogen packing" by some storage food suppliers, this method is largely ineffective for either insect control or food preservation since it only replaces a portion of the oxygen in the container, allowing both insect growth and oxidation to continue to some degree.

2) Vacuum packing is the first step without the second. A vacuum is drawn and held while the container is sealed. No nitrogen is added. This process does remove the oxygen, and will be effective as long as the vacuum is maintained. While sometimes useful for household food preservation, this method is not used by storage food suppliers.

CO2 Flushing

Many storage food items keep well with a minimum of treatment. Most grain and beans are stable and will keep their nutritional value for years if kept cool and dry. They are, however, prone to harbor insects, such as weevils and moths. The use of carbon dioxide, in a simple process called "CO2 flushing" or "CO2 packing" will very effectively eliminate these pests. The principle here is not the elimination of oxygen, but elevating the levels of CO2 in the storage container. Slightly higher-than-normal CO2 levels will kill insects, rodents, and incidentally people (if they are in a sealed environment). Simple equipment is used, consisting of a bottle of pressurized CO2, regulator, hose, and injection wand. The cans or tins are filled with the food product and a specified volume of gas is injected into the container before it is sealed.

Unfortunately, plastic buckets cannot be flushed with carbon dioxide. While nitrogen is inert and non-reactive (one of the "noble gases"), CO2 is not. It reacts over time with the interior walls of the plastic bucket, chemically bonding with them. In doing so, the gas vacates the spaces around the food product inside the bucket, and in fact will cause a powerful vacuum to be created inside the bucket. This wouldn't't be so bad, except that it causes the buckets to "suck in" from the

vacuum effect. This distortion can weaken the bucket's stacking ability, and often is so severe that the walls crack, admitting air and insects.

Diatomaceous Earth

Buckets used alone, without CO2 or nitrogen, serve the admirable purpose of inhibiting rodents and moisture, and are far superior to unprotected bags of grains or beans. But there is an additional way to control insects in stored grains and beans, which is the simplest of all methods.

Diatomaceous Earth (DE) is a marine deposit which is mined from the earth, ground up into a fine, talc-like consistency, and used by many farmers to prevent insect infestations in silo-stored grain. Microscopic marine creatures called diatoms formed these deposits millions of years ago. The silica shells of diatoms are extremely abrasive to insects, scratching off the protective waxy coating on their shells, thereby killing the insects from dehydration of body fluids.

DE works as well in buckets as it does in grain silos and is completely non-toxic to humans and animals. Favored by natural farmers over chemical fumigation of silos, it is an inexpensive method for those packaging their own food. Few commercial suppliers of storage foods are aware of DE and none are selling products treated with it.

Commonly Asked Questions

- ### How long will dehydrated food store if I do not open the can?

This is not an easy one to answer. It depends on how the food was packaged and the conditions under which it was stored. Very little scientific testing has been done to determine shelf life. It appears nevertheless that a storage life of 5 years can be expected for 95 percent of the storage foods, with minimal care. Ideal storage is in a dry area that can be maintained in the 40 degree range. Since this is impossible for most of us, 50-70 degrees is acceptable but will shorten shelf life. Any higher temperatures are sure to eat rapidly into the shelf life. It is good to store your food on north walls, close to the floor, in root cellars, in basements without heaters, or even under the house if it is always dry.

Some foods are more sensitive than others. Leavening agents — yeast and baking powder — may need to be replaced after a few years, and oils even before that. If liquid vegetables oils are stored, they should be checked for rancidity after one year. Mono-saturated oils with high oleic acid content, like olive oil and canola oil, have a naturally higher resistance to rancidity than other oils. Dehydrated oily products also have a relatively shorter shelf life. These include butter and margarine powders, and to a degree, powdered milk.

- ### How long will the food keep after I have opened the can?

The information here is more certain then unopened shelf life. Freeze dried foods (meats, fruits or vegetables) can be good for up to 2 or 3 months, if care is taken to always keep a plastic lid on, and if the can is stored in a cool, dry place. Regular air dried food, because it shrinks when dried, keeps much longer then freeze dried. If the plastic lid is kept on the can, minimum storage even for items such as powdered butter is 4 months.

- ### What is the difference between freeze dried and air dried food?

Freeze drying is a process in which a raw or cooked food item is flash frozen, then placed in a vacuum chamber in which the moisture in the food is drawn off. This is a sophisticated method that relatively few processors have the equipment to accomplish. It is a super way to preserve food, but the products are more expensive than air dried foods. Some dried foods are only available in freeze dried form, including meats.

Freeze dried foods cook much faster than air dried. Freeze dried green peas, for example, need only boiling water poured over them and they are ready to eat in about 5 minutes. Air dried peas must be cooked in boiling water for about 20 minutes. That's a lot more cooking fuel!

Air dried dehydrated foods are less expensive than freeze dried. Air drying causes the foods to shrink, making it possible to put more food in a can. Hence, air dried foods take less storage space — as much as 75 percent for some items. Air drying is something that a homemaker can do oneself. It is fun but time consuming.

Example of a Seven Month Emergency Food Supply.

Item	Net Wt.	Sample Daily/Weekly Usage
Meat		
Turkey breast, diced, freeze dried	1 lb.	two 3 oz. servings monthly
Beef hamburger patties, freeze dried	3 lbs.	six 3 oz. servings monthly
Vegetables		
Carrots, diced	2.5 lbs.	two 4 oz. servings weekly
Broccoli, freeze dried	1 lb	one 4 oz. serving weekly
Potato Flakes	6 lbs.	four 4 oz. servings weekly
Tomato Powder	2.75 lbs.	2 to 4 oz. weekly as paste, sauce or juice
Stew Blend ,	9 lbs.	three 4 oz. servings weekly
Grains and Seeds		
Hard Red Wheat	100 lbs.	five slices of bread daily
Rolled Oats	16.5 lbs.	one 4 oz. serving daily
Rye	16.5 lbs	five slices of rye bread weekly
Barley	11 lbs.	2 oz. daily in soup or stew
Millet	4.75 lbs.	three 3 oz. servings weekly
Corn	5.25 lbs.	two slices corn bread weekly
Popcorn	5 lbs.	1.25 cups, popped, daily
Buckwheat, hulled	5.5 lbs.	one 4 oz. serving weekly
White Rice	25 lbs.	four 4 oz. servings weekly
Alfalfa Seed	2 lbs	1/2 cup, sprouted, daily
Eggs and Dairy		
Egg Mix for scrambling and omelets	6 lbs.	1 egg daily Instant Nonfat
Dry Milk	22.5 lbs.	two 8 oz. servings daily
Cheddar Cheese Powder	3.25 lbs	4 or 6 servings weekly, sauce or spread

Fruit

Applesauce Flakes	3 lbs.	three 4 oz servings weekly
Raisins	5 lbs.	one 1 oz. serving daily
Whole Blueberries, freeze dried	1 lb.	one 2 oz serving weekly

Beans

Lentils	5.25 lbs.	one 4 oz. serving weekly
Pinto Beans	5.25 lbs.	one 4 oz. serving weekly
Kidney Beans	5.25 lbs.	one 4 oz. serving weekly
Navy Beans	5.25 lbs.	one 4 oz. serving weekly

Fats and Oils

Butter Powder	2.75 lbs.	1.5 tbsp daily
Olive Oil	1 gallon	1.5 tbsp daily

Misc.

Elbow Macaroni	10 lbs.	four 4 oz. servings weekly
Salt	7 lbs	1/2 tsp. daily
Herbs and Spices,	1 lb	varies
Corn Starch	4.25 lbs.	1.5 tbsp. daily, soup or gravy thickener

Nutritional Guidelines

Considerable research has been undertaken in recent years to develop guidelines and recommendations for storage food programs. Shown above is a typical seven month emergency food supply that would provide a comprehensive, nutritionally balanced program consisting of a broad variety of dehydrated fruits, vegetables, proteins, grains, seeds, and legumes. This program should provide a hearty diet for a normally active adult male for seven months.

Depending on your personal dietary preferences, you may want to vary different food quantities listed above, as well as supplement the program with additional vegetables, freeze dried meat, sweeteners (honey or sugar), oil and condiments.

In spite of all the innovations in food preserving and long-term storage, we unfortunately still have not seen the development of the freeze dried pizza, but the M.R.E. is the next best thing.

M.R.E.

MREs or meals ready to eat were originally developed for the military.

They are fully hydrated and precooked meals with the individual portions vacuum sealed in a foil laminate. You just drop the foil laminate packages in boiling water to heat or open the packaging and eat cold. M.R.E.s are especially applicable for grab and run emergency packs and situations where cooking and meal preparation is not feasible. They only last for about three years. M.R.E.s are loaded with sodium. So plan on drinking lots of water with them. Autopsies of soldiers in Viet Nam who subsisted on M.R.E.s for extended periods of time indicated that the high sodium content can contribute to serious health problems, but they do have their place.

Other Considerations

Most people don't have enough variety in their storage program. The typical food storage program consists of the four basic food items, wheat, powdered milk, honey and salt. Wheat can be considered a backbone of an inexpensive storage program because of its long shelf life, nutritional balance and utility. Statistics suggest that many people can't survive very well on a diet in which whole wheat is the main staple. The big problem is wheat allergies. People with wheat allergies can many times tolerate refined wheat products (made with unbleached or white flour), but not whole wheat products. Many people are allergic to wheat and don't realize it because at this time they are only consuming a small amount of whole wheat in their daily diet. Many people can tolerate wheat in small amounts, but once it becomes the main staple in their diet, they start experiencing serious health complications. This is especially true with young children. The other aspect of a limited food storage program is what is called appetite fatigue. This occurs when people get tired of eating the same foods every day and many times will end up preferring not to eat at all. Again children are especially susceptible to this phenomena. One solution is to purchase less wheat and purchase a variety of other whole grains. A food staple which has good long-term shelf storage is beans. Beans are relatively inexpensive and they can add variety and flavor to a diet. A vast portion of humanity survives quite well on beans and rice.

Nutritional Analysis Comparisons (%)

Grain	Protein	Fat	Carb.	Fiber	Ash
Barley	8.2	1	78.8	0.5	0.9
Buckwheat	11.7	2.4	72.9	9.9	2
Corn	3.5	1	22.1	0.7	0.7
Millet	9.9	2.9	72.9	3.2	2.5
Oats	13	5.4	66.1	10.6	3
Quinoa	**16.2**	**6.9**	**63.9**	**3.5**	**3.3**
Rice	7.5	1.9	77.4	.9	1.2
Rye	9.4	1	77.9	.4	.7
Wheat	14	2.2	69.1	2.3	1.7

**Alpine Aire
Gourmet Reserves**

Another less known option to wheat is Quinoa. Quinoa is an indigenous South America grain like food which has been cultivated in the Andes since at least 3,000 B.C. and was the mainstay of the Inca culture. Quinoa is high in protein, calcium and iron. Quinoa has a 15 to 20 year storage life when nitrogen packed and can be cooked and served like rice or added to baked goods.

Another more expensive option for making your storage program more appetizing is to purchase some Alpine Aire freeze-dried, dehydrated and instant gourmet meals. These don't have an exceptional shelf life, but they are very tasty and easy to rotate into your daily diet. (Some may last up to ten years.) The entrees include thirty-one vegetable, seafood, chicken, turkey and beef meals, fourteen side dishes, ten desserts and three soups. Unlike M.R.E. entrees, they are not loaded with sodium, preservatives or white sugar. They are easy to prepare, by adding hot or cold water, and in many cases can be eaten as is.

Apline Aire Gourmet Reserves can be ordered from Yellowstone River Trading Co., P.O. Box 206, Emigrant, MT 59027, (406) 333-4707 or (800) 327-7656.

When designing your food storage program start from the bottom up instead of the top down. By bottom up I mean look at what your family likes to eat and try to approximate this in your program. Obviously you are going to be handicapped due to shelf storage life, but some of this can be overcome by continually rotating, consuming and replenishing items in your food storage program which have a shorter shelf life. These items could be mayonnaise, canned tomatoes, ketchup, pickles, spaghetti sauce, etc. One help to bottom up food program designing is purchasing a good food storage cookbook. Use the cookbook and your family's favorite recipes to put together a rotating 30-day menu. One excellent cookbook is, *Cooking With Home Storage*, by Vicki Tate. This book can be obtained by calling (801) 835-8283 or writing to: Vicki Tate, 302 E. 2OO N., Manti, UT 84642.

Calculate the quantities of each item needed to fulfill the menu for the full 30 days, including spices, oil and sweeteners, etc. Then multiply the totals times the number of months worth of food you want to have. Finally go out and buy food.

Top down designing is when you shop for a generic one-year food supply without any consideration about how you are going to use the stuff.

One aspect to breaking the monotony of a limited diet is spices and seasonings. Include into your food storage program sufficient quantities of flavorings such as bouillon, onion, garlic, cinnamon, items necessary for baking such as baking powder, soda, yeast, powdered eggs and any other spices you usually cook with. Having these flavorings and spices available will help you to be creative when you are limited to cooking with only those items in your food storage program. Make sure you stock enough cooking oil.

One thing which long-term food storage tends to be deficient in is vitamins. Vitamin C, D and multi-vitamins are important supplements to a long-term food storage diet. This is an especially important consideration for children since children's bodies don't store reserves of nutrients in their bodies like adults do. Also, the situation or environment which would cause you to have to rely on your long-term food storage in all likelihood would bring with it considerable stress. Stress depletes vitamins and minerals in the body and is a major contributor to sickness and disease. Vitamins are another item which lose their potency but they can be rotated into daily use.

Psychological foods and desert goodies such as jello, pudding, tapioca, and pie ingredients may seem frivolous, but can help normalize an otherwise drab diet. Again, this is an especially important issue with children. Also, you can spruce up a storage food menu with dehydrated or freeze dried fruits and vegetables and home or commercially canned food.

When a family or shelter group has their food storage program fully organized and rehearsed they can reap the tremendous advantage of not having to work out so many things during the actual crisis when they are under stress. The more survival procedures, including food preparation, that can be worked out, rehearsed and prearranged ahead of time, allows for the maximum continuity of normal life and attention can be directed to critical tasks. A period of stress is not an effective time to make a radical change in diet.

In Europe during and shortly after World War II, most people were familiar with hunger. One quart of vegetable oil was a highly prized

Country Living Grain Mill

commodity because of its scarcity, the flavor it added to food and its high calorie content. Honey and sugar have always been items of high barter value during food shortages.

I personally am not too optimistic about the idea of living out of #10 cans and 5 gallon buckets. Above and beyond appetite considerations, they are not renewable food sources. Gardens and small livestock should not be overlooked in terms of their importance. Goats can pretty much live off the land. They can stay alive eating the bark off trees. I am not saying you will get the highest production and flavor of milk under these conditions, but they will survive when the cow has died for a lack of hay. For similar reasons, other renewable food sources such as sheep, chickens, rabbits turkeys, and bees might be considered as part of your long-term food program.

Hiding Food

Don't have all your eggs in one basket. In the event of a national emergency, Executive Order 10998 would empower the president to take over all food resources and farms. This means the National Guard could come and confiscate your food reserves with the exception of whatever they considered to be a two-week food supply. Reports from some U.S. military participants in Fort Polk 1995 Partnership for Peace exercises indicated that one of the training scenarios involved an operation against a group of people who had committed the crime of hoarding food. Hide a substantial portion of your food in a secret, secured location. If they can't find it, they can't take it!

The government previously maintained food reserves to feed the population in the event of a national emergency, but these reserves for the most part have been liquidated. Much of our wheat reserves was sold to the Soviets, paid for with loans from New York banks. As was mentioned in Chapter 1, in 1977 the Soviet government completed a five-year food storage project which provides the entire population with 2 pounds of grain per person per day for 300 days. Most of that wheat came from the U.S.

All things considered, I would not in any way feel guilty or unpatriotic about not wanting to give up my food reserves. Most people make great sacrifices and do without certain luxuries in order to obtain food reserves. At the same time, the rest of the population was spending their money on trips to Las Vegas, vacations and new cars.

Comprehensive Food Storage Programs for home or shelter can be obtained by contacting Yellowstone River Trading Company, Safe-Trek Outfitters, Nitro Pack or any of the other companies listed in the Resource section in the back of this book. This list includes phone numbers and addresses.

Passport To Survival, by Ester Dickey, is one of the standards of survival food storage and meal preparation. *Essentials of Home Production and Storage,* published by the L.D.S., can be obtained through the Welfare Services, 7th Floor, 50 East North Temple St., Salt Lake City, UT 84150. These are books everyone who is starting to put together a food reserve should read.

Equipment

Most one-year food programs come with an inexpensive handmill and a cheap can opener. Try grinding wheat in one of these Corona handmills. You won't have to pump iron anymore. If you can afford it, invest in a good handmill. Since the bulk of most long-term food programs is made up predominately of grain, it is very important to have some sort of non-electric means of turning grain into flour and meal. The hand-powered grain mill shown on the previous page is probably the best one on the market today. It can be ordered from Yellowstone River Trading Company and other sources listed in the back of the book. Don't forget a good can opener!

A pressure cooker is a must! Pressure cooking cooks food with the least expenditure of energy. It also cooks things faster and it is one of the best ways to rehydrate dehydrated foods. Pay a little more and get one made of stainless steel. The established medical and scientific community is just starting to acknowledge the long-term health dangers of cooking in aluminum pots.

Forget the microwave. It is a nice clean and energy efficient idea but it destroys the value of the food. The U.S. Navy did a study on a ship where half the crew was having noted increase in sickness. What the study revealed was that the half of the crew that was prone to sickness ate in the second meal shift which had all of their food reheated in microwave ovens.

Chapter 17
Communication

The purpose of this section is to explore the communications options of shelter occupants during catastrophic events.

Having the ability to receive communication will significantly improve the psychological health of the shelter occupants. Also, having radio communication capability in a shelter during a disaster can be critically important to physical survival.

When confined in a shelter underground for a period of time, the only viable source of knowledge about what is happening on the surface, locally, nationally and internationally, is a radio. The unknown, resulting from being cut off from what is going on elsewhere, lends itself to wild imagination, doubt and great insecurity. Every shelter should at least have a good battery operated AM/FM radio and a shortwave receiver with spare batteries. These should be stored in a steel container with a tight fitting lid, such as a garbage can, to facilitate E.M.P. protection. Antennas, pre-positioned on the surface, will also be necessary to insure reception.

Not only can communication have psychological benefits, it can have definite strategic importance also. Being able to actually carry on two-way communications with people in other shelters or people outside of the area affected by the weapon detonation would be extremely valuable.

Radio communication is controlled by the government in almost every nation in the world. The U.S. government assigns certain frequencies for certain purposes. Specific bands are allocated for specific types of communication; citizen, emergency service, commercial broadcasts, military, cellular, satellite telemetry, marine and amateur, to name a few.

Basic Radio Technology

Radio communication operates through the use of electromagnetic energy. When electrical power flows through an antenna, the antenna transmits an electromagnetic radiation in all directions. This electromagnetic broadcast is referred to as a radio wave. Radio waves travel best through outerspace and air, but they can also to a lesser extent transmit and penetrate through earth and water. The radio communication spectrum is comprised of a wide range of different radio frequencies.

There are three basic terms which relate to the properties of a radio wave. These are amplitude, wavelength and frequency. Radio waves have a crest (high point) and a trough (low point). The strength or amplitude of the radio wave is the distance between the crest and trough.

Y = Amplitude

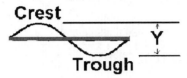

The wavelength of the radio wave is the distance between the crests of the wave. A short wavelength is a radio wave with a short distance between the wave crests and a long wavelength is a radio wave with a long distance between the wave crests.

Frequency is the number of complete waves which occur in a given radio wave in one second. A higher frequency radio wave indicates a radio wave with a shorter wave

X = Wavelength

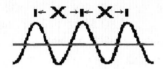

length.

Low and high frequency radio waves are measured in kilohertz (KHz) which relate to thousands of cycles per second and very high frequency radio waves are measured in megahertz (MHz) which relate to millions of cycles per second.

Short Wave

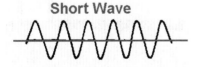

Long Wave

The AM radio band is referred to as a medium frequency. It starts at 540 KHz and increases in frequency up to 1600 KHz or 1.6 MHz. AM stands for amplitude modulation and FM stands for frequency modulation. This relates to two different ways of connecting sound with a carrier wave. The FM band is in what is called the very high frequency range (VHF). It starts at 88 MHz and increases in frequency up to 108 MHz.

AM radio broadcasts long distances but is susceptible to static interference. FM does not have the extended range of AM but it produces a clearer signal. Lower frequency radio waves transmit their signals farther because they bounce off the ionosphere and reflect back down to earth thus enabling them to go over the horizon. Higher frequency radio waves are limited to line of sight transmissions.

One advantage to higher frequency is that it penetrates buildings and earth better than lower frequency. I have personally noticed that I can transmit and receive with my UHF radio (ultra high frequency in the 450 MHz range) while in a shelter under 12 feet of earth, inside a steel tank. Under the same circumstances, a VHF radio (very high frequency 150 MHz range) would not be able to receive or send a signal through such an earth and steel barrier. For this reason, emergency medical services in many cities use UHF instead of VHF since they can receive and transmit better inside buildings with UHF.

Frequency / Meter Designation

One confusing factor in communications is when someone refers to trans-

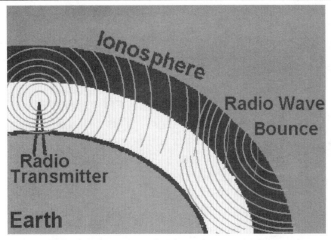

Ionosphere

Radio Wave Bounce

Radio Transmitter

Earth

$$\text{Frequency in MHz} = 300 \div \text{Meter}$$

$$\text{Meter} \overline{\smash{\big)}\ 300} \ \text{Frequency in MHz}$$

$$40\ \text{Meter} \overline{\smash{\big)}\ 300} \ \ 7.5\ \text{MHz}$$

mitting on a certain frequency and another person, usually a licensed amateur, talks about communicating on a certain meter band. It is basically comparable to one person designating a given length in inches, another in centimeters. Above is a conversion formula which shows how to convert frequency in MHz to meters.

Receiving

In terms of receiving broadcasts, there are many sources of signals that can provide a variety of information. In the case of the listen only mode, an ordinary AM/FM household radio will work but with limited effectiveness.

The recommendation would be to obtain an extended range receiver. This range should be from about 155 kilohertz (KHz) to 30 megahertz (MHz) and be capable of detecting signals which are either amplitude modulated (AM), frequency modulated (FM), single side band modulated (SSB) (either upper or lower side band) or continuous wave (CW) telegraphy. The receiver usefulness is greatly enhance if it also can receive signals in the very high frequency (VHF) band from 120 MHz to 180 MHz FM. A separate small vertical VHF antenna is desirable for this band.

Radio Communication Spectrum

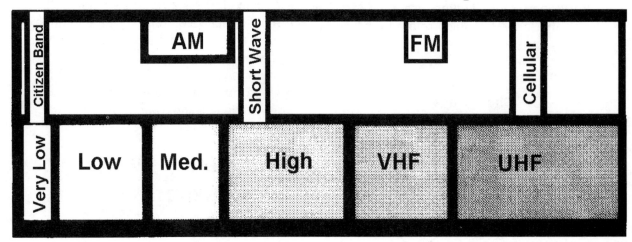

Citizen Band | AM | Short Wave | FM | Cellular

Very Low | Low | Med. | High | VHF | UHF

Very Low Frequency	VLF	10 - 30 KHz	30 - 10 Meter
Low Frequency	LF	30 - 300 KHz	10 - 1 Meter
Medium Frequency	MF	300 KHz - 3 MHz	
High Frequency	HF	3 - 30 MHz	Short Wave
Very High Frequency	VHF	30 - 300 MHz	
Ultra High Frequency	UHF	300 - 3,000 MHz	

Commercial AM / FM Broadcasts

Depending upon the disaster situation many commercial radio stations may not be operating. The government, however, utilizes commercial stations as part of its emergency broadcast system. Considering the widespread geographical distribution of these stations, the probability would be quite high that some stations will maintain broadcasting and that those who have higher quality receiving equipment will be able to receive the signals.

Foreign Shortwave Broadcasts

Shortwave radio broadcasts between 3 and 30 MHz are capable of long distance transmission. This frequency of radio waves bounces off certain portions of the ionosphere. The ionosphere itself is made up of layers of charged particles which start at about 30 miles high and go on up to 260 miles in altitude. Short wave signals propagate better at night because during the day certain layers of the ionosphere actually absorb radio wave energy.

There is a sizable volume of English language shortwave DX radio stations on foreign soil. By far the larger number are in Europe. The BBC in the United Kingdom, the Swiss International Radio and the U.S. Armed Forces Radio network are some of the more well known. Good reception requires more sophisticated radio and antenna equipment but it is probably well worth the expense since information from foreign broadcasts may be very valuable during a national emergency. Fairly inexpensive short wave receivers can be purchased from Radio Shack or any of the radio equipment supply houses.

The better grade of short wave receivers have what is called BFO. BFO is an option which enables the radio to receive single sideband transmissions. This means being able to receive and monitor transmissions from ham operators around the world. In the event of a

Grundig World Receiver
AM, FM, Long Wave & Short Wave

217

national or international crisis, amateurs (Hams) are the best source of direct and unprocessed news. Hams are real people who will generally tell you what is actually going on in their area without the mass media's editorial filtering.

Government Radio Services

These services are of both broadcast (for many listeners) and point-to-point (two-way radio) in nature. In either case the ability to eavesdrop can be fruitful if unsecured voice transmissions are used. Typical of these are the weather radio service, air traffic control, marine radio service, etc.

Most amateur radio supply houses also sell a fairly inexpensive circuit board that, when installed in a PC and coupled to an antenna, will allow a person to receive satellite weather pictures.

Military Radio Services

The information carried by the military signals (both US and those of an invader) is potentially the most useful and probably the most difficult to understand. During a conflict, information regarding military planning, movements and positions could be vital to survival planning and actions. However, these communications tend to be of a secure nature either by intent (coding and/or scrambling) or by the use of other languages in the case of a foreign invader.

Scanners

Scanners are listen only devices which scan the VHF and UHF frequencies. You can program into the scanner the frequencies you

want the scanner to monitor. This is a good way to listen to emergency services. Also, even if you have a multiple channel VHF or UHF transceiver, it is good to have a scanner so you can hear what is going on other channels while you are transmitting on a given channel.

Transmitting

When considering transmit/receive mode the selection of equipment becomes very broad. The options can be narrowed by determining who you want the capability to communicate with. To start out, a shelter should definitely have whatever radio is compatible with other shelters in the area. This could be inexpensive C.B. radios, commercial VHF or UHF, amateur 2 meter ham and even marine band radios during an emergency. If there is another shelter very near, communication by a buried landline would be ideal. For transmitting longer distances, the only option will be amateur ham equipment which operates in the lower frequency range. This requires that someone in your group be a licensed operator and that your shelter have the right equipment. For most shelters, transmitting options may be limited to CB and commercial VHF and UHF due to the lack of licensed operators for amateur equipment.

Citizens Band

Citizens Band (CB) radio service is in the 27 Megahertz band. CB radios are inexpensive and common nationwide. CB radios are limited by law to 5 watts of power which effectively limits them to local communication. If you look in the newsletters which advertise used radio equipment, occasionally you will see advertisements for100 watt linear amplifiers. These are quite illegal but if you had one on the shelf, it could give you inexpensive medium distance (across the country) communication in the event of a national disaster. CB has longer range capability because it is a low frequency wave which can be bounced off the ionosphere when transmitted at a high enough wattage.

There is a more expensive line of CB radios called a CB Side Band radio. These can transmit up to 12 watts legally and thus can transmit farther than a conventional 5 watt CB.

All CB radios have the same channel frequencies and therefore all CB radios can talk to each other. This is not the case with commercial VHF and UHF radios which have specifically assigned and licensed frequencies. Also, for this reason, CB communication is less secured in terms of privacy of communication. Many people have them and everyone has access to the same channels.

Commercial VHF & UHF

State, city and county emergency services have designated frequencies within the VHF and UHF frequency ranges. The VHF also includes a low band of 32 to 49 MHz. Emergency services used to be in this area,

Citizen Band Transceiver

Relm WHS 150

but most have moved up to high band VHF. There is a lot of used low band equipment floating around which is inexpensive. This low band equipment will potentially transmit farther because of the lower frequency but it is just about all crystal equipment. The expense of re-crystalling the radio is three-quarters of the cost of a good synthesized radio. The only way to make effective use of this old equipment is to be able to get a license for the frequency which the radio is already crystalled for.

The commercial VHF and UHF frequencies are licensed. You can apply for your own licensed frequency which can then be programmed into radios in neighboring shelters. Whatever you do, don't transmit on a commercial frequency unless it is licensed to you or you have formal written permission to do so. If someone complains, you would be in hot water with the FCC.

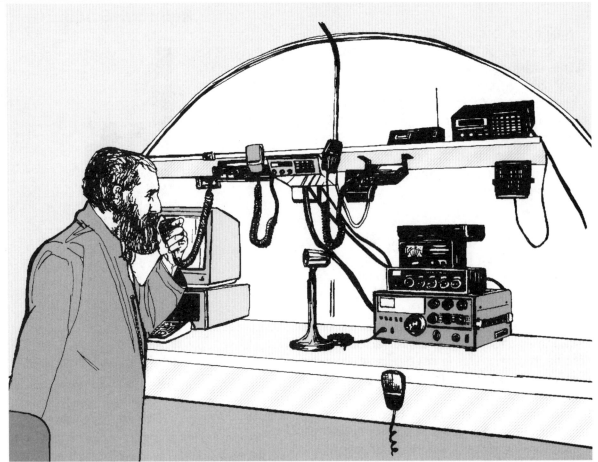

Commercial VHF and UHF radio equipment is readily available and reasonably expensive. This equipment is characterized by its simplicity of operation for people with minimal telecommunication knowledge and experience. These radios are generally not field programmable like similar amateur radios. Field programmable means you can program frequencies into the radio channels with the radio itself. Commercial equipment usually

Relm VHF 25 Watt 16 Channel Mobile Radio (Field Programmable)

does not have this capability and must be taken to a service shop to have new frequencies installed. The exceptions are King radios and the Relm WHS-150 VHF (commercial). Relm also makes a UHF version. This radio is made by Alinco for Relm and is one of the best buys on the market today. This 16 channel radio can be purchased for $349 from Falcon Direct, #36 20th Avenue N.W., Birmingham, AL 35215, telephone (800) 489-2611. Recently both Motorola and Icom have come out with field programmable handheld VHF radios.

Field programmable is an important feature. When confined to a shelter in an emergency, a person can't just run down to the local radio shop to have a new frequency programmed into the radio.

Amateur Radio License Levels		
License	**Test**	**Transmission Privileges**
Common	No Morse Code w/ written test	2 meter band (in VHF range)
Novice	Morse Code 5 words per minute w/ test	Limited portions of all bands
Technical	5 words per minute w/ harder written test	Less restrictions on bands
General	13 words per minute w/ harder written test	Even less restrictions on bands
Advanced	13 words per minute w/ harder written test	Very little restriction on bands
Extra	20 words per minute w / hardest written test	Full use of all amateur bands

Relm also makes a 16 channel 25 watt VHF field programmable mobile unit. Other than size, the main difference between a hand held and a mobile radio is power. Hand-held radios have minimal transmission power (5 or less watts) and thus have a shorter effective distance. Mobile VHF and UHF radios usually start at 25 watts and go up to 40 and 100 watts. Falcon Direct also sells a unit that will boost a 25 watt radio up to 100 watts for about $300.

Amateur radios are less expensive than commercial radios. Some amateur radios can be modified to transmit and receive in the commercial VHF and UHF bands. This is against FCC regulations but it is frequently done. With amateur equipment you can also purchase dual band radios, which can transmit and receive on both VHF and UHF.

FRS Band

Another new option is the Family Radio Service Band which was recently set aside by the FCC specifically for families and sportsmen (no license required). These radios are low power in the UHF (460 Mhz), are inexpensive and have a range of up to 2 miles depending on terrain.

Marine Band Radios

Marine Band radios are another option. It is against Federal Communication Commission (FCC) regulations to use them on land-based applications, but during a major disaster the FCC might not be around to put up a fuss. Don't plan on using them during non-disaster times. The big advantage to marine band is that the radios are very inexpensive.

Amateur Radio Service

This is the oldest, most flexible and therefore most useful of the radio services available. Amateur radio frequency bands are spotted throughout the radio spectrum. In order to transmit on the designated amateur bands, one must pass a FCC exam and apply for an amateur license. There are six levels of amateur licensing, beginning with Common on up to Extra (see the chart). Amateurs with the proper license and equipment can communicate locally, nationally and internationally. This service offers the possibility for a number of communicating modes, such as voice, telegraphy, teletype and packet.

In the lower frequency range (30 Mhz or less) most amateur radio equipment makes use of what is referred to as single side band transmission. In essence this concentrates the power of the wattage and gives the signal a greatly increased distance capability. Amateurs using single sideband radios can transmit farther with less wattage.

Amateur radio capability in combination with a local (shelter-to-shelter) VHF, UHF and CB capability facilitates a broad "look" at the world while keeping in touch with the local area. In a case where a number of shelters are clustered in an area, one shelter with an amateur capability can serve the entire group by relaying information over a local network.

Amateur radio operators (hams) have a wealth of knowledge. In almost every reasonably sized city in the United State there is an Amateur radio club. These clubs put on courses for people who want to get their Amateur licensing. Local hams are a good source of information and advice and they can be a great help in locating used equipment.

It is illegal to encrypt radio transmissions over the designated amateur bands. Packet communications are not considered encryption. An exceptional book for anyone interested in learning more about communication is the ARRL Handbook. This book costs $30 and can be obtained by contacting The American Radio Relay League, 225 Main Street, Newington, CT 06111.

FCC

Be aware that the FCC swings a big club. Anyone caught operating without a license gets an $8,000 fine. Anyone using an unauthorized frequency gets a $5,000 fine. Even hindering an inspection by a FCC agent is a $7,500 fine. The FCC does not need a search warrant. Above and beyond fines, the FCC can confiscate any equipment involved in the violation. The local hams monitor radio traffic on most of the frequencies and commonly turn in unauthorized users to the FCC.

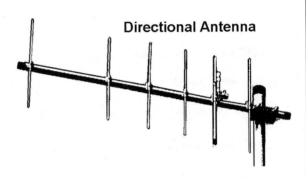

Directional Antenna

There are some exceptions to required licensing for transmitting on some amateur frequencies in an emergency. In the event of a real national emergency, the regulatory people will be covering their heads like everyone else and if the disaster is large enough, regulatory agencies will be a memory from the past.

Contact your local County Disaster and Emergency Services Coordinator (DES) and ask him if you can get his permission to communicate with him in the event of an emergency on his designated state frequency. Also ask him if you can get permission to transmit on your state's "common" mutual aid channel. There is a good chance they will give you permission to transmit to emergency service agencies via the state common mutual aid frequency. Almost all law enforcement, fire, emergency medical services and DES radios scan this frequency.

Antennas

With any form of radio there must be an antenna system. This may take many forms; a long horizontal wire attached to two vertical supports, a vertical rod or mast fixed to a ground support or a tower with a directionally controlled beam array. In many cases the mast antenna is the best for a shelter because it is omni-directional and it presents the least visible image when installed. This type of antenna is portable, reliable and easily installed. A mast type antenna can be quickly stored or erected as conditions require.

Directional beam antennas allow a radio operator to concentrate his receiving capability in a given direction and thus receive with more clarity a distant, weak transmission which would otherwise be drowned out by competing signals from other directions. This type of antenna requires a rotating device and position indicator.

Along with any antenna, a feedline is needed to connect the antenna to the radio equipment. Long feedline runs produce signal loss. Be sure to use high gain cable so as to reduce the effect of this signal loss.

Antennas are tuned for limited frequency ranges. A broad band antenna will do a mediocre job of receiving and transmitting on a number of different frequencies where a spe-cifically tuned antenna will give excellent transmission and reception on a limited spectrum of frequency. A separate antenna is required for every radio.

Radio Nets

In a real emergency everything that can go wrong most likely will go wrong. Organization, training and regular drills are needed to insure operator proficiency. The ideal disaster communication structure is a pre-established radio net (group of radio stations). Such a net should have regularly scheduled on-the-air meetings in which all stations check in with a net control station. During these drills, simulated disaster situations can be held to sharpen communicating skills.

Shelter Radio Room

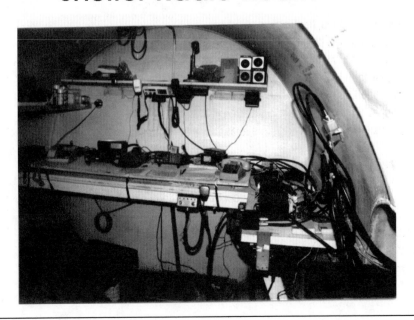

223

Chapter 18
E.M.P. Shielding

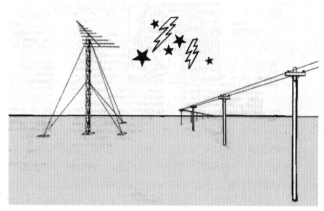

Electromagnetic pulse is a serious problem which any shelter builder should understand and make at least basic contingencies for. When a nuclear weapon is detonated in the high atmosphere, it produces an effect called E.M.P. In simplified terms, this is an electrical charge which collects on wires, cables, antennas, etc. and produces very high voltage for a fraction of a second. It is so fast that surge and lightning protectors will not stop it and it will damage any electronic equipment it travels to, even if the equipment is turned off. A basic solution is to keep sensitive equipment unplugged and stored in sealed metal containers. The sealed metal container will act as a shield against the pulse.

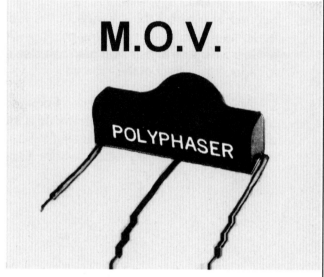

M.O.V.

More sophisticated protection can be implemented by installing special EMP rated gas tubes on antenna wires and EMP rated metal oxide varistors (MOVs) on power lines leading to equipment. The gas tube and MOV protection will not work if it is not properly connected to a special grounding system.

Metal oxide varistors are used on power supply lines which feed sensitive electronic components. MOV surge arresters divert dangerous electrical surges created by E.M.P. to ground. MOV surge arresters come in various sizes. The exact size needed for a particular piece of equipment will depend on

Polyphaser Gas Tube

the type of electrical components being protected. The manufacturers rate the MOVs.

Gas Tube Installation

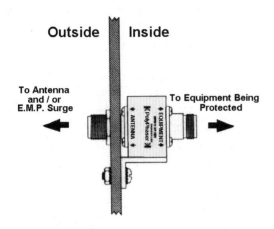

Outside   Inside

To Antenna
and / or
E.M.P. Surge

ANTENNA   PolyPhaser   EQUIPMENT

To Equipment Being
Protected

Gas Tubes

Gas tubes are a type of surge arrester which is typically used on antennas to divert E.M.P. which has collected on the antenna to ground. Gas tubes are also used on telephone lines.

A typical gas tube application for a shelter antenna would involve a gas tube and grounding system on the surface at the base of the antenna and another gas tube and grounding system where the connecting antenna cable meets the radio unit. The gas tube and grounding system on the surface would divert the pulse collected on the antenna to ground and the second system next to the radio would divert any E.M.P. which collected on the connecting cable running from the antenna to the radio.

In general, it is good to have a very short connecting wire between the MOV and the equipment being protected. For example, take a situation where you have a power line coming into a shelter. If you placed an MOV right where this line enters the shelter, it will divert the E.M.P. which has collected on the

outside line. E.M.P. can still collect on any internal wiring in the shelter downline from the MOV and destroy the connected electronic components if the run is too long. Therefore, if possible, have each piece of sensitive equipment individually protected with the shortest possible lead between the MOV and the piece of equipment. This whole E.M.P. subject is somewhat based on theoretical knowledge. Some authorities believe that a 1/4 inch lead, under some circumstances, can accumulate enough E.M.P. to destroy the piece of electronic equipment it is attached to. Other authorities say that a typical handheld VHF or UHF radio with a short whip antenna would not collect enough E.M.P. In general, any wire or antenna over 36 inches is going to attenuate a damaging E.M.P.

Grounding

Grounding is one of the most important factors involved in E.M.P. protection. If you are going to try to divert E.M.P. with an MOV or gas tube, you have to have somewhere to send it. E.M.P., like water, tends to follow the path of least resistance. If the grounding system is inadequate, the MOV will not divert the E.M.P. to ground. Thus, the E.M.P. will go by the MOV and damage the attached equipment.

A typical grounding system involves a lead or strap going from the MOV to a ground rod system or some other buried metal object which has adequate surface exposure to facilitate grounding to the earth. The effectiveness of grounding varies depending on the compaction, moisture content and soil type.

EMP Grounding System

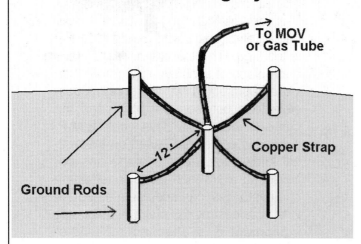

To MOV or Gas Tube

Copper Strap

12'

Ground Rods

The connecting leads between the MOV and the ground rod should be a metal strap, preferably copper, braided or solid. The strap should not have bends in it, being as straight as possible.

The ground rod should be at least a #2 copper rod 8 feet long. Multiple ground rods should be used for each MOV. Typically, 5 ground rods are used in a single MOV or gas tube application. These are attached in a star pattern (see drawing) with the lead or strap coming from the MOV connected to the center rod and straps connecting the center rod to the four other outer rods. The outer rods should be placed about 12 feet away from the center stake.

Ground rods should not be driven into disturbed soil. Undisturbed, virgin soil with maximum compaction will produce the most effective grounding.

If the shelter is made of a steel tank or culvert, this shell can be used as a substitute for a ground rod system. In this situation, at least 20 percent of the surface of the shell must be in contact with the earth and this surface area cannot be treated with a coating, such as tar or epoxy, which would inhibit electrical conduction. It is important to have good connections between the connecting straps and the ground rods. The best way to do this is through a cad weld. Many commercial electricians will have this cad welding equipment for fusing connections onto grounding rods. A typical MOV protection array for a 120-volt wire would utilize three individual MOVs as follows:

One MOV connects the hot wire and the ground wire. Another MOV connects the hot wire and the neutral wire. The last MOV connects the neutral wire and the ground wire. All three MOVs are connected by a lead to the ground rod.

MOVs and gas tubes can be purchased from Poly Phaser Corp. Their toll free number is 1 (800) 325-7170. Nevada residents call 782-2511.

FEMA has a free publication available on shielding E.M.P. It is called CPG 2-17, Volumes I, II, and III. These can be ordered by calling (202) 646-3484.

Integral Shielding

Integral shielding involves the use of a faraday cage. A faraday cage is a closed metal container which will collect the E.M.P. on its outer surface area, thus shielding the electronic equipment inside the container from the damaging effects of E.M.P. One simple application of integral shielding would be to store unplugged electronic equipment inside a steel garbage can with a tight lid on it.

A more complex application of integral shielding would be a steel tank shelter with tight closing metal hatches outfitted with metal mesh gaskets. In order to maintain the integrity of this integral shield or faraday cage, all of the penetrations into the steel tank, including incoming wires, air supply and exhaust pipes need to be configured to eliminate pathways for E.M.P. penetration into the inside of the tank.

E.M.P. is a large spike wave. For this reason a small hole in the side of a steel tank will be less apt to allow the entrance of a tall E.M.P. wave spike, whereas on the other hand, an elongated slot through the wall of a tank could allow the tall wave pattern entrance into the inside of the tank.

Earth cover, in general, over the top of the shelter provides little protection from E.M.P. Pipes entering the shelter penetrating through the integral shield must be no larger than 8 inches in diameter. Also, these pipes need to be welded to the tank where they penetrate through the surface and the pipe needs to be a minimum length of six times its diameter.

Vehicles

Some authorities suggest that most automobiles with electronic ignition and computerized control systems would only be temporarily affected by E.M.P. Pre-1975 vehicles with non-electronic ignitions will generally not be affected by E.M.P.

Notes:

227

Chapter 19
Radiological Monitoring

There are two deadly types of radiation — neutron and gamma — but for the most part, radiological monitoring involves measuring only gamma radiation. Neutron or initial radiation occurs for a few seconds in close proximity to the blast. Gamma radiation is the more deadly radioactive element found in radioactive fallout particles. Gamma radiation is the long-term (two- to six week) problem which shelter occupants will have to contend with.

Equipment

Radiological monitoring is a twofold activity. The first aspect involves measuring the amount of radiation received per hour, both inside and outside of a shelter. This aspect is referred to as determining radiation exposure rates and is accomplished with the use of a survey meter. The second aspect of radiological monitoring involves measuring the total amount of radiation which people have been exposed to during a given period of time. This aspect is referred to as accumulated exposure and is measured with the use of a dosimeter and a dosimeter charger. Every shelter should have these three pieces of equipment!

Survey Meter

Radiation is an unseen, unfelt danger that can only be detected and identified with a

Autonnic Research Survey Meter

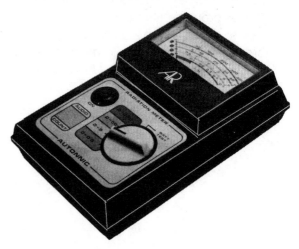

survey meter. Survey meters are also referred to as dose rate meters and fallout meters. A survey meter is so named because it's used to scan, or "survey," an area or surface to determine the radiation exposure rate present. Survey meters measure the rate that people are being exposed to gamma radiation in terms of roentgens per hour (r/h). In short, a survey meter is somewhat like the speedometer in your car, but instead of telling you speed in miles per hour it tells radiation strength in roentgens per hour. It is kind of like a thermometer which tells you how hot the radioactivity is. Survey meters measure gamma radiation and some can detect beta radiation.

Rad Gun Survey Meter

A survey meter to be used in a shelter should have a capacity of 0 to 500 roentgens per hour. Do not purchase a survey meter, like the government civil defense issued CD V-700, with a capacity of 0 to 50 milli-roentgens. Such low range units were made for training purposes but can be used in the decontamination process; detecting and dealing with radioactive contaminates on people and things. Survey meters are powered by batteries. Over time, batteries have a tendency to corrode and leak. The batteries should be removed from survey meters which are not in use and both should be stored in a dry place.

The National Academy of Sciences' Advisory Committee on Civil Defense in 1953 concluded: *"The final effectiveness of a shelter depends upon the occupants of any shelter having simple, rugged, and reliable dose rate meters, [survey meters], to measure the dose rate, [rate of exposure], outside the shelter."*

Most survey meters have a multiple range reading capability. In other words, they have some sort of selector knob which will change the scale to register and measure lower or higher levels of radiation exposure in roentgens per hour. For instance, a Autonic AR model 81 survey meter has four ranges, 0 to 0.5 r/h, 0 to 5 r/h, 0 to 50 r/h and 0 to 500 r/h.

If you took a reading at the 0 to 500 r/h and it hardly registered on the meter scale, you would turn the knob down to a lower scale which is in closer proximity to the exposure rate. You are not going to be able to get an accurate reading of radioactive fallout with a field strength of 4 r/h if you have the survey meter selector knob turned to the 0 to 500 r/h range. An accurate reading can only be obtained by turning to the 0 to 5 r/h range. If the field strength is stronger than the range you have selected, the needle will peg itself or run off the scale. This is an indication that you need to turn the selector knob to higher ranges until you get an accurate reading.

Outside - Inside Exposure Ratio

Radiological monitoring involves getting both inside (sheltered) and outside (unsheltered) readings of the radiation field strength in r/h with a survey meter. By making an initial survey, both sheltered and unsheltered, preferably no more than 3 minutes apart, you can calculate an outside-inside exposure ratio. For instance, the unsheltered outside radiation exposure rate might be 1,000 r/h and the sheltered inside radiation exposure rate 2 r/h. You would divide the unsheltered exposure rate of 1,000 by the sheltered exposure rate of 2 and this would give you an outside-inside exposure ratio of 500. Later that evening you might take another inside or sheltered reading which registered 1.5 r/h. By using your pre-determined outside-inside exposure ratio of 500, you can determine the unsheltered exposure rate without having to subject yourself to the hazard of going outside and getting another reading. The outside unsheltered exposure rate can be fairly accurately estimated by multiplying the sheltered exposure rate of 1.5 r/h, by the outside-inside exposure rate of 500, which would mean the unsheltered exposure rate would be 750 r/h, (1.5 r/h X 500). The outside exposure rate will generally change proportionately with the inside exposure rate according to the outside-inside exposure ratio.

Be advised that if you are getting sheltered exposure rate readings of over 2 r/h, you have a problem. Different parts of the shelter are apt to have different protection factors. Survey other parts of the shelter to see if there are other areas in the shelter which provide better radiation exposure protection where people can be moved to. If there is not enough room for everyone in the areas of better radiation shielding, rotate people in and out of protected areas to minimize, spread out, and balance out the overall radiation exposure rates. If you register an inside sheltered exposure rate reading of 10 r/h or more you are dealing with a potential life and death situation.

The outside-inside exposure ratio will change with time because the energy level or penetrating capability of gamma radiation changes with time. Also rain and weather can physically shift concentrations of fallout material off the roofs of buildings and coverings over shelters and onto the ground. This rearrangement of fallout concentrations could affect the ratio for better or for worse. This is why it is important to continually survey different parts of the shelter and recalculate ratios as much as is feasible and safe.

External Probe

Most, if not all, of the civil defense grade survey meters can be ordered with external probes. An external probe is a 25- or 30- ft. extension cable with a monitoring head which

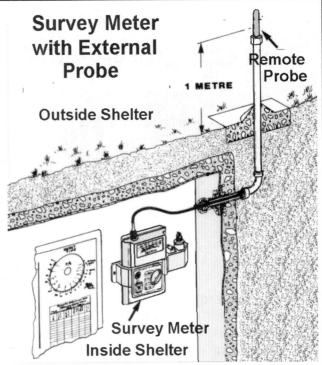

Survey Meter with External Probe

Outside Shelter

1 METRE

Remote Probe

Survey Meter Inside Shelter

can be positioned outside with the cable running back into the shelter and connected into the survey meter. This allows you to take a reading of the roentgens per hour outside of the shelter without having to go outside (See pictures). You may be starting to realize that having a survey meter with an external probe could prove to be a valuable asset.

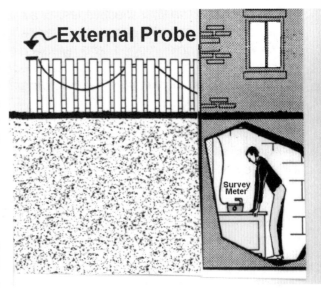

External Probe

Survey Meter

230

Maximum Allowable Accumulated Exposure

Length of Time		Maximum Dose in Roentgens
24 hours		75 r
100 days or more	(over 35 years in age)	150 r
100 days or more	(under 35 years in age)	100 r

The Seven-Ten Rule

Another useful method for calculating the outside, or unsheltered rate of radiation exposure is called the seven-ten rule. For every sevenfold increase in time after a nuclear weapons detonation, there is a tenfold decrease in the radiation exposure rate. For example, if an initial reading is made one hour after a detonation with a survey meter showing a radiation exposure rate of 1,000 roentgens per hour, seven hours later (1 hour times 7) the exposure rate would be 100 r/h, forty-nine hours later (7 hours times 7) would be 10 r/h, and 343 hours later (49 hours times 7), which is a little more than 14 days, the radiation level would be 1 r/h. So in 14 days the outside radioactive fallout with an initial strength of 1,000 r/h would diminish to 1 r/h.

Numerous companies make survey meters. One of the best Survey Meters on the market today is the American made model 80M 300 made by Nuclear Research, 125 Titus Ave, Warrington, PA 18976, telephone (215) 343-5900. Another excellent unit is the Universal Survey Meter model RD-10 made in Finland by Alnor Oy which is manufactured to Finnish army specifications and is built to withstand E.M.P. effects. The Alnor is sold by Yellowstone River Trading Co., PO Box 206, Emigrant, MT 59027, (406) 333-4707 9\or (800) 327-7656.

Yet another excellent unit is the AR model 81 made by the Atomic Research Limited, Tollesbury, Essex, CM9 8SE, England, telephone (0621) 869460.

Older and more economical Radgun units like the model AGB 10KG-SR can be purchased from Jordan Nuclear Company, 3244 Arroyo Seco Ave, Los Angeles, CA 90065, telephone (213) 222-8141.

Dosimeter and Dosimeter Charger

Dosimeter

A survey meter should not be confused with a Dosimeter. They are two different things. A Dosimeter is like your odometer, but instead of telling you how many miles you have traveled, it tells you the accumulated radiation you have

Ends of a Dosimeter

231

Dosimeter Charger / Calibrator

been exposed to in roentgens. A dosimeter is an instrument that tells you what quantity of accumulated radiation you have been exposed to or have absorbed over a period of time. This is valuable in terms of delegating duties in exposed areas. Having a person wear a dosimeter during emergency conditions allows the shelter manager to make sure individuals don't get exposed to too much radiation. If the dosimeter indicates the individual has been exposed to the maximum safe accumulated radiation exposure, then the shelter manager should assign any task involving additional radiation exposure to another person who has not yet exceeded their safe radiation exposure limit. A dosimeter can detect and register accumulated radiation exposure over a given period of time. Anyone going on an outside mission or doing work in a poorly shielded area should wear a dosimeter so their accumulated exposure to radiation can be monitored.

Be advised that no one should go on an outside mission unless the outside radiation exposure rate has reduced to 25 r/hr or less. Any outside missions when radiation rates are in excess of 25 r/hr could endanger the lives of those involved. Every outside mission should be evaluated in terms of critical importance, the exposure time required to accomplish the mission and the rate of radiation exposure which the participants will be sub-

jected to. The decision to undertake an outside mission should not be clouded or based on emotion but should be based on a logical evaluation. Such a logical evaluation would include calculating the potential accumulated exposure of any individual involved in the contemplated outside mission and comparing it against the accumulated exposure which potential candidates for the mission have already received. If a potential candidate will not exceed his maximum safe accumulated exposure and if the mission is deemed critically important, then the mission could proceed as long as it is well defined and time completion deadlines are specified to prevent excess exposure.

Note: Any amount of radiation exposure will have an effect on the human body. People can have different reactions to the same dose of radiation. The absolute "safe" limit may be as low as 2.4 r for 24 hours. The human body has the ability to daily repair or compensate for a radiation exposure of 10 r. Pregnant women and children under 18 years of age should avoid any unnecessary radiation exposure. The fetus or unborn child is very susceptible to injury from radiation exposure.

A dosimeter is a pen-like object about 1/2 inch in diameter and about 4 1/2 inches long. Dosimeters have a clip on them, similar to the clip on a pen, so they can be clipped onto clothing. One end of a dosimeter has a magnifying glass that you look through and the other end has a contact which plugs into the dosimeter charger.

Before a dosimeter can be used it must first be recalibrated. While in storage, dosimeters tend to accumulate a false reading or calibration drift. Stored dosimeters should be periodically checked for calibration drift and recalibrated back to zero when necessary. Recalibrating is also referred to as zeroing. First remove the cap off the charging receptacle on the top of the dosimeter charger and then press the contact end of the dosimeter firmly down into the charging receptacle. While looking into the magnifying end of the dosimeter, turn the control knob which is on top of the charger adjacent to the charging receptacle until the dosimeter reads zero.

After using a dosimeter to check the accumulated dose of radiation a person has received, a reading of the dose needs to be taken and recorded. Once this exposure rate is recorded, the dosimeter needs to be zeroed out again.

Reading a Dosimeter

Looking Into a Dosimeter

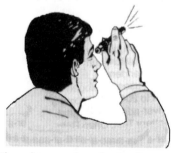

If the charger has batteries you should be able to look down into the magnifying glass end of the dosimeter while it is being held gently in the charger and as you press, a light will come on which will enable you to read the scale. After you have recorded the reading, zero out the dosimeter.

A dosimeter can be read without the use of a battery powered charger light source. This is done by looking through the magnifying glass end of the dosimeter and pointing the other end of the dosimeter up toward a light source. The light behind the charging end

Dosimeter Scale

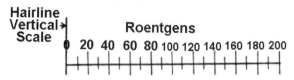

of the dosimeter will illuminate the scale inside the dosimeter and facilitate a reading.

Dosimeters and dosimeter chargers can be purchased from: Dosimeter Corporation, 11286 Grooms Road, Cincinnati, OH 45242, telephone (800) 322-8258, and from Jordan Nuclear Company, 3244 Arroyo Seco Ave., Los Angeles, CA 90065, telephone (213) 222-8141.

Dosimeters can be purchased with different ranges. The best range for civil defense purposes should probably be 0 to 200 r.

In a pinch, a dosimeter can be expediently used to survey exposure rates and a survey meter can be expediently used to determine accumulated radiation exposure. This is, of course, contrary to everything we have previously explained.

During a nuclear crisis you might end up in a shelter that does not have a survey meter but there is a dosimeter and a dosimeter charger in the shelter. You can use the dosimeter to take a reading by calibrating the dosimeter back to zero with the dosimeter charger as previously explained. Put the dosimeter in the area that you want to survey, (for example, let's assume you want to determine the outside rate of radiation exposure). Put the dosimeter outside, leave it there for 15 minutes and then retrieve it. View the scale inside the dosimeter as previously explained. If the scale registers an accumulated exposure of 50 r, multiply 50 r by 4, because 15 minutes is 1/4 th of an hour and a survey is always done in roentgens per hour, r/h. The result will be 50 r X 4 = 200 r/h. The unsheltered rate of exposure is 200 r/h.

In a different situation, you might be in a shelter during a nuclear crisis and you have a survey meter but not a dosimeter. You can use the survey meter to determine the accumulated exposure which people are receiving while inside the shelter by using the survey meter and taking a reading every hour of the radiation field strength. Then total the readings at the end of each 24 hour period. This will give you a total of the accumulated radiation exposure in roentgens for that 24 hour period.

It is important that each shelter have a person assigned to be responsible for radiological monitoring who understands the concepts involved and knows how to use the equipment available. This person should take scheduled readings, calculate the outside-inside exposure ratio, determine which areas in the shelter have safe and unsafe radiation shielding, supervise the maintenance of individual accumulated radiation exposure records and closely monitor the exposure of those working in poorly shielded areas or who are involved in highly exposed outside missions. This radiological manager should also oversee personnel job assignments. Those who have received the greatest dose accumulation of radiation should be assigned to a task in the most shielded portion of the shelter. Those who have received minimal exposure should be assigned

Accumulated Radiation Exposure Record

Name_____ Age _____ Male___ Female___

If female are you pregnant _____or possibly pregnant_____

Date of Exposure	Amount of Exposure	Total Exposure to Date

to the tasks which involve greater exposure to radiation.

Nomograms

In this section we will discuss the use of the "exposure rate nomogram" and the "entry time - stay time, total exposure nomogram". In order to make use of either of these nomograms, you must have an accurate recording of what time a nuclear weapon detonation took place. Also, neither of these nomograms will work if your shelter is being affected by radioactive fallout from multiple weapons detonations.

Exposure Rate Nomogram

Outside exposure rates decrease in a mathematically predictable manner with the lapse of time due to the natural process of radioactive decay. The "exposure rate nomogram" is used to pre-determine or estimate future unsheltered exposure rates based on certain known factors. These necessary factors are: the time which has elapsed since the detonation (burst) of the nuclear weapon and an unsheltered (outside) exposure rate taken at a determined time after the burst.

Let us now examine the "exposure rate nomogram" (Example A). Note in the upper left hand corner a heading on top of a vertical scale labeled "exposure rate at H + t". The readings on this scale indicate the unsheltered exposure rate that would exist at a given time after a detonation if all the radioactive fallout material produced by the burst was already on the ground. The "H" represents the hour at which the weapons detonation took place and the "t" represents the time which has lapsed since the deto-

nation. Now examine the "time after burst (H + t)" heading half way down in the center of this same page. The scale under this heading indicates the time, ("t"), which has elapsed since the detonation of the weapon. At the upper right hand corner of this same page is the heading "exposure rate at H + 1". The readings on the vertical scale under this heading represent the theoretical calculated exposure rate for one hour after the detonation of a weapon. The "exposure rate nomogram" uses this one-hour-after-detonation exposure rate figure as a basis, because when estimating future exposure rates you need to go back to an exposure rate which is as close as possible to the time of the detonation.

One hour after the burst of the weapon (H + 1) is a calculated theoretical standard because in reality you couldn't get an exposure rate reading one hour after a burst. This is due to the fact that most of the radioactive fallout would still be airborne and would not yet have fallen down and settled onto the ground.

Example of "Exposure Rate Nomogram" Calculation

Let us now take an example and work it out on the exposure rate nomogram. The hypothetical situation is that it is 3 hours after the detonation of a nuclear weapon and we just took an outside exposure rate reading which indicated 200 rh. Now we will transfer our two known factors onto the exposure rate nomogram. The time, 3 hours, will be marked on the vertical scale under the "time after burst (H + t)" which is in the center of the page. The exposure rate reading of 200 rh will be marked on the scale under the heading on the upper left, "exposure rate At H + t". Lay a straight edge on the exposure rate nomogram so that the straight edge passes across the entire page, from left to right, with the upper edge passing through the 200 r/h mark on the scale under the "exposure rate at H + t" and passing through the mark at 3 hours on the scale under the heading "time after burst (H + t)". Now that the straight edge is positioned so that it intersects our two known factors, follow the top edge of the ruler as it slopes down the page to the right and mark where it crosses the scale on the right under the heading "exposure rate at H + 1". If you remove the ruler you will see that the intersecting line from the other two known factors crossed the "exposure rate at H + 1" at 750 r/h. This 750 r/h mark now becomes a straight edge pivot point for all estimates of future exposure rates.

Now under the same situation, let us say that we want to estimate what the unsheltered exposure rate is going to be 20 hours after the detonation. Keeping the ruler's right upper edge on the 750 r/h mark on the "exposure rate at H + 1" scale, lower the left end of the ruler until the upper edge of the scale intersects the 20 hour mark on the "time after burst (H + t)" scale in the center of the page. With the ruler fixed at these two points follow the top edge of the ruler up and to the left to the point where it crosses the "exposure rate at H + t" scale on the left hand side of the page. This intersecting point on the scale reads 20 r/h., which indicates that 20 hours after the burst the outside exposure rate will have decreased to 20 r/h.

Entry Time - Stay Time Total Exposure Nomogram

This nomogram is a tool which can be used to calculate ahead of time when the exposure rates will have reduced enough that someone can safely enter an exposed area, how long someone could safely stay in this exposed area and the total exposure someone would receive as a result of being in this exposed area for a given period of time.

As you examine example B, "entry time - stay time, total exposure nomogram" you will notice five vertical scales across the page with headings.

The first heading, moving from the left to the right, is "total exposure". The reading on this scale represents the radiation exposure someone would receive under a given set of circumstances.

The second heading is "exposure rate (1 hour) (R-1)". The information on the scale under this heading represents the exposure rate one hour after the detonation of the weapon, "H + 1". This figure represents a known factor which you would obtain from your calculations on the "exposure rate nomogram" under the "H + 1" heading on the right. The third heading is "E/R-1". The information on the scale under this heading represents the radiation exposure divided by the exposure rate one hour after the detonation, "H + 1". The E/R-1 scale is simply a common pivot point for the two scales on either side of it. When using the "entry time - stay time, total exposure nomogram" you will always use this center scale as part of making a calculation with either the two scales to the right or the two scales to the left.

The fourth heading is "stay time (hours)". The information on the scale under this heading represents the amount of time, in hours, that someone stays in an area exposed to radiation.

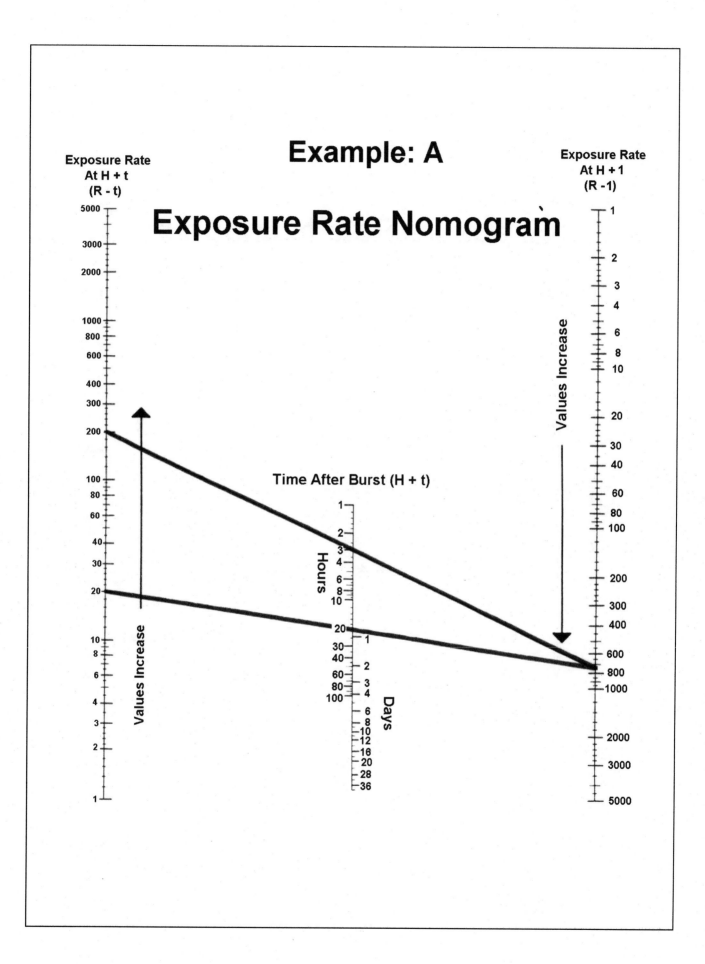

Example: A

Exposure Rate Nomogram

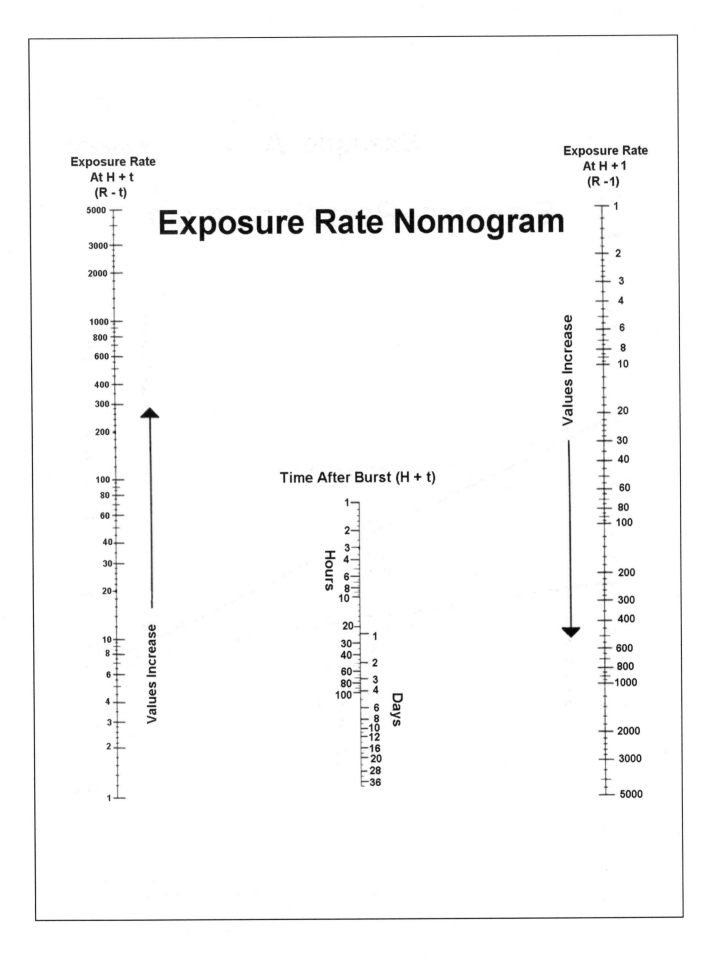

The fifth heading is "entry time". The information on the scale under this heading represents the time, after a burst, when someone enters an area which is exposed to radiation.

To use the "entry time - stay time total exposure nomogram" you need to know three factors or quantities. First, you mark the two known quantities on the appropriate vertical scales. Both of these factors or quantities need to be either on the right or left side of the center E/R-1 scale. Connect both of these points with a straight edge and locate the point where the established angle of the straight edge crosses the E/R-1 scale. The next step is to connect this established focal point on the E/R-1 scale with the third known factor on the other side of the E/R-1 scale. Extend this line from the E/R-1 focal point to the third known factor. Using a straight edge continue the line from the angle established by the connection of E/R-1 and the third known factor, beyond the point representing the third factor, if necessary, so that the remaining scale is intersected. The reading taken from this intersection determines a previously unknown factor.

Here is an example for the entry time - stay time total exposure nomogram calculation. Imagine a hypothetical job which needs to be done in an area exposed to radioactive fallout. The exposure rate at this unsheltered area was 20 r/h 5 hours after the burst, H + 5. The job needs to be done at H + 10, or 10 hours after the burst. One person will have to do the job and it will take 2 hours to complete. What is the total exposure which the individual will receive as a result of doing the job?

The first thing we must do is use the "exposure rate nomogram" and determine what the exposure rate was at H + 1. As we previously discussed, this is established by joining the two known factors, 5 hours after burst and the H + t exposure rate of 20 r/h, and extending the line until it crosses the H +

1 scale. The resulting figure is H + 1 = 140 r/h. Now we take this H + 1 figure of 140 r/h and mark it on the "entry time - stay time, total exposure nomogram" on the second scale from the left entitled "exposure rate at H + 1 hour". Next we go to the fourth vertical scale over from the left, headed "stay time (hours)", and we mark the point indicating 2 hours. Continuing, we go to the fifth vertical scale from the left entitled "entry time (H + t)" and we mark on the scale the point indicating 10 hours after the weapons detonation. We now join these two points on the "stay time" scale and the "entry time" scale creating a line which we extend with a straight edge to the left and intersect the focal point scale headed "E/R-1". Continuing, we will now extend a line to the left from our established focal point on the "E/R-1" scale to the third known factor previously marked on the "exposure rate at H + 1 hour" scale as 140 r/h. We will extend a line with a straight edge from the angle created by the connection of these two points, continuing to the left until it intersects the scale headed "total exposure". If we read the point where the line crosses this scale we will determine that the person doing this 2 hour job will receive an accumulated dose or exposure of 16 r.

The use of this nomogram will help you evaluate the safety of a contemplated mission in terms of exposure to those doing the work.

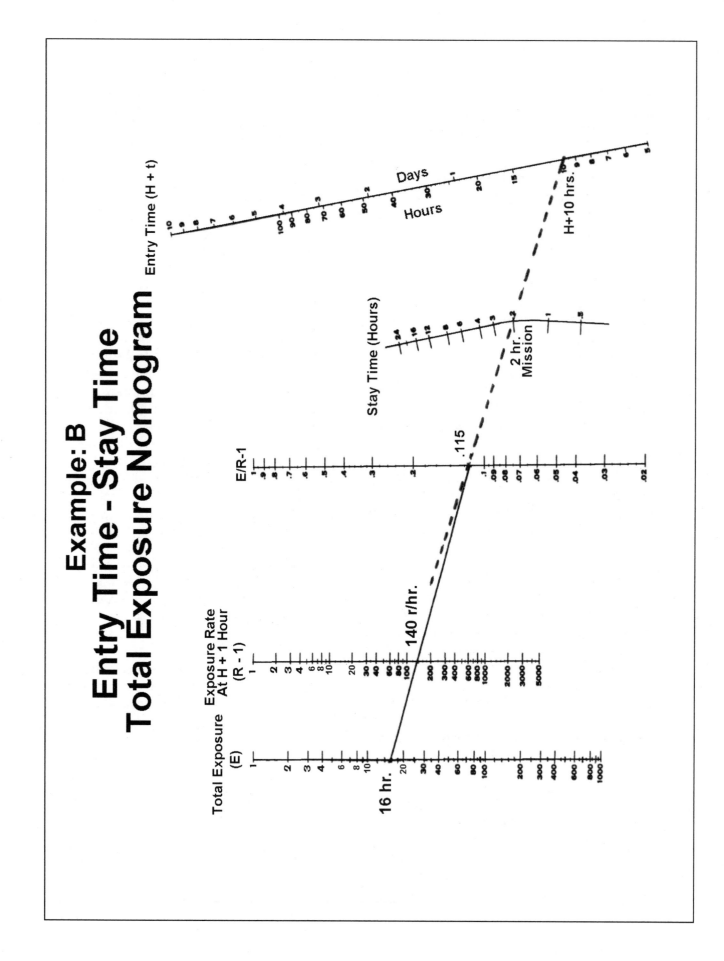

Example: B
Entry Time - Stay Time
Total Exposure Nomogram

Entry Time - Stay Time
Total Exposure Nomogram

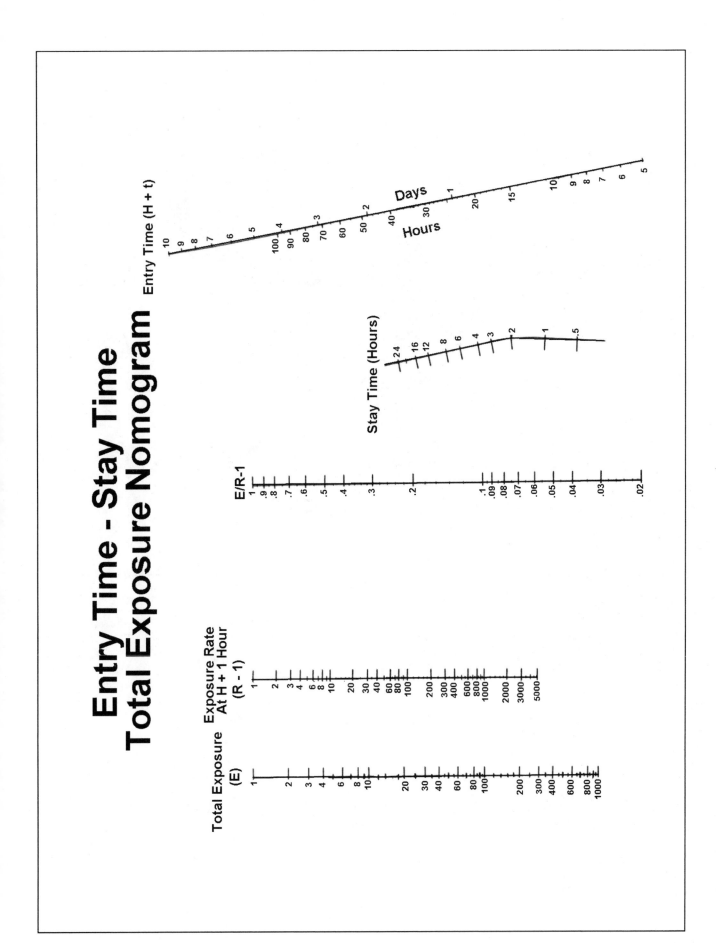

Notes:

Chapter 20
Decontamination

The best solution to the decontamination problem is to live near your shelter. Decontamination removes radioactive, biological or chemical contamination from the body surface of individuals who have been exposed to contaminates. The main purpose of decontamination is to prevent injury due to the presence of contaminates on the body surface by removing the contaminates as quickly as possible to minimize the potential of internal contamination through the penetration of broken skin, inhalation or ingestion. An additional purpose of decontamination is to prevent the spread of contaminates to adjacent body areas or to other people.

Radiation is the easiest contaminate to remove. Decontamination of an individual exposed to radioactive contaminates usually involves removing and discarding clothing, cutting off hair, scrubbing down with soap and water, and rinsing off the body surface. After this is done, the individual can be checked with a radiation meter to see if any radioactive material has been missed in the decontamination process. It should be noted that if the individual has ingested or inhaled radioactive particles, there is no practical means of internal decontamination. Generally, anyone who has ingested radioactive particles but has been decontaminated will not be a serious contamination threat to other people around them.

General Principles of Decontamination

Start Decontamination As Soon As Possible

Decontamination procedures should be instituted as soon as an individual enters the shelter. It is very important to have a decontamination area set aside in the entryway of the shelter, away from any further radiation exposure but before entering into the living quarters of the shelter. Contaminated areas of the body should be identified and differentiated from non-contaminated areas as early as possible.

Triage

Decontamination should proceed based on established priorities of need. The highest priority areas are those with the highest levels of contamination. The next highest priority are wounds, since these might allow contaminated substances into the body. The third priority are body orifices — nose, mouth, ears, etc. for the same reason.

Protective Gear for Decontamination

Only Do What Is Necessary

It is unnecessary to go into complex decontamination procedures if the initial simple ones are successful in eliminating the radioactivity. Use the more complex procedures only if they are warranted by persistent contamination readings. Be aware that it is virtually impossible to completely decontaminate the skin. Thus, don't go overboard in an attempt to attain absolute decontamination and try to use highly abrasive substances that could damage the skin.

Adequately Survey Victims

It is important to adequately survey victims both before initiating decontamination procedures and after completing them. A radiation level reading with a survey meter should be taken over the whole body when the victim arrives. It is also important to use a Q-tip or cotton swab on the nasal passageways and take a reading on this to determine the possible extent of inhalation of radioactivity and thus internal contamination. Before discontinuing the decontamination proce-

dure, recheck the whole body with the survey meter as well as the nasal passages.

Use Cool Water

To minimize the likelihood of further absorption of radioactivity, tepid to cool temperatures for water, the body and the environment during decontamination are desirable. Fever or excessive heat could cause greater absorption of radioactivity into the blood stream, deep body tissues or lymph nodes. Hot water showers, warm irrigations and overly heated rooms should be avoided.

Prevent Infection

Once the immediate effects of radiation exposure and acute life-threatening injuries are past, infection is the greatest concern for the victim. High doses of radiation tend to suppress the immune system and increase an individual's susceptibility. Thus strict aseptic procedures should be followed.

ISI Battery Powered Respirator

Give Psychological Support

The radiation victim needs special psychological support. In addition to the stress produced with any disaster, victims of a radiation accident must cope with the fears and anxieties of the unknown aspect of their situation. Providing information, frequent communication and reassurance for victims and their families is most helpful in minimizing reactions and improving coping mechanisms.

Basic Procedures for External Contamination

Decontamination Personnel

It is important to protect the people who are helping with the decontamination work. The main concern is that these people are protected against being splashed in the face or eyes with contaminated water. A rubber hazmat suit, like the one shown on the next page, with face protection, rubber gloves and rubber boots are ideal. A respirator like the ones shown are good for face and eye protection, but be advised that the threat of inhaling fallout particles at this stage is minimal. The fallout particles are fairly big and should not be airborne once a person reaches an inside decontamination facility. The protection equipment makes it easier to clean up the decontamination personnel once the process is finished.

Remove Clothing

Since clothing and shoes often serve as effective barriers to radiation exposure, simply removing and isolating the victim's clothes in a plastic bag may be all that is required. Label the plastic bag with the victim's name and a radioactive warning label. Radiation surveying should be performed and the levels recorded. As we mentioned before, nasal swabbing should also be done at this time.

Shower the Victim

If most of the skin surfaces have been contaminated, the victim should take a shower. He should use abundant amounts of tepid to warm water, never hot. Soft soaps, preferably acid-based, and very mildly abrasive materials, such as terry washcloths, soft brushes, or sponges minimize the chance of skin abrasion yet give effective cleansing action. Soaps and detergents emulsify and dissolve contamination. By using highly abrasive or alkaline soaps, organic solvents and hard brushes, the skin may tend to become abraded and increase its permeability to radioactivity. Discontinue washing the skin if it becomes reddened, thinned or abraded. The skin is actually capable of cleansing itself and, depending on the area of the body, the outer layers are shed and renewed every 12 to15 days. Begin with 2 to3 minute washes and repeat these 2 or 3 times.

Be Cautious With Washing Procedure

The Latest Fashion In Decontamination Attire

Avoid getting water into the ears, eyes, nose or mouth since these openings are potential routes for the entry and spread of contamination. Wash all areas thoroughly, including body folds. Wash from the top down to minimize recontamination of previously cleansed areas. Dry thoroughly.

Only if these measures prove ineffective should the use of other chemicals be considered. Some potential agents used for residual skin contamination include aqueous agents such as a mixture of 50 percent powdered detergent and 50 percent cornmeal made into a paste, scrubbed onto the skin and washed off. Waterless agents, such as mechanic's hand cleansing cream, can also be used. These procedures can be repeated if necessary.

It would be a rare instance that would not respond to the above attempts at decontamination. Some last-resort options include sanding the skin with emery cloth (not coarse sandpaper), using sticky tape, and wrapping the affected areas in plastic wrap, occlusive dressing, or using spray or liquid dressings if the contaminated areas are well defined. By wrapping the area, the spread of contamination is limited until the skin can naturally replace itself. Wrappings should be removed after 24 hours and the area thoroughly washed.

Isolate and Dispose of Decontamination Wastes

Wastes such as the drainage water from all decontamination procedures, discarded materials, hair, dressings and wrappings must be safely disposed of in an area separate from the shelter's septic system.

Partial Body Decontamination

If the skin has been contaminated only in certain localized areas, these areas may be individually decontaminated using the washing procedures just described. It is important to cover all adjacent uncontaminated areas with plastic and tape during the cleansing process to protect against the spread of the contamination.

Certain body areas require special attention when they are contaminated. The hair and scalp, natural body orifices, nail beds and wounds all merit particular attention during decontamination.

Hair and Scalp

Shampoo with liquid soap for three-minute periods. Repeat several times if needed. The victim's eyes, nose and mouth must be protected. Use goggles, if available, to shield the eyes, manually close the nostrils and insert swimmer's ear plugs in the ears during the washing and rinsing process. Thoroughly dry the hair before resurveying for contamination.

If repeated shampooing does not decontaminate these areas, the hair may have to be clipped. Do not shave the head as it is not worth taking the risk of cutting or nicking the skin and thus creating an avenue for radiation to enter the body.

Eyes

If the eyes show contamination, they should be irrigated with abundant quantities of water or saline. The water or saline flush should be directed to flow from the inner corner toward the outer corner of the eye. This process should be repeated as often as necessary to provide decontamination.

Ears

The ear canals may be irrigated repeatedly with normal saline or water, preferably using an ear syringe. Suction can be used to remove the water that does not drain naturally. If the eardrum has been perforated, irrigation should not be used.

Nose

The nasal passages may be decontaminated by having the victim blow his nose on gauze compresses and cleaning the nostrils with saline-soaked cotton swabs. The mouth and throat may be decontaminated by repeated gargling with normal saline or water.

Fingernails

If the fingernails or toenails show contamination trim them carefully, discarding the clippings as you would any other radioactive material. Cleanse the nails using liquid soap on a soft scrub brush, toothbrush or cotton swabs.

If a wound area is giving a persistent reading of contamination, it is possible that radioactivity has entered the internal tissues of the body through the wound and will need to be dealt with as internal contamination.

Decontamination Procedures for Internal Contamination

Internal radioactive contamination exists when radio-nuclide become incorporated within the body. In general, the major problem associated with chronic irradiation is that the organ affected may become cancerous. The following are the four steps radiation takes in the process of internal contamination.

The radio-nuclide gain entry into the body through the respiratory tract (inhaled through the nose or mouth), or through wounds on the surface of the skin and also possibly through the skin itself. This is called entry deposition. Next the radioactivity moves from its site of entry into the body either to the bloodstream or the lymphatic system. This is referred to as translocation. Once radio-nuclide gain access to the general circulatory systems, they tend to concentrate in specific target organs of the body.

Radio-nuclide are removed from the body by the kidneys. They enter the bloodstream from the target organs or from the lymphatic system. As the blood circulates, it passes through the kidneys. There it is filtered and the radioactive particles are excreted into the urine with other body wastes.

The quicker the process of internal decontamination can begin, the greater the chance of success, as damage from internal contamination is somewhat proportional to the length of time that the contaminant has been present internally. Whenever there is reasonable suspicion that internal contamination has occurred, begin treatment as early as possible.

The radiation can be most effectively stopped during entry deposition and translocation. There are very few forms of treatment available for preventing target-organ deposition or enhancing normal clearance mechanisms.

If it has been determined that radio-nuclide are present in an open wound, three principles should be followed:

1) Be sure to protect adjacent areas from cross contamination during decontamination of the wound. Cleanse these areas with saline and then dry them. Cover them with adherent, nonabsorbable wraps, such as an occlusive dressing.

2) Decontaminate the wound to minimize further contamination by repeated irrigations using normal saline or sterile water. Hydrogen peroxide can also be used as an irrigation. If the wound is allowed to bleed freely, this will flush out the contaminant, if necessary. Closure and dressing of the wound with a thick, absorbent dressing to prevent infection is the final step.

3) Initiate therapy aimed at reducing absorption and internal deposition and/or enhancing the excretion or elimination of absorbed nuclide.

Potassium iodide can be used as a thyroid blocking agent. (See chapter on nutrition and radiation)

A decontamination facility should be a room or space isolated from the main part of the shelter. This room is usually situated at the end of an entryway where radiation shielding exists. The space should have a drain in the floor and preferably be a grated floor. The space should have running water, preferably warm, and handheld shower sprayers. You need electricity for hair clippers and ideally a vacuum hose tied into a central vacuum system or a rainbow type vacuum. It is important that the central vacuum discharges the resulting air outside. The rainbow type vacuum will also work because it can contain the contaminates in the filter and water.

The decontamination room must have a survey meter and protective clothing and respirators for the decontamination staff. Plastic bags with twist ties should be stocked for containing the contaminated clothing and belongings. Tags should also be available to identify the bag's owner. The bagged contaminated items can be reutilized once the radioactive contaminants have decayed. Soap, towels and cleaning agents should also be stocked. Basic robes should be available to clothe the people who have gone through the decontamination process.

Decontamination Room

Notes:

Chapter 21
Medical Considerations

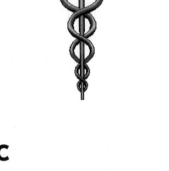

It is inevitable that during or after a national crisis, traumatic injuries and medical emergencies will occur. The nature of the disaster will limit the availability of conventional medical services as we know them today. The extent to which medical services will continue to remain available depends on the scope and severity of the disaster.

As part of being prepared to face disaster one should make some sort of medical preparation. The extent to which one can provide medical services in a disaster is going to be proportionate to the level of training acquired and the available equipment.

Due to limited medical help, during a disaster many of the routine injuries and emergencies which are successfully treated by today's medical facilities, will move into a terminal category. The general health of the population will be weakened due to breakdowns in sanitation and malnutrition caused by food shortages. Malnutrition weakens the body's immune system which increases susceptibility to uncontrolled infections. If the disaster is nuclear, immune systems will be further suppressed by exposure to dangerous levels of radiation.

Triage, or the sorting of patients, will become a more common practice in this type of disaster environment. Triage is usually done when medical resources are limited, and thus sorting concentrates available resources on the most salvageable patients.

Allopathic Considerations

Pre-hospital Care

For the sake of preparedness it would be good to go out and get some training in pre-hospital care. There are a number of levels to what is called pre-hospital care. The first level is standard first aid (8 hour course). The next level is called first responder (46 hour course), then emergency medical technician, E.M.T. (120 hour course) and finally paramedic (one year of training). All of these levels require a formal training program. In most states there are various levels in the First Responder program which step up to E.M.T. level. Likewise the E.M.T. program starts with basic life support and has various levels which move toward advanced life support. Paramedic is the highest level of pre-hospital care and it involves advanced life support including the administration of certain drugs.

It is very hard to keep your E.M.T. certification unless you are working as an employee or a volunteer with an emergency medical services (E.M.S) organization and most people tend to lose

Typical lst Responder Kit

1 stethoscope

1 sphygmomanometer (adult)

1 sphygmomanometer (child)

1 sphygmomanometer (pediatric)

10 butterfly closures

I kelly forceps

1 splinter forceps

I disposable penlight

1 utility scissors

1 Airway Set

6 dressings 3" x 3"

6 dressings 4" x 4"

3 2" kling bandage

3 3" kling bandage

20 elastic strip dressings 1"

2 surgipads 8" x 10"

6 surgipads 5" x 9"

6 sponges 3" x 3"

10 non-stick dressings 2" x 3"

1 mult-trauma dressing

6 cover sponges 4" x4"

10 non-stick dressings 3" x 4"

2 triangular bandages

4 eye pads

10 alcohol swabs

1 bandage 2"

4 rolls 3" gause

4 rolls 4" gause

1 wire splint

1 air splint arm

1 air splint leg

2 tape 1" x 10 Yds

2 tape 2" x 10 Yds

2 cold packs

3 abdominal pads

10 cleansing wipes

10 knuckle bandages

10 fingertip bandages

1 burnsheet 60" x 92"

1 rescue blanket

1 15 liter per min. oxygen requlator

1 "D" cylinder

6 non-rebreathing O2 masks

6 non-rebreathing O2 masks child

6 nasal cannula (adult)

4 nasal cannula (child)

2 nasal cannula (pediatric)

1 hand powered suction unit

1 pocket rescue valve mask set

1 box disposable ruber gloves

1 O.B. kit

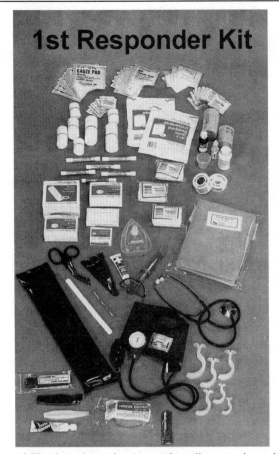

1st Responder Kit

skills that they don't continually practice. It is virtually impossible to maintain paramedic certification unless you are employed in the E.M.S. field and practicing advanced life support.

Every prepared individual needs to know the basics of patient assessment. Patient assessment involves checking a person's vital signs. When a serious injury or condition occurs, the body automatically makes an attempt to compensate for the problem or deficiency. Any sign of this irregular compensation effort indicates the presence of a problem in the body.

If a person is suffering from bleeding or dehydration this will produce a shortage of fluid (ie. blood) in the circulatory system. Blood is the carrier of oxygen and nutrition to the vital organs and all of the living tissue of the body. When a shortage of blood occurs in the circulatory system and thus parts of the body start receiving inadequate supplies of oxygen, the brain gets signals from different parts of the body that pressure is dropping due to a lack of fluid volume. The first thing the brain does is to try to compensate for the problem by constricting all of the blood vessels which reduces the overall volume of the vascular system. This keeps the plumbing system full by reducing its size and the reduced blood volume is initially able to meet the basic demand functions of the vital organs. This is what is referred to as compensated shock. If the problem persists, be it dehydration or bleeding, and the blood volume continues to decrease then the brain shifts into plan "B".

Once the brain has exhausted its capacity to compensate by reducing the size and volume of the vascular system, then all it can do is turn up the speed of the central pump, (the heart). By speeding up the pump and moving the reduced volume of blood at a faster rate, the brain can temporarily service the demands of the vital organs. Since the lungs oxygenate the blood and work in sync with the heart, as soon as the heart rate increases, the rate of breathing also increases. The increase in the heart rate, or pulse, and the increase in respirations are the first two signs of decompensated shock. The third sign is the actual reduction of pressure in the vascular system.

The pulse rate can be checked by pressing one's finger on a vein, like the jugular (neck), or the brachial (upper arm) or radial (wrist) arteries. The pulse rate can also be determined by listening directly to the heartbeat with a stethoscope. With the use of a stethoscope and a blood pressure cuff, a person can measure blood pressure.

Vital Signs

Age	Weight Pounds	Pulse	Respirations	Blood Pressure Systolic	Diastolic
Newborn 0 - 2 mo.	5.5 - 8.8	94 - 145	30 - 60	60 - 90	
Infants 2 - 12 mo.	8.8 - 22	124 -170	30 - 60	74 - 100	
Toddlers 1 - 3 years	22 - 33	98 -160	24 - 40	80 - 112	
Preschool 4 - 5 years	33 - 44	66 - 132	22 - 34	82 - 110	
School Age 6 - 12 years	44 - 88	70 - 110	18 - 30	84 - 120	
Adolescent 12 + years	over 88	55 -105	12 - 16	94 - 140	
Adult		60 - 80	12 - 20	90 - 180	60 - 104

Blood Sugar Levels: Low = less than 60, Normal = 70 to 180, High = above 240

Note: Due to the fact that circulatory systems of children and adolescents have a great ability to compensate for shock (blood volume loss), any sign of drop in blood pressure for children and adolescents indicates a critical situation. When adults start going into shock the vital signs will generally show a gradual drop in blood pressure and a corresponding increase in respirations and pulse rate. When the circulatory system of a child or adolescent exceeds its capacity to compensate for blood loss, through constriction of the blood vessels, an immediate crash of vital signs occurs. This usually happens so rapidly that the patient does not survive without immediate intervention of advanced life support. Be aware that dehydration can also cause shock. The best indicator of shock in children and adolescents is the appearance of delayed capillary refill. Capillary refill can be observed by pressing your finger against the fingernail bed or the leg of an infant. Once you remove your finger after firmly pressing for about four seconds the area should look lighter in color than the surrounding area due to a lack of blood in the tissue. In a healthy person this lighter colored tissue should return to its pinkish color within 2 seconds. This is called capillary refill. If it takes much longer than two seconds you have a serious problem, compensated shock. Treat the patient immediately for shock and obtain the best available medical treatment as soon as possible. Other signs that indicate the onset of shock in infants and children include pale skin color, a blue color to hands and feet and the presence of a lethargic or altered state of consciousness.

For adequate instruction on reading vital signs and patient assessment, the reader should order a copy of *Brady Emergency Care*, by Grant, Murray, and Bergeron, Sixth edition, published by Brady-Prentice Hall, ISBN 0-89303-155-0, telephone 1-800-638-0220. It is not within the scope of this book to adequately cover pre-hospital care. Also a

Locatin of Brachial Artery

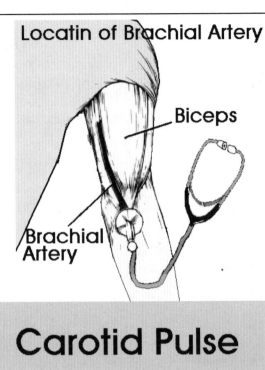

Biceps

Brachial Artery

Stethoscope

Sphygmomanometer

Blood Pressure Cuff

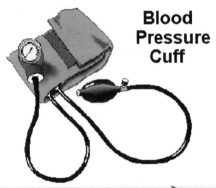

Carotid Pulse

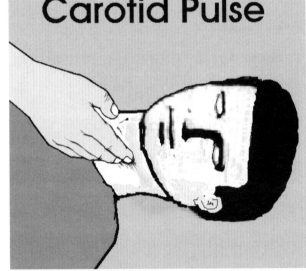

Checking Blood Pressure

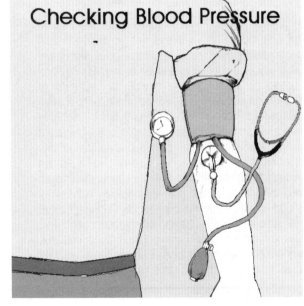

Radial Pulse

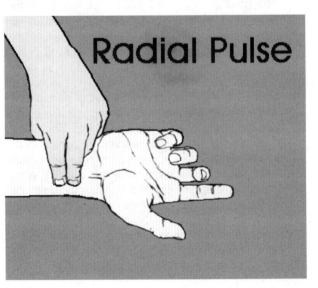

Nasal Cannula

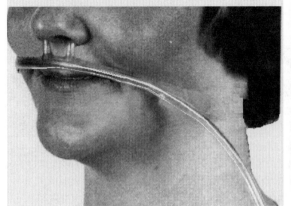

Non-rebreathing Mask

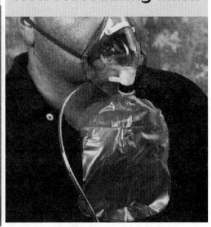

book is not good enough. A person needs to get training and hands-on practice.

Oxygen is the most powerful drug available to those practicing basic life support. Medical grade oxygen works wonders on patients suffering from shock and numerous medical complications. Unless the patient has a history of chronic obstructive pulmonary disease (COPD), you cannot overdose them with oxygen. *Brady Emergency Care* gives full instruction on the administration of oxygen. Generally speaking, an adult suffering from severe trauma or shock should receive 8 to 15 liters of oxygen per minute via what is called a non-rebreather mask. If the patient's condition is not critical or oxygen is in short supply, oxygen can be administered via a nasal cannula at rates of 1 to 6 liters per minute.

The problem with oxygen therapy is it uses up compressed oxygen fairly quickly. For this reason the use of oxygen may be restricted unless you can stockpile a quantity of "H" size cylinders. A small "D" cylinder will last about 28 minutes when being discharged at a rate of 10 liters per minute and a "H" cylinder will last about 9.5 hours when being discharged at the same rate. You will need the appropriate regulator to use the oxygen out of a cylinder and the large "H" cylinder will also require a flowmeter.

Like any other compressed gas cylinder, oxygen bottles need to be secured in position with adequate clamps or brackets. If a bottle tips over and breaks off the exposed regulator, the cylinder becomes a torpedo like projectile which has the capacity to pass through masonry walls.

Oxygen administration equipment and supplies for pre-hospital care can be purchased from the following sources: Moore Medical Corp., 1-800-234-1464; Dixie USA, 1-800-347-3494; or Dyna Med, 1-800-854-2706.

All of this prehospital training is valuable, but in a major disaster we are talking about a situation where there is not a hospital to take the patient to. The

Open Drop Mask
For Administration of Ethyl Ether

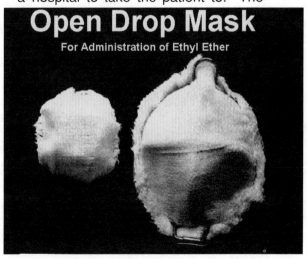

situation suddenly involves more than just stabilizing and transporting the patient. It now involves the actual treatment of the patient. The only person certified to do this is the doctor. But if it means the patient dies or you try, it is much better that you try!

Beyond Pre-hospital Care

There are two excellent books which should be in every shelter and emergency medical kit. One book is titled, *Where There Is No Doctor, a Village Health Care Handbook*, by David Werner, and the other is titled, *Ditch Medicine*, by Hugh L. Coffee. Both of these books are written for unskilled personnel. *Where There Is No Doctor* deals with instruction geared toward improvising medical services in remote third world situations. This is exactly the type of environment we might find ourselves in during a war or a major national disaster. *Ditch Medicine* picks up where *Where There Is No Doctor* leaves off. This book deals with serious trauma, gunshot wounds, intravenous therapy, surgery and more.

In a major disaster, antibiotics will be in high demand and in short supply, due to their lifesaving capabilities. For this reason, any available antibiotics will be extremely expensive and will probably be one of the most highly valued barter items. *Where There Is No Doctor* contains a comprehensive list of antibiotics, the diseases they treat, doses, administration, precautions and side effects. I will not reproduce this information here because I feel it is absolutely critical that you acquire a copy of this book.

Antibiotics should not be used on infections which the body's own immune function can successfully contain. There are several reasons for this. A certain number of people will have serious allergic reactions to the an-

tibiotic. Antibiotics also tend to kill beneficial bacteria in the intestines. With repeated use of antibiotics, resistant strains of bacteria can develop. Antibiotics should only be considered a last resort in a life-threatening situation involving an uncontrolled bacterial infection.

It is difficult to obtain and stockpile prescription medications without a prescription. This can be overcome by purchasing through the mail from foreign sources. According to a February 1994 article written by Zane Binder which appeared in *American Survival Guide*, the U.S. Postal Service, the FDA and U.S. Customs are not obstructing small shipments of prescription drugs mailed into the country from foreign suppliers to American residents. The following are three foreign sources for ordering prescription drugs through the mail:

Medicine Club International, POB N7108, Nassau, Bahamas. 1(800) 358-2765

Masters Marketing Co., Masters House, No.1 arborough Hill, Harrow Middx. HA1 1TW, England, Tel. 081-427-9978

Qwilleran, POB 1211, Birmingham B18 6NW, England

Be aware that if they wanted to, the FDA could use the newly expanded property seizure laws to take any piece of property on which they found a stockpile of unprescribed medications.

Another legal option is the acquisition of veterinary drugs. Many of the antibiotics used for humans are also used for livestock. Do not order an antibiotic which is not listed for use in the back of the previously mentioned book

Where There is No Doctor. For those with advanced life support skills, be advised that the veterinary supply houses also sell epinephrine.

Jeffers Vet Supply is a source of mail order veterinary drugs, telephone 1-800-641-2836. Your local agricultural / farm supply store will also sell a lesser variety of drugs over the counter.

One of the really hard things to get in a disaster will be pain killers and anesthesia. Non-prescription analgesics (painkillers) should be stockpiled. This starts with aspirin (acetylsalic acid) and Tylenol (acetaminophen). If you can come by it, codeine is the next level of painkiller, but obtainable only with a prescription. Lidocaine (Xylocaine) is very useful when it comes to deadening pain before suturing a wound. This is a local injectable painkiller which also requires a prescription. Morphine is the top of the line painkiller for serious trauma. Morphine is a wonder drug when it comes to shock, pain suppression and other medical emergencies, but it is highly controlled because of illicit use. It is unfortunate that some legal vehicle does not exist for the stockpiling of medications for disasters.

The old fashion anesthesia was ethyl ether. This is not the stuff you start your diesel truck with when it is cold outside. Squibb makes this in 1/4 lb., copper-lined cans. If your group has the expertise and equipment to perform surgical procedures, ethyl ether should be in your medical kit.

Modern anesthesia is too high tech and expensive for most disaster preparedness programs and requires highly skilled personnel. Ethyl ether can be dripped onto a cloth over the person's nose and mouth to facilitate the inhalation of this anesthesia. The fabric covering the nose and mouth should permit the free passage of air and anesthetic vapor and the unobstructed elimination of carbondi-oxide. This fabric covering can be made of 8 to 12 layers of surgical sponge gauze.

Ethyl ether is highly flammable and overdoses can cause respiratory arrest. The antidote is artificial respiration and the administering of oxygen or fresh air.

All drugs sold in the United States have expiration dates. Expiration dates are conservative estimates of the date at which the drug's stated potency, as listed on the label, can no longer be guaranteed. This does not mean that after this said date, the drug becomes toxic. It means that a prescribed dosage might not work due to a degeneration in the drug's potency.

There are three factors to consider when storing perishable items. These are temperature, sunlight and humidity. If the drugs are in a dry capsule form, try to keep them in a freezer or refrigerator. Ultraviolet ray (sunlight) exposure tends to degenerate many products. Keep it out of sunlight. High moisture contributes to the degeneration. Medications in sealed, tamperproof containers probably will not be affected by humidity, but medications in powdered forms should be carefully repackaged in airtight containers.

Naturopathic Considerations

A person can't cure serious gunshot wounds, open fractures or a ruptured appendix with natural remedies, but many infections can be effectively controlled with the use of herbal preparations and homeopathic remedies.

Typical Homeopathic Medication Kit

Preparation Name	Symptoms
Aconitum Napellus	Fever, inflammation, pain
Apis Mellifica	Insect bites, bee stings
Arnica Montana	Trauma, exhausted, septic conditions
Arsenicum Album	Vomiting, diarrhea, pains, hay fever
Aurum	Depression, headache, photophobia
Belladonna	Fever, headache
Bryonia Alba	Pains, cough
Calcarea Carbonica	Swollen glands, cough, chest pains
Calcarea Fluorata	Hemorrhoids, varicose veins, delayed dentition
Calcarea Phosphorica	Fractures, caries, headache, sore throat
Calcarea Sulphurica	Wounds, thick, vertigo, nausea
Cantharis	Cystitis, burns
Carbo Vegetabilis	The great reviver
Causticum	Rheumatic or arthritic pain
Chamomilla	Sleeplessness, nervousness, colic and teething
China	Debility, diarrhea, anemia
Cina	Worms, cough
Drosera	Cough, vomiting
Eupatorium	Influenza, cough, fever
Ferrum Phosphoricum	Inflammations, fever, wounds, hemorrhages
Gelsemium Sempervirens	Flu, headaches, colds
Hepar Sulphuris Calcareum	Inflammation, croup, sore throat, colds
Hypericum Perforatum	Nerve injury, puncture wounds, spasms

Preparation Name	Symptoms
Ignatia Amara	Pain, chronic spasms, relieves bed wetting
Ipecac	Cough, asthma, stomach disorders
Kali Carbonicum	Pains, palpitations, cough, nausea.
Kali Muriaticum	Glandular swellings, acne, eczema, bursitis
Kali Phosphoricum	Brain fatigue, depression, blurred vision, asthma
Kali Sulphuricum	Inflammation, diarrhea
Ledum Palustre	Puncture wounds, stings, bites, poison oak
Magnesia Phosphorica	Cramps, spasms and pain
Natrum Muriaticum	Digestive disturbances, headache, colds
Natrum Phosphoricum	Rheumatism, jaundice, colic, headaches
Natrum Sulphuricum	Fever, vomiting, asthma, flu, jaundice
Nux Vomica	Heartburn, nausea, vomiting, headache
Phosphorus	Inflammations, bleeding, colds, cough, pain
Pulsatilla	Stomach upset, colds
Radium	Pains, depressed, nausea, arthritis, fever
Rhus Toxicodendron	Strain, rheumatic pains, skin eruptions
Rumex Chrispus	Pains, cough, itching, diarrhea
Ruta Graveolens	Strain, periosteum injury, eye strain, headache
Silicea	Chronic pus formation, rheumatic affections
Spongia Tosta	Cough, goitre
Thuja	Ill effects of vaccination, otitis, wart
Urtica Urens	Urticaria, neuritis, diarrhea, burns
Veratrum Album	Fainting, pains, cramps, diarrhea, shock
X-Ray	Anemia, burns, psoriasis, nausea, rheumatism

Natural measures will help maintain health and improve the body's immune resistance. In extreme life threatening situations, natural treatments cannot substitute for synthetic drugs and surgical intervention. There is a fine line between the legitimate boundaries of allopathic medicine and naturopathic medicine or the health field. I have seen abuses by both camps when they overstep their boundaries.

There have been amazing results from the use of a Bach Flower preparation called "Rescue Remedy". This preparation helps bring people back under emotional control after sustaining a serious injury. Arnica is another effective homeopathic remedy for injuries. I have also used essential oil of Lavender topically to effectively reduce the pain caused by both minor and major closed fractures. Bach remedies can be ordered from the Bach-Nelson Company, 1(800) 314-2224.

The best and most comprehensive book I have seen on the subject of homeopathy is a paperback titled *Family Guide to Homeopathy*, by Dr. Andrew Lockie. This book is published by Fireside / Simon & Shuster, ISBN:0-671-76771-2. Homeopathic kits are fairly inexpensive and can be purchased from Boiron, 1208 Amosland Road, P.O. Box 54, Norwood, PA 19074, telephone 1-800-258-8823 or Standard Homeopathic, Los Angeles, CA 90061

Another branch of naturopathic medicine involves the use of herbs and herbal preparations. Dr. Michael Murray has some very good books regarding medical applications for herbal preparations. These books include: *Healing Power of Herbs* ($17.95), *Encyclopedia of Natural Medicine* ($21.95) and *Natural Alternative to Over the Counter Drugs* ($30.00). These books can be ordered from Impact Communications 1-800-477-2995. A company called Phyto Pharmica,

1-800-553-2370, sells herbal preparations according to Dr. Murray's formulas.

Traditional Chinese medicine has made use of herbal preparations for thousands of years. One in particular, Yunnan Paiyao, was used successfully by the Vietcong for treating gunshot wounds. The preparation is taken both topically on the wound site and orally. It is also purported to be effective in treating traumatic hemorrhaging, bruises and contusions. It is recommended for the treatment of infections to the skin that produce puss, redness, swelling, abscess and boils.

A more western version of the Chinese gunshot preparation is the mixture of goldenseal and cayenne. This can be put directly on open wounds. It facilitates clotting and suppresses infection. **Goldenseal** and **Echinacea** have a natural antibiotic effect and can also be taken orally for the treatment of bacterial infections. According to Dr. Christopher, a noted herbologist, cayenne, when taken orally in substantial quantities causes the vascular system to dilate. This has shown to be very helpful for an individual suffering acute chest pain caused by an obstruction to the coronary artery, i.e. heart attack.

Nutrition

Certain nutritional factors can greatly help protect the human body against damage from radiation and increase immune resistance to disease and sickness. Most of these nutrients can be eaten ahead of time to strengthen the immune system and then taken in larger doses after radiation or disease exposure. Nutritional factors can boost the effectiveness of the body's natural protective mechanisms and reduce the hazards of radiation and other environmental toxins. Each of the major protective nutritional factors are introduced and discussed briefly in the following:

Vitamin E

This vitamin is the most important free radical trap which can counter harmful effects of radiation. Vitamin E is especially high in wheat germ and some unrefined vegetable oils (see oil precautions below). Good levels are found in oats while all leafy greens, most whole grains and nuts are reasonable sources. The major reason that nitrogen or carbon dioxide are used to store foods is to displace oxygen and therefore protect vitamin E and the polyunsaturated fatty acids from oxygen damage (and to kill insects). Unless you spend lots of time figuring out and maximizing your intake from food sources, a daily supplement is prudent to maintain your system in the best possible condition and to build your stores for radiation protection.

The most effective form of vitamin E supplement is called alpha-tocopherol-acetate. The acetate form is synthetic but it stores much, much better than natural vitamin E sources. This is one case where the cheapest, drug store sale vitamins are the best buy for daily use or for long-term storage. 200-400 IU (international unit) sizes are the best for most people on a daily basis. Larger amounts (about 1000 IU per day) are desirable after radiation exposure.

Carotenoid

The importance of this class of compounds as free radical traps is only beginning to be appreciated by researchers. Carotenoid has long been known to be efficiently converted to Vitamin A and Vitamin A is very important to oppose infection. Green and yellow vegetables and fruits are good

sources, especially pumpkins, yellow squash, sweet potatoes, yams, carrots, egg yolks, and yellow fruits. Many sea weeds are excellent sources of carotenoid and have other important anti-radiation properties as will be discussed later. A desirable level of supplementation has not been established and supplementation may not be essential if attention is paid to including sufficient highly colored yellow and green foods in the diet.

Sulfur Amino Acids

Sulphur Amino Acids are technically called methionine and cysteine. These trap free radicals directly and are also essential to detoxify lipid peroxide compounds which are secondary products of radiation damage. The sulfur amino acids may be the hardest single nutrient to get enough of in our diets. Eggs are a superb source of sulfur amino acids (as well as a source of well-rounded amino acid nutrition). Most grains are rather weak in sulfur amino acid content although millet is excellent, brown rice and barley are pretty good and whole wheat is a modest but useful source.

Many types of fish are pretty good sources of sulfur amino acids including haddock, tuna, cod and catfish. Sesame seeds and Brazil nuts are good sources. This is one of the nutrients which should be stored because radiation exposure optimally requires more than can be obtained practically from nutritional sources.

Vitamin C

Vitamin C is essential for radiation protection and for good health in general. It may be possible to get enough Vitamin C from foods if great attention is paid to this point. It is prudent to take supplements because many other considerations come into daily meal planning and the best foods for vitamin C content are not always available (especially in a long-term food storage supply). Vitamin C supplements should certainly be stored.

If we are to make optimum use of our own recovery machinery we should greatly increase our vitamin C intake in times of severe stress, infection or injury. Vitamin C is essential in wound healing and in repair of any damaged tissue including joints and ligaments.

Time release forms are best for daily nutritional intake. 1,000-2,000 milligram/day is the minimum dose that should be taken at two different times a day. The powdered forms can be taken at 10-30 grams a day to combat severe infection (1 gram = 1,000 milligrams).

Selenium

Selenium is a trace mineral which detoxifies peroxide products of free radicals and is an immune stimulant. Only about 100 micrograms are needed per day. It is prudent to take a 50 microgram per day supplement daily since it is hard to be sure how much is in the food we eat. Wheat is a major dietary source but the selenium in the wheat depends on the amount of selenium in the soil on which the wheat (or other plant) is grown. Other whole grains are useful sources as are vegetables. Ocean fish and garlic are very good sources.

Selenium has been in the news for its toxic properties which are certainly a reality at high levels. Special attention should be given to keeping dietary supplements secure and out of the reach of little children.

Polyunsaturated Fatty Acids

It is common medical practice to recommend lots of polyunsaturated fatty acids in the diet to lower blood cholesterol. Followed blindly this recommendation can present problems, because polyunsaturated fatty acids are the compounds in our bodies that are most sensitive to damage by free radicals.

Olive oil is the most healthful type of oil. It is rich in a mono-unsaturated fatty acid (called oleic acid) which is effective in lowering cholesterol but is very resistant to damage by free radicals. There are some new strains of safflower and sunflower (which grow best in northern climes) that mimic olive oil in desirable properties but these may not be in production as yet.

Proteins

Balanced protein nutrition is required for optimum operation of almost all defense mechanisms in our bodies including resistance to infection, radiation and environmental toxins. These defense mechanisms require the synthesis of fresh, new protein enzymes which can only be accomplished if a balanced ratio of the essential amino acid building blocks are eaten within about a 3-5 hour period. The key essential amino acids of concern to balance in nutrition are called lysine, isoleucine, tryptophane, and the two sulfur amino acids already mentioned (methionine and cysteine). Obtaining this balance is the goal of combining grains with beans in the diet. Rice-soy, corn-beans, legume soup with bread are examples of good combining.

Sanitation and Disease

Sanitation is not a luxury, it is a survival necessity. More civilians die in wars from diseases resulting from the breakdown in sanitation than from flying bullets and bombs.

Modern sanitation, not immunization, is responsible for eliminating most major diseases. During a time of disaster, public sewer and water facilities will not function. As a result, diseases such as cholera, typhoid and scurvy will reoccur if good sanitation practices are not maintained. Close confinement and a lack of medical care will compound the effects of poor sanitation. Following is a description of several of the major diseases that are of potential concern to people in shelters during a disaster.

Typhoid Fever

Typhoid is an acute infection of the intestinal tract. The main symptom is a continual but erratic fever that can last up to four weeks. The beginning of this disease may pass unnoticed with a general malaise (bodily weakness), anorexia (lack of appetite) and headaches. During the first week, the patient's fever can rise to 104 degrees and he becomes helpless and submissive. Frequent nosebleeds are common and diarrhea and constipation are often present, accompanied by abdominal tenderness and distention. Discrete rose-colored spots may be seen on the body. The death rate is only 5 percent if treatment begins in the early stages.

Cholera

Cholera is an acute infection of the intestinal tract. Symptoms are a moderate diarrhea which increases in severity and develops into sudden violent diarrhea and vomiting. The diarrhea is followed by diffuse water vomiting. Unless the dehydration is checked, the disease is fatal. Within a few days, the patient either recovers or dies. The death rate, if untreated, is 20 to 85 percent.

These two diseases, typhoid and cholera, are caused by the water, food or people being contaminated by human sewage. Common modes by which these diseases are spread between people include soiled clothing, skin and mouth contact. Once having contracted either typhoid or cholera, the person can be a carrier for days or weeks, spreading it to others through skin contact.

Hantavirus

Every so often I see a new article in the newspapers about someone else contracting Hantavirus or mice being found which are infected with the virus. Hantavirus is transmitted to people through airborne particles of mouse droppings and nesting materials. The virus grows in cells lining the lungs and when the immune system attacks the infection, small holes develop. As a result, liquid fills the lungs and death occurs. It takes about two weeks for an infected person to get sick. Symptoms are similar to flu. Two to three days after initial symptoms the body collapses and 24 to 48 hours later, death occurs.

Keep your shelter and your food storage areas liberally distributed with bar bait or De-Con type mouse poison. Dark, uninhabited areas with improperly packaged food will result in eventual mouse infestations.

If you end up with an infestation and you have to clean it up, do not sweep up or vacuum droppings and nesting material until it has been disinfected. Disinfect with 10 parts water and one part household bleach. Spray the area down and let it soak for 15 minutes before sweeping or wiping up. Carpets can be cleaned with a steam cleaner.

Hepatitis

Hepatitis is usually very slow to get started and begins with anorexia (lack of appetite), vague abdominal discomfort and nausea. Fatality ranges from 1 to 20 percent even when treated medically. The form of hepatitis that may occur during shelter occupation is Hepatitis A, which is transmitted by personal contact and from handling contaminated water or food. Food is a more common mode of transfer than water. If food is washed off with contaminated water, it becomes infected also. This disease is commonly transferred by food handlers who have the disease in the incubation period when the symptoms are not yet noticeable. The prevention of this disease is basically cleanliness — washing hands after using the toilet, washing before preparing food, keeping soiled clothing away from food, and washing soiled clothing with plenty of soap and water.

In addition to a strict adherence to all cleanliness procedures, it is advisable for people in shelters to rinse their eating dishes, cups and silverware in rinse containing one tablespoon of Clorox-type bleach to a quart of water.

Chapter 22
Security

Security concerns are a very sensitive issue. The social trend today is toward disarming the public. Anything relating to gun ownership or shelter defense considerations is looked upon as politically incorrect. If the local media picks up on any such activity, they will be sure to accuse you of making plans and building fortifications with the intent of having some sort of confrontation with the government. These types of legitimate shelter security concerns do not have their source in an intent to conflict with the government, but a concern that during a national crisis there won't be enough government to insure law and order.

Shelter Observation Tower

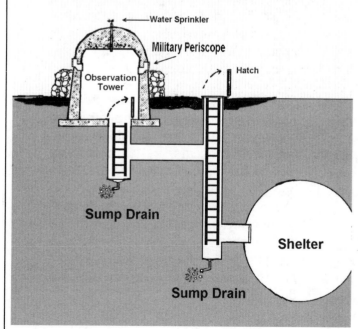

←— Water Sprinkler

Military Periscope

Observation Tower

Hatch

Sump Drain

Shelter

Sump Drain

Social Security

If you are going to build a shelter, it would be wise to make adequate contingencies for its defense. Shelters are a big investment and we all value our families. Generally, people without shelters will not just sit home and perish from radiation poisoning. Once the rains started, Noah discovered a lot of last-minute converts to dry-land boat building. Those who for years mocked Noah and his sons made a last-minute effort to survive the flood, ran to the ark and tried to get in. I'm sure the ark was filled to capacity and if the people outside could have gotten in, they would have either swamped the ark or thrown Noah and the animals overboard.

Expect those who hindered and mocked you to be the first ones to pound on your door. I am not suggesting that anyone should be hardhearted or unforgiving to such individuals. The point is that the shelter air supply system will only support life for a specified number of people. One of the most stressful and emotionally trying situations you may be

Oleoresin Capsicum

faced with in the hour of occupying your shelter is what to do with the flood of people who are not involved in your shelter project and want to be let in. Know the safe capacity of your shelter's life support systems and when capacity is reached, you will have to close the door even with people still outside begging to be let in.

If you don't make that hard decision, then you aren't fulfilling your ultimate responsibility to the members of your shelter, because overloading the capacity of the shelter's air system will cause everyone in the shelter to perish. This is just another form of triage. If you try to save everyone you won't save anyone. Ultimately, you are only responsible to do your best at what you can do and know how to do with the number of people your air system can handle. Once that number is reached, additional people will endanger the lives of all the occupants. Everyone has a moral obligation to fill their shelter to the safe capacity of the life support system, but once the shelter occupancy has reached the capacity of the air supply system, the door must be closed. This is a very impersonal issue, but any people remaining outside may very well take it personally. This type of situation can quickly become irrational and hostile.

A shelter is generally a very vulnerable thing. It is only as good as its air supply. Shelter air supply systems, by their nature, are very susceptible to tampering. Make contingency plans to deal with the unfortunate nature of this potentially life threatening problem. If someone, out of spite, cuts off the incoming air source, the occupants will at some point have to either open the door and surrender or perish from CO_2 poisoning. It would be good to have a substantial reserve of compressed air, the capacity to scrub CO_2 and a good means of defending your air supply.

In regards to this issue be creative. Try to have some non-lethal alternatives which you can try first. If a life threatening incident occurs, try to contact the local authorities by radio and ask them how they would like you to deal with the situation. Document and record if possible any such urgent communications. A major disaster is not necessarily going to bring the end of the world and you might have to answer for rash actions after the dust settles.

If you are going to own firearms, you need to be well trained in how to use them. In terms of semi-auto rifles, an SKS is the most inexpensive thing you can buy. An AR-15 or a .308 M1A are expensive $1,000 to $2,000. The .308 caliber has longer range than the .223, but it is bigger and the ammo weighs more. For close-in combat, a 10mm or 45 ACP semi-auto hand gun seems to be one of the better options. These calibers have good knock-down power. If you can, try to purchase from a private party. Someday soon these may be illegal.

Good night vision equipment is expensive. American-made third generation night vision equipment can cost up into the $3,000 range. The market is flooded with Russian equipment which is poor. American-made second generation night vision equipment performs a bit better than its Russian counterpart.

The real performance of night vision systems is determined by a number of factors. These include gain, photosensitivity, resolution, signal to noise ratio and operational life. Gain indicates the number of times an image intensifier amplifies the incoming light. Photosensitivity is measured in microamps per lumen and it represents the ability of the unit's photocathode to produce an electrical response when subjected to light waves. Resolution is measured in line pairs per millimeter and it represents the ability of an image intensifier to distinguish between objects close together. Signal to noise ratio is a measure of the light signal reaching the eye from a particular night vision unit divided by the perceived noise as seen by the naked eye. The higher the ratio, the better the equipment. Operational life represents the total rated number of hours an image intensifier will continue to convert light energy into electrical energy. It is important to realize that inexpensive Russian night vision equipment ends up costing more in terms of cost per hour when expected operational life is taken into consideration.

Night vision requires the presence of some ambient light in order to work. Thus night vision will not work in a totally dark room or cavern without the use of an infrared illuminator. Infrared illuminators greatly enhance the effective capacity of night vision equipment. The naked eye cannot see infrared light, but an infrared illuminator creates a noticeable flag waving signature for anyone else out there who might be looking for you through a piece of night vision equipment.

American made 3rd generation equipment will allow a person to see 800 yards with the light of a full moon. The same equipment will only allow you to see 200 yards on a moonless night with an overcast sky. American made 2nd generation equipment will allow a person to see 675 yards with the light of a full moon. The same 2nd generation equipment will only allow you to see 100 yards on a moonless night with an overcast

45 ACP Caliber Colt

.223 Caliber AR-15

Night Vision Comparison

Night Vision Unit	Operational Life in Hours	Resolution (Line pairs Per Millimeter)	Signal To Noise	Photoresponse (Microamps Per Lumen)	Tube Gain	Field Of View	Cost Per Operational Hour
U.S. 3rd Gen (ITT)	> 10,000	45	18 : 1	> 1,000	30,000 - 50,000	40^0	$.15 - .35
U.S. 2nd Gen (ITT)	> 5,000	36 - 40	13 - 14 : 1	> 300 - 500	30,000 - 50,000	40^0	.20 - .40
Russian (Night Spy / Moonlight)	400 - 800	32 -36 Center Only	6 - 10 : 1	< 150	< 10,000	< 12^0	1.00
Japanese (Fujinon)	~ 1,000	32 -36	10 : 1	~ 200	10,000 - 15,000	40^0	1.90

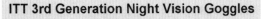

ITT 3rd Generation Night Vision Goggles

ITT 3rd Generation Night Vision

sky. In my experience the Russian equipment and some of the American-made 2nd generation equipment give poor resolution at distances. It becomes difficult to discern between a rock and a man at a distance greater than 50 yards.

American-made 3rd generation night vision equipment can be purchased from Yellowstone River Trading Co., P.O. Box 206, Emigrant, MT 59027, (800) 327-7656, (406) 333-4707. A good source of publications in regards to security issues is Paladin Press, P.O. Box 1307, Boulder, CO 80306, (303) 443-7250. *Soldier of Fortune Magazine* is another good resource. A wholesome source of preparedness, self defense and security training is The Center For Action (702) 723-5266.

Litton 3rd Generation Night Vision Scope

Chapter 23

Psychological Considerations

Psychological Responses to Disaster

It is important to understand some of the ways people react during emergencies, disasters and unusually adverse conditions. It is very difficult to predetermine what different individual reactions will be. This is something that can only be determined through actual experience. There are, however, fairly common reactions which can be predicted. Understanding these reactions will help the individual and the group to maintain some sort of objectivity and implement remedies and prevention.

The natural psychological and emotional reactions which people have as a result of a disaster include fear, terror, panic and emotional containment.

Fear

Fear is a normal reaction. Fear fulfills a function within certain limits. Fear serves as a warning and an aid. Fear mobilizes the body's reserved strength and increases its physical abilities. Fear allows people to ac-complish feats which otherwise would be unthinkable. Fear can be a useful driving force.

The symptoms which people show when under the effects of fear are an increased need for contact (conversation provides a sense of security) and increased inner tension, which can result in the involuntary shaking of the hands, sweating, flushing, palpitations of the heart and pressure on the bladder. Fear is usually short term in duration and does not have long-term adverse side effects.

Terror

Terror is fear which has intensified and is usually provoked by an unexpected danger. Suddenness and surprise are essential ingredients for producing a terror reaction. Terror is a storm of emotion which express itself violently and irrationally and can increase the danger in a situation. Terror is an emotional obliteration of the will, reason and common sense. The mind is completely flooded and it is unable to clearly evaluate the real level of danger. The power to deliberate is put to flight by the absolute emotion of terror. Terror overwhelms the ego and paralyzes certain mental facilities.

Terror manifests in two phases. The first reaction is apathy. This will be seen as an absolute inhibition and immobility, in other words, terror stricken. Typically terrorized people in dangerous situations can become indifferent to their own safety and the destruction going on around them. The second phase of terror which follows apathy is the impulse to flee. This is usually indiscriminate, without common sense. Both phases can involve a dazed condition and a loss of memory.

Terror is an emotional reaction which generally does not benefit the individual but increases his vulnerability to injury and disaster. Also, terror is an emotional reaction which has the very dangerous quality of being explosively contagious.

One significant indication of when normal fear is on the verge of dissolving into terror is when the need to talk increases and people begin to scream. The raising of the level of the voice indicates an acceleration of emotions which are beginning to overwhelm the individual's faculties of logic and control. When people are showing signs of trembling, walking back and forth and doing unnecessary things, this indicates a buildup of inner tension. The only thing you can do for a person suffering from terror is isolate them. If such an individual is violent, they need to be brought under control for their own good and to prevent infecting others. Methods of force should be limited to keeping them quiet with a gag and tying them up. If the means is available, sedate the individual.

Panic

Panic is a group emotion. Panic, like terror, is explosively contagious and spreads quickly. Panic is a reaction to a suggestion of danger which may or may not be real. When the individual loses his power to exercise common sense, he can easily be swept away by group emotions. Panic is a mass psychosis which can drive a whole group to irrational acts.

Panic is grounded in man's herd instinct and his need for security. People fulfill their needs for security through group contact and interaction with other people. This security and herd instinct outpictures in day-to-day life as people find places in various social groups. People feel protected by the similar behavior of those around them. This is seen at sporting events. What happens is that the individual in the group lets the imitation instinct take over individual thinking and thus satisfies or fulfills a need for community feeling.

Panic usually starts with one weak-willed person and spreads lightning fast throughout the group by way of suggestion. Those in the group who usually have the ability to control their inner anxiety can also be swept along with this herd reaction of abandoning one's ego and dissolving into the group. Strength of character and a well exercised willpower are the only deterrents to being swept away in such group reactions.

Emotional Containment

Emotional containment is a positive, involuntary reaction which some people experience in the face of a dire emergency. In this situation, the dominating emotional impact of the incident is totally suppressed. This allows for a better possibility of control and common sense. Under emotional containment one can think clearly and act purposefully. An extremely focused consciousness accompanies emotional containment, which shuts out anything going on outside the person's immediate surroundings.

Depression

Many factors can cause depression. A natural or nuclear calamity and shelter confinement are sufficient potential causes in themselves. Couple this with the unknown, concerns for the fate of close friends and

relatives, a perceived inability to improve one's personal situation, and feeling out of control of one's future and such a situation could easily become overwhelming for many individuals. This could result in a breakdown of normal psychological defenses, and thus a downward spiral of mental and emotional despair, ultimately resulting in emotional depression.

Depression Symptoms

Dysphoric mood is a common symptom associated with depressed individuals. This includes an all-pervasive sense of doom and gloom. As this symptom progresses, individuals will lose interest in things which were previously important to them or gave them enjoyment or gratification.

Social withdrawal is another symptom of depression associated with the loss of pleasure and meaning in life. The depressed individual crawls into a cocoon in an attempt to protect himself from a falsely perceived threat from people around him and the outside world.

A significant decrease in someone's energy level is another depression indicator. This will manifest as an inability to do a job effectively and efficiently.

Thoughts of low self-esteem and self-worth clearly indicate depression. Depressed individuals revolve and are dominated by feelings of inadequacy, negative evaluations of their self-worth or their worth to others. This low self-worth is definitely based on a distortion of reality. The depressed individual usually looks for signs confirming his negative sense of self-worth.

Depressed individuals tend to suffer from distorted thinking and faulty information processing. This manifests as overpersonalization. Such an individual tends to illegitimately relate everything that happens in the environment around him to himself, even if there is no logical connection. This symptom also manifests as an overgeneralization and a tendency to magnify or minimize events out of proportion. In advanced cases of depression, the individual loses the ability to mentally focus and engage in effective problem solving.

The last symptom category of depression is the presence of suicidal ideas, thoughts of death and suicidal gestures. Suicidal and death thoughts indicate a desperate need to escape the turmoil and pain of depression. People who lose their sense of hope and thus their purpose for life see death as a way out — a way out of the guilt, despair and their loss of control of their surrounding environment. This suicidal fixation can manifest either in an obvious fashion or in a disguised form. The obvious fashion would be an attempt at suicide. The disguised form is more subconscious, which results in accidents. The individual's inability to focus, reduced reaction abilities, mental preoccupation (which results in a lack of vigilance), unnecessary risk taking and recklessness all lead to the depressed individual endangering himself and others around him.

As you can see, the individual suffering from depression should not be delegated critical responsibilities involved in operating the shelter because his ability to function in a safe and accurate manner is impaired, not only endangering himself but possibly everyone in the shelter. Depressed individuals cannot be counted on to accurately monitor their own accumulated exposure to radiation. These individuals must be closely monitored so that they don't unnecessarily expose themselves to radiation.

Treating Depression

Early recognition and treatment is vital.

It is important that you make contact with the person suffering from depression. Don't attempt to approach him in an aggressive or demanding manner, such as telling him to "snap out of it." Try to make a gentle, understanding contact with the objective of securing rapport. On the other hand, don't indulge in pity. Get him to tell you what happened or what is going on.

Vigorous exercise increases the production of the neurotransmitters norepinephrine and serotonin in the brain. These naturally occurring chemicals act as natural antidepressants and tend to elevate moods.

Control the diet. Eliminate stimulants, including all sugar (sweets), caffeine and nicotine intake. Stimulants are a roller coaster ride. They temporarily elevate and then quickly lower the individual to a crash in depression.

Do not give a depressed individual alcohol or other depressants, such as tranquilizers, antihistamines, codeine cough suppressants or any other similar acting medications.

Rearrange the individual's environment if possible. Try to alleviate burdens of stress on the individual. If his position is causing the stress, give him a different, less stressful responsibility in the shelter.

Keep people busy so that they don't have too much time to dwell on their supposed problems. Make or create work in the shelter if necessary to keep everyone occupied. There is a certain amount of truth in the old saying "idle hands are the devil's tool." Preoccupy individuals confined in shelters with tasks, objectives and goals.

During one of the major World War II Pacific sea battles, a U.S. aircraft carrier was badly damaged, almost sunk and took massive casualties. The captain opted not to abandon the burning, listing ship and managed to stabilize the vessel. En route back to San Francisco, he intentionally kept the surviving crew working around the clock repairing what damage they could, cleaning up the devastation, recovering bodies and body parts and giving them sea burials. The captain did this intentionally so that the remaining crew, who had lived through a nightmare, would not have time to dwell on the tragedy and carnage they had experienced.

Predictable Psychological Reactions to Disaster

A disaster can be segmented into seven general phases and each of these disaster phases can be associated with predictable, human psychological reactions.

Warning Phase

During the warning phase the predictable psychological reactions which people affected by the disaster experience include anxiety and denial.

Alarm Phase

During the alarm phase the psychological reactions experienced by those affected by the disaster include increased activity and panic.

Impact Phase

During the impact phase of a disaster people tend to become emotionally stunned, overwhelmed and shocked. This is typified by victims aimlessly wandering about.

Inventory Phase

During the inventory phase those stunned by the impact of the disaster are confused and disoriented as they start to regain their senses. The first thing people do is take inventory of their situation by beginning to assess their losses and the current level of danger in an attempt to figure out how to cope with the situation. Once people feel assured that things are reasonably under control and that they are safe again, some of these people are able to start helping and assisting other victims of the disaster. Research indicates that about half the people in disasters will move quickly through the warning, alarm and impact phases into an adaptive mode and thus be able to help other victims while the other half will have difficulty adjusting to the situation, stay stunned and thus remain in a victim mode. Those remaining in the victim mode will require the most help.

If you are involved in a disaster, the first thing you should do once the inventory phase starts is to enlist and organize those who have made it through the warning, alarm and impact phases and start assisting less fortunate victims.

The second step is to set up a triage program, sorting victims according to a hierarchy of need (who needs the most service right now, who needs the least) because you have limited resources and you need to pour your initial effort into those who have the most need. It's not only the most need but it is also a practical matter of efficiency and effectiveness. Some people get triaged out of the first response assistance because they are too far gone and their condition can't be improved.

This is a practicality and a reality that rescuers and medical-aid personnel have to face in a disaster. You can't waste time and medical resources on people who have mortal injuries when there are numerous people suffering from serious but salvageable injuries waiting for treatment. This involves making hard decisions. In a typical major disaster there aren't enough immediate medical services available to treat all the injured at once. You concentrate on the seriously injured which have a real potential for recovery.

Rescue Phase

During the rescue phase victims are elated, they are happy to be alive and they are euphoric, but this reaction is short lived. Very quickly reality sets in and the victims begin to realize the seriousness of the situation. Thus euphoria will often turn to anger, depression or desolation.

Recovery Phase

During the recovery phase victims start experiencing a sense of relief. The immediate danger seems to be over, but bouts of anxiety and fearfulness tend to reoccur. This insecurity and anxiety is not over concerns for here-and-now situations, as is the case in the rescue phase, but is caused by concerns about the future. In the recovery stage, the victim starts thinking and focusing on the future. This is a key indicator that the recovery phase has started. Even though victims are having a negative response to the future, they are moving through it. This is a positive psychological sign.

Reconstruction Phase

During the reconstruction phase the victims have adapted to change and they have essentially returned to a functional level. The victims are not necessarily the same people they were before the disaster happened, but they are now at a level where they can mobilize their internal resources, assess external resources and begin to function in a helper/responder role. All the human faculties essentially return to a minimally productive level but the one thing that needs to be watched for in this last phase is the potential for relapse.

Some victims will assume that they have put the disaster behind them and that they have dealt with it, but they don't realize that resolution and recovery is going to be an ongoing process. This involves dealing with reemerging feelings, reverting back and repeating some of the phases. Thus some victims may relapse back into a deeper state of any one of these phases.

Wherever you have personal wounds in your psychology or your history, then those may match up with any one of these phases. You don't necessarily go back to the beginning and run through all the phases again. You may go back to a specific phase.

Disaster Myths

There are a number of commonly accepted myths in regards to people's reactions to disasters. Examining these myths will help us better understand, prepare for and cope with disasters.

People will heed warnings.

The first myth is that people will heed warnings. People generally do not heed warnings. Historical research on the Pompeii disaster indicates that there were many warnings from nature as well as warnings from people in society which were not heeded. The truth is that even though people generally ignore warnings, they will listen if they are repeated and repeated by trusted leaders, close neighbors, friends and family.

People always panic in a crisis.

This is not true. Mass panic only happens in about 10 percent of the disasters and then only when there is no clear evidence of an escape route or protection from the disaster. The existence of a prepared shelter or other preparedness becomes a key element in helping people psychologically deal with the crisis when it arrives. When people know that they have a potential escape, such as a shelter, and that the shelter will protect them to a reasonable extent, they will be less likely to panic. The only other question is are they in a position to get to the shelter. The key element is that a person has a safe place to go and a way to get there under the worst circumstances.

In a disaster, the entire social structure of the community will remain psychologically devastated.

People have a miraculous resiliency and the ability to respond to difficult situations. History has shown that people do have the ability to rise to the occasion and deal with what needs to be dealt with. This is not to say that it happens without pain, mistakes or difficulty. The truth is that communities restore themselves to normal functioning and only about 50 percent of the people will need help coping with depression or anxiety.

Factors That Affect Human Reaction to Disaster

Escalating Factors

There are certain factors which can escalate the individual's negative reaction to a disaster. These factors include:

1. **severity of injury,**

2. **the intensity of the incident,**

3. **the length of time which the crisis lasts,**

4. **and the occurrence of more than one disastrous event within the same time frame.**

The duration factor can be very difficult for people to deal with because they may not know how long the disaster is going to continue. The simultaneous occurrence of multiple disastrous events compounds the intensity of the overall disaster.

Mitigating Factors

There are certain factors which can reduce the individual's negative reaction to a disaster. These factors include:

1. **A quick, well-organized rescue effort.**

The speed, comprehensiveness and organization of the disaster relief and rescue effort obviously mitigate the negative impact of a disaster.

2. The Existence of a Family or Community Network

It is important to recognize who is a stranger and who isn't in the disaster situation. Try to get people matched up with people they know. If they don't know anybody, get them introduced and start that networking happening.

Variable Factors

Age of the Victim

The young and the elderly are the most vulnerable in a disaster. They are the most concerned about security. They feel the least capable of coping because their internal resources are usually more limited. Their physical capabilities are more limited and particularly with children, they haven't developed the capacity to really understand what is going on. The very elderly may have gone back to that state in terms of dementia of some kind where they don't fully understand what is going on.

Objectives for Managing the Disaster and the Victims

The primary basic objectives of emergency personnel managing a disaster should include the following:

1. **Safety first!**

2. **Bring the disaster situation under control**. The obvious purpose is to prevent additional people from getting hurt or killed.

3. **Organize and provide basic services for displaced persons.** This should include food, clothing, shelter and security.

4. **Organize and provide family quarters in shelters as soon as possible.**

5. **Assist displaced victims in finding relatives or friends who can better help them through the crisis.** If relatives or friends are not available, try to match them up with someone they know. If no familiar persons are available, then pair the displaced victim up with someone who is outgoing. A person or family who is willing to take this displaced person under their wing and make them part of their family group is ideal.

6. **Encourage victims to return to "normal" schedules.** In a disaster situation there will be no real normal schedule, but try to approximate one. Try to get people on a regular eating and sleeping schedule. Get people working on activities that utilize their skills and assets as soon as they can be moved through their own traumatic response to the situation.

7. **Protect victims as much as possible from further stress.** Provide psychological counseling services to the extent that trained personnel are available. If trained personnel are not available then find someone who seems to have a natural listening ear for those having trouble. Provide a listening ear for people to vent their problems, emotions and anxieties. Counselors should avoid making statements such as "Well, it could have been worse," or "you're lucky to be alive." Such statements do not provide people with comfort.

When acting as a counselor, you need to be very perceptive. This means not only listening to the words but also listening to what is the underlying message, the hidden message, and what is the behavior saying. If there is an incompatibility, the behavior is usually giving you a truer message than the words. A gentle physical touch, a calm soothing voice will go a long way in helping victims recover from the trauma of a disaster.

What has been learned about disasters is that people should contain their feelings. It is not the time to help people cry and express their feelings. It's not a problem if someone cries or wails, if that is what is going on with them. However, what you want to do is get them to a point where they feel more control over their emotions. They have enough lack of control going on in their lives without feeling personally, emotionally out of control. You need to help them reconstitute. There will be plenty of time later to set up some groups to help people work through their feelings and begin to express these feelings. This will help them purge the negative emotions that are there and get into a healthier emotional state.

Containment is a good by-word all the way through. You want to contain. You want to protect. You want to organize. You want to bring things back into focus in people's lives to help them move into the coping phase. You want to encourage the defense mechanisms. You don't want to encourage a lack of reality about what's happening but you want people to feel that, internally and externally, the resources that are necessary for them to feel safe are there.

Reassure people that what they are experiencing is normal. Whatever they are experiencing, there is no such thing as abnormal in a crisis or disaster situation. Reassure them that "there is nothing wrong with you experiencing what you're experiencing." The victims are reacting to something that is out of the ordinary human experience, so how can anyone say what is normal and not normal. Reassure people that they have it within themselves to survive.

Critical Incident Stress

Some of the conditions that cause critical incident stress are events, emergencies or disasters which result in sudden or unexpected death. These can burden rescue personnel, family and bystanders with great amounts of stress. This stress is often magnified if the victim was a child, if the emergency involved carnage or great human suffering, if a death occurred after a prolonged rescue attempt, or if a victim at the incident was expressing uncontrolled emotion. These stressing situations are experiences outside of the usual range of human experience and would be distressing to almost anyone.

This type of stress is called critical incident stress. The individual's reaction to stress can be either acute or delayed. The acute stress response (immediate reaction) to the incident can include becoming physically ill, the inability to function, becoming immobilized, and mental or emotional breakdown.

The delayed stress reaction is referred to as post traumatic stress disorder (PTSD). PTSD is a delayed psychological reaction to severe stress which results in behavioral problems. PTSD has been a serious problem with fire, E.M.S. (Emergency Medical Services), and law enforcement personnel nation-wide. It is the major cause for people leaving the E.M.S. field today.

In one particular incident in Billings, Montana, firemen responded to a mobile home fire. When they arrived the mobile home was completely engulfed in flames. After suppressing the fire, the firemen found the bodies of five children and two parents in the rubble. This is the type of situation a person may deal with in a real disaster.

Many of these particular firemen later had nightmares, mainly due to the visual impressions they experienced at this incident. This is the same problem which Viet Nam war combat veterans have experienced. Some of the typical symptoms are nightmares, intrusive thoughts and disturbing memories. These disturbing memories can work their way into the person's daily thoughts in an uncontrollable way. The burden of this becomes so overwhelming that the individual starts to fear and avoid any additional exposure to stressful events.

Thus, these types of inordinately stressful experiences may result in critical incident stress. If this stress is ignored or suppressed, it can lead to physical and emotional symptoms as well as behavioral problems which incapacitate valuable people.

Common Stress Reaction Symptoms

Common reactions to stress include depression, apathy, defensiveness, energy loss, compulsive drinking, smoking, or eating, repetitious behavior, organizing one's things beyond the point of productivity, illness, nausea, disorientation, palpitating heart, hyperventilation, and difficulty in sleeping.

Ways to Reduce Reactions to Stress

Prior To An Incident:

1) **Pre-incident training.** This is one of the most effective things you can do. Prepare yourself the best you possibly can with reading, discussion and training.

2) **Get information** before the event as to what you might expect in the various different possible scenarios.

3) **Teamwork.** Spend time with the people who are in your disaster response group ahead of time to develop a good working relationship.

4) **Communication skills** There are some very specific things you can learn that will help you communicate more effectively. One of the best resources is parent training books which give instruction on how to talk to children and teenagers. They discuss communicating in a combat situation.

During Or After An Incident:

5) **Support groups.** Organize support groups within your group, agency or shelter project. The family is the basic and most natural support group.

6) **Be as physically active as you can**. Take breaks. Don't get into the self-defeating situation of "they can't do without me" because if you don't take any breaks, they'll have to do without you. You will collapse at some point. So take control of when others have to do without you.

7) **Discriminating humor.** Humor is a very effective tool in breaking tension

and there is a time and a place for it, but be careful that it is not heard by those who will misinterpret the reason you are using it.

8) **Eat well**. One of the important tasks in firing up the shelter is getting someone in the kitchen. Get the soup kitchen moving because people are going to expend a tremendous amount of energy.

9) **Get some sleep.** There are such things as power naps (15 minutes to get the energy back), but if you want to get sleep that is regenerating you're going to have to get between 1 1/2 and 3 hours of sleep minimum in order to move through the REM sleep cycle, which helps you regenerate. There is a period of sleep where the eyes move rapidly. The sleep clinics that study this find that this period is where the recharging goes on. If you are going to send someone off to "get some sleep," you are going to have to provide about three hours for them to do that. Four-hour watches work well. Normalize your schedule as soon as possible. Develop what is the new "normal" cycle. Get the routines put in place. People like ritual and they like routines. It feels comfortable and safe.

10) **Maintain a sense of hope**. Remember that your survival effort is important. It is important in terms of the survival of your family and loved ones. Remind yourself and remind one another why you are doing this. It may seem so chaotic at first that you'll wonder, "Why are we doing this?" Some may even wonder, "Wouldn't it have been better for us to have just stood under the bomb and let it drop on us?" You may imagine that no one will think this but your shelter group will have the same problems of any other group in the world that has gone through traumatic experiences.

Critical Incident Stress Debriefing

It is important in a crisis to maintain emotional control in the interest of stability, but it is also very important for people to be encouraged and assisted in expressing pent up emotion as soon as it is opportune after a traumatic incident. This process is referred to as critical incident stress debriefing. There is a whole science regarding how to deal with critical incident stress. Psychologists have developed a model that helps first responders get through the emotional stress that comes with their job.

Most people don't like to show weakness, but it is very unhealthy from a psychological standpoint to suppress the expression of emotions. Denial or dismissal of one's own emotional pain will eventually lead to depression or worse. Suppression can lead to depression and post-traumatic stress disorder. Typically, those affected by PTSD become subjective, irrational and unreliable. Because of this problem, national emergency organizations, including police, fire and E.M.S., have implemented CISD or critical incident stress debriefing programs. Whenever people are involved in a traumatic incident it is important to get them to sit down and say what they saw and what they experienced. This process enables individuals to vent an experience which, if otherwise suppressed, would emotionally incapacitate them or become a mental scar.

Patrick M. Wolberd, M.S.W., was a major contributor to this chapter on Psychological Conditions. He has a private counseling practice in Billings, Montana, and is also a member of the Critical Incident Stress Debriefing team there. He can be contacted at (406) 657-0611 or by writing to P.O. Box 20601, Billings, MT, 59104

Psychologically Preparing Oneself for Disaster

People can condition and prepare themselves psychologically for dealing with difficult and stressful situations. The author has had to consciously develop a calm and composed nature in emergency medical response situations. The logic I personally used, in combination with a great desire to change, was that having an emotional reaction to an emergency, being flustered, in a hurry and excited caused me to fumble, make mistakes and not follow established procedures. Such a reaction in no way helps the situation or the patient. It is better to consciously be in control of your emotions, slow down and take your time doing things effectively the first time.

In the same manner, one can condition

oneself to be in emotional control under other contemplated emergency situations. This potentially includes not being overwhelmed

by carnage, dead bodies and people who are seriously maimed. If you allow yourself to be swept away by floods tides of emotion, you will not be able to help those around you who are in distress and seriously need assistance.

Anthony Robins, in his book *Awaken The Giant Within*, affirmed this concept in what he refers to as *"The Law of Reinforcement"*....."*Any pattern of behavior that is continually reinforced will become an automatic and conditioned response. Anything we fail to reinforce will eventually dissipate"*.

Cataclysmic vs. Localized Disaster

There are two categories of disaster. The first and most common category includes localized disasters, such as flood, fire, earthquake or large accidents where the emergency services in the area are strained or incapacitated but help from the unaffected surrounding areas is able to respond and stabilize the situation. The second category of disaster involves a catastrophic or cataclysmic disaster which impacts a wide geographic area. This is a disaster which is so massive that little or no disaster relief aid or help can be expected from any outside source.

What this means is that the individuals involved in such a cataclysmic disaster will have to wear two different hats — that of the victim and that of the rescuer. This seriously complicates the recovery and rescue process. During common disasters, like the earthquake in Northridge, California, most of the rescue and relief help came from outside the Northridge area. These rescuers did not have to worry about the fate of family members, friends and relatives, their houses or their belongings and thus they were able to concentrate their entire attention on helping victims.

During a cataclysmic disaster, the rescuers are going to be in the position of having been affected by the disaster as well as having to help others. Preparedness and rehearsal can significantly improve one's ability to function and cope with such a dual role disaster situation. The more prepared a person is, the more they have thought through a potential disaster scenario, the more likely it will be that they will be in that 50 percent group which moves quickly through the traumatic response to the disaster and is thus able to move into the mode of helping other victims. The better you are prepared, the better you are informed to anticipate the worst case scenario, the more likely you are going to move through the disabling phases of the crisis into the helper phase.

Dealing With Children's Reactions to Disasters

This section is intended to help parents deal with children's fears and anxieties during and following a disaster.

During a disaster the primary concern of any parent is the physical safety of their children and the family as a whole. As a result, the disaster's emotional impact on the child is frequently neglected.

Understanding Children

The average child's environment consists of certain regularities. For most school age children, regularity involves the presence of parents, awakening in the morning, preparing for school, meeting with the same teacher, the same children, playing with friends and sleeping in their own bed. Essentially, the child is able to depend on a series of

predictable events. The child expects dependability from adults and certainly from the forces of nature. For the preschooler, life is much the same. He spends his day within the familiarity of

281

his world — be it at home, with baby-sitters, a nursery school, etc. His family life remains more or less constant. When there is an interruption in this natural flow of life, the child experiences anxiety and fear. How the adults help the child to resolve these "problem times" may have a lasting effect on the child.

Fear and Anxiety

Fear is a normal reaction to any danger which threatens life or well-being. What is a child afraid of during and after disaster? The child is afraid of injury or death. He is afraid that the disaster might reoccur. He is also afraid of being separated from his family and possibly being left alone.

The child, who is dependent on adults for love, care, security and food, fears most the loss of his parents and being left alone. In a disaster, even the child who is usually competent and unafraid may react with fear and considerable anxiety to an event which threatens the family. Since most adults also react

emotionally with normal and natural fear to disaster, the child becomes terrified, taking parental fears as proof that the danger is real.

There are certain things that parents can do to help their children during and

after a disaster. It is of great importance for the family to remain together. Being together with the family provides immediate reassurance to a child. Fears of being abandoned and unprotected are alleviated. For example, children should not be taken to the shelter and left there alone while the parents go out again to bring in more supplies, possessions, etc. Immediately after a disaster, parents should not leave the child in a "safe" place while they themselves go elsewhere to inspect possible damage. Without the opportunity to experience the fear of being left alone, the child is less likely to develop clinging behavior.

Realistically, parents are also experiencing fear. However, they have the maturity to cope with the stresses upon them. A demonstration of strength should be apparent to the child, who will feel more secure and reassured. However, it will not harm the child to let him know that you are also afraid. As a matter of fact, it is good to put these feelings into words. This sharing will encourage him to talk about his own feelings or fears. Communication is most helpful in reducing the child's anxiety and also the adult's anxiety. The child may then express some fears which are not real and the parents will have an opportunity to explore these fears and reassure the child.

It is important to listen carefully to your children. Listen to what the child tells you about his fears, how he feels and what he thinks about the situation. Explain to the child, as well as you can, about what is happening and what to expect in the near future. A child may express his fears in play or in actions. If these fears are unrealistic, explain and reassure him. You may have to repeat yourself many times. Don't stop explaining just because you have told him this once before.

The most frequently reported problem that parents encounter with their children occurs at bedtime. The child may have difficulty falling asleep. He may wake up often during the night or he may have nightmares. It is necessary to be somewhat flexible regarding bedtime in a shelter situation. Bedtime may be delayed when the child is anxious and wants to talk longer, but a limit should be set. It is natural for a child to want to be close to his parents and for the parents to want to have their children near to them.

There will be important concerns and things to do during and after a disaster. A child can and should be included in these activities. It is reassuring for a child if he is involved with the parent in these jobs. Establishing a familiar routine as soon as possible is very helpful. For the parents of a very young child, the task is more difficult. Such a child may need more physical care and more holding. This makes it harder for parents to attend to the other things that should be done. Unfortunately, there is no shortcut. If the child's needs are not met, there will be continuing problems, even after the disaster.

The content of this section on children and disasters was drawn from a pamphlet titled "Coping With Children's Reactions to Disasters." This pamphlet was developed by and is available free from the San Fernando Valley Child Guidance Clinic, 9650 Zelzah Avenue, Northridge, CA 91325, phone # (818) 993-9311.

Death and Dying

This is not a popular subject but it is one that needs to be covered. All of us could be faced with an inordinate exposure to death in the coming years. I personally experienced a near-death situation when I was 19 years old. The one thing I recall distinctly from the experience was the thought that no one had, in any way, prepared me for this.

Death is the ultimate threat. Many people today view death as the opposite of life and the complete end of everything. The idea of the extinction of self is a terrifying concept.

Members of present day society are generally uncomfortable in the face of death. It is a taboo and people don't usually have open and frank discussions about it. In many people's minds, death is an overwhelming disaster which is too terrible to think about.

The way most people deal with death is to avoid it or deny its existence. Consequently, many people have not developed a realistic concept of death nor a means of dealing with it. Very few people face death and admit that it is an

283

inevitable part of life. This is especially true in the case of those who are young and full of life. They view death as something very distant, something that happens to other people. The general social practice for dealing with death is to try to limit its appearance, to avoid talking about it and to spend as little time as possible with dying people.

In less developed third-world countries and agrarian cultures, people are better equipped to deal with death. In rural environments, children grow up with a closer relationship to nature. As a result, they are exposed to the cycles of life and death in the changing of the seasons and in the shorter life spans of pets and livestock. Modern urban living tends to separate people from exposure to the basic cycles of nature and fosters false attitudes of immunity from the effects of nature and the inevitability of death. The advance of modern medicine and the subsequent increase in the average life span has also contributed to an illusion that man has mastered death.

Fifty years ago, a major portion of American culture was centered around a more rural, agrarian structure which included a close-knit family life. This family structure usually included a large circle of aunts, uncles, cousins and grandparents living in the same house or in close proximity. If a parent died, other family members were readily available to assume the position of the deceased parent.

Today's modern family unit does not usually include family members other than father, mother and children. As a result, family members tend to have stronger emotional investments in each other. In the event of the death of a family member, the sense of loss seems to be far greater and there is not necessarily someone readily available to fill the void.

Grief and Loss

The unexpected death of a close friend or loved one can cause a person to feel lost and uneasy and have thoughts which are intangible and frightening. Grief invariably follows death. Grief is an emotional reaction which occurs when a tie of love is severed or we experience a significant loss. Thus, grief tends to occur following any significant loss and will vary in intensity according to the nature and strength of the attachment which is broken. The loss created by death can interrupt a person's sense of security and continuity and threaten their psychological well-being.

Loss is an inevitable part of growth and everyday life. The experience of loss begins with birth and the separation from the warm, safe environment of the womb. Throughout a child's early development, he discovers new objects in his environment, develops attachments to some, and subsequently experiences loss when these objects are removed. Examples of this phenomena might include weaning from the breast or bottle, the move from a crib or the separation from a blanket or stuffed toy. Homesickness is a mild grief reaction to a separation from or loss of familiar surroundings.

As a person matures they experience separation from or loss of familiar persons and objects. Learning and maturing involves the gradual development of adaptive responses and effective mechanisms for coping with losses. But when a loss is traumatic, normal coping mechanisms may not have the capacity to deal with the crisis. This results in a disruption of the individual's normal emotional and behavioral patterns.

The Grieving / Healing Process

Grief is the emotional suffering which results from a loss. The grieving process has been compared with the healing of a tissue wound. Successful grieving, like successful tissue healing, follows a predictable sequence. The healing of a tissue wound cannot be hurried; it takes time. This rule governs the emotional wound. Grieving also takes time.

The grief process has a generally predictable course with fairly consistent symptoms, but the way each individual experiences grief will vary depending on their age, sex, ethnic background, past experience with death and individual coping patterns.

Severe grief causes a breakdown in normal coping patterns, yet some semblance of habitual response is usually retained. For example, the person who meets all of his life crises with denial will inevitably use denial when faced with sudden loss. Those who are openly emotional in every phase of their lives will be highly emotional in grieving.

Some responses to loss and grief may not conform to stereotypes and standards of conduct which most of us have inadvertently acquired during our lives. For example, persons who do not cry as the result of the death of a loved one are sometimes perceived as not being very much affected by the loss. The person who jokes even when faced with severe loss is felt to be behaving inappropriately and others around this individual may disapprove. In reality no judgments should be made in this type of situation. People will cope with death in the best way they can and the individual's surface appearance does not necessarily reflect the depth of their emotional grief. If someone seems to be reacting to death in an inappropriate manner, always try to be accepting and supportive.

Shock and Denial

The first reaction to news of a death is usually shock and denial. This denial reaction is often followed by a numb or dazed state of mind whereby the person attempts to shut out conscious acknowledgment that the death occurred. The affected person may start carrying out automatic activities as though nothing happened. Another reaction is for the person to sit motionless, in deep introversion, making it difficult for anyone to communicate with them.

After the initial shock response, a certain percentage of people will carry on as if nothing had happened. These people immediately begin to take care of the necessary arrangements, comfort and support other members of the family, and seem to accept the death as a reality. It is important to realize that while this person may have accepted the loss intellectually, they have suppressed the emotional impact of the death and are still in a state of partial denial. Very often this state persists through the funeral preparations and burial. These people are described as "taking it well" or "holding up well."

Perhaps what is really going on is that the grieving person is attempting to protect himself from overwhelming grief through denial. This is done by either blocking out the reality of the event or the painful emotions aroused by it. Shock and denial are more intense when the death is sudden and unexpected. The grief-stricken survivor may move into shock and denial within a matter of minutes of learning of a death or it may take as long as 14 days or more.

Crying

Crying is a typical response in this stage and should be encouraged, as it plays an important role in the grieving process. Crying is a legitimate release in the presence of death for both males and females. One should not only accept crying but expect it. No one should be shocked at the sight of a man's tears under these circumstances. When men lose their customary control of emotions, they need reassurance and support because they feel ashamed of their inability to regain their composure. It is important to accept the crying without any hint of disapproval.

Anger

Another very important emotion that may be openly displayed in the second stage of the grieving process is anger. This anger may be irrationally directed at anyone who the grieving person imagines must have allowed or contributed to his loved one's death. This anger is more a manifestation of a person's feelings of frustration and helplessness and their inability to do anything about it. Anger may also be felt toward another family member who somehow failed in an obligation toward the deceased. Anger may also end up being directed against the self if the person feels himself to be somehow at fault.

It tends to be easier for people around a grieving person to tolerate crying than it is for them to tolerate anger. It helps to understand where the anger is coming from. Do not react to this anger or any rash statements which the individual might make when he is grieving. Remain calm and compassionate until the anger subsides. Do not take it personally or be offended.

Occasionally a grieving person will have the fear that they are going insane. Those who express this fear need to be quickly reassured that the feelings they are having are a natural part of the grieving process. It is normal to have such feelings under these circumstances and they will eventually go away.

Guilt

Parents, especially, tend to feel guilt over the death of a child and may berate themselves or each other. Parents may even injure themselves in an impulsive gesture of aggression or self destruction. Guilt is probably felt to some degree by all bereaved persons as they search their minds for ways in which they may have failed the loved one. Persons mourning a loss should be allowed to express their feelings of guilt. These expressions frequently begin with the phrase, "If only I had done this (or had not done that), the situation might not have happened." Such expressions are a normal component of the grief response and need to be expressed.

Do not judge or be too quick to reassure the person who is experiencing guilt. Any quick reassurance you give may cause the individual to stop expressing his guilt. It is important to let the person express his guilt feelings freely. Listen sympathetically and after they have had a chance to express their feelings then offer reassurance that will help to relieve their guilt. One possible response would be, "I understand how you feel. It's natural for you to feel this way right now, but you may feel differently about it later."

Mourning

The third stage is the process of grieving is referred to as mourning. This may begin during the funeral rites or shortly after. During this final stage, which may go on for a year or more, the bereaved person works through his feelings toward his loss and is able to resume a normal life that revolves around new interests.

Supporting Family in the Event of a Death

A surviving family member should never be left completely alone during the grieving process. Any person who is alone in the world, with no relatives, friends or neighbors who they can stay with needs to be put in the care of someone who is warm and compassionate until they are able to cope with the loss. During acute states of anguish people can become temporarily suicidal. Also try to provide grieving family members with some sort of privacy so they can feel free to express their emotions.

Supporting Children in the Event of a Death

When a death occurs in a family, the adults can become physically and emotionally overwhelmed. Because of this, sometimes the emotional needs of children are not dealt with. This is compounded by the fact that many adults find it difficult to talk to children about death. Children cannot be protected from the knowledge of death. Children are no different then adults in that they are not immune to the effects of death and they also suffer from the loss. Thus children need to be included in the family's grieving so that they can heal their emotional wounds. A child's response to grief may be different than that of an adult. Often children will express their grief through their emotions and actions. This may be misinterpreted by adults to be behavioral problems. Unresolved grief in children can manifest as regressive behavior which can continue on into adulthood.

A child's understanding and perception of death depends on his age and previous experience. Children, nine and older, have developed an understanding of death. They view death as a real event that can happen to them too. In general, children under the age of three years have little or no comprehension of death. But children in this age bracket do understand separation and they suffer seriously from the loss of a parent.

Children need assurance of the continuing love and protection from other family members. Their major concern is that some one person will still be there to take care of them and give them the unconditional love and security they need. Their questions should be answered honestly but simply, in terms they can understand.

It is a mistake to tell a young child that a loved one is sleeping or gone on a trip. This may make the child afraid of going to sleep, and in relation to going on a trip, the child might interpret this as being abandoned. The child can deal with the truth, simply stated by a family member,

such as the loved one was so sick or so badly hurt that the doctors couldn't save them and so they died.

Guilt feelings always accompany grief. The child, in a past fit of anger, may have once wished ill upon the family member who is now dead. This can result in the child perceiving the death as a fulfillment of their secret wish. Even if the child does not verbalize this concern, he should be reassured that such a thing could never happen.

The question will have to be addressed of whether or not the children will be allowed to see the body. School-age children should certainly be allowed, but should never be forced. The decision should rest with the adults responsible for the child who best know his maturity and emotional status. If the death resulted in head or facial trauma or dismemberment, it may not be advisable.

Giving Support to A Dying Person

All of the previous information and advice given in relation to supporting those in grief due to the death of a loved one also applies to the dying individual who is going through his own personal grieving process. Dying individuals can become overwhelmed with anxiety. You can help these people come to grips with this situation by assisting them in the process of separating out and dealing with some of the individual strands which make up the overwhelming ball of anxiety. Try to determine if the person's major concern is for their own welfare or for that of another family member. Help this person sort out smaller worries. If some of these can be alleviated, the situation may not be so overwhelming.

In the first stage of dying, the individual mourns the impending loss of self. Shock and denial constitute the first phase, followed by an anger stage. The anger arises from feelings of frustration, injustice and resentment over one's situation. The third stage is bargaining with God or with some person who is perceived as being able to grant a temporary reprieve. This is an emotional respite, an attempt to postpone the inevitable. The fourth stage, depression, is a time of mourning past losses or failures, a time to grieve and cry. The last step is the stage of acceptance — a time of peace and contentment but also of sadness. It is important to keep in mind that the stages of grieving are not clear-cut. One stage merges into the other and a regression into an earlier phase may occur at any time.

It is a mistake to assume anything from the facade that the dying person presents. The person who appears to be calm and in control may be just as frightened and unable to cope with his anxieties as the person who openly expresses his feelings. Communication is the key to dealing with grief. Good communication involves conveying a warm, receptive attitude; meeting the person's eyes with concern; offering a touch, a gesture. Supportive measures in themselves are not necessarily helpful or comforting; it is the way in which they are administered that makes a difference. Be compassionate, honest and sincere. Listening is one of the more important and effective things you can do.

If a dying person is alert and oriented, they need realistic support at all times. Encourage the dying individual to express his concerns. Answer his questions as honestly as possible and always maintain as much hope as is reasonable. Dying individuals usually fear for the worst and many times honest answers can alleviate their worst fears.

The individual who is not able to talk but who is conscious or semiconscious should be assured that those taking care of him are aware of his concerns and the person should be given a realistic assessment of their situation. This individual has a basic need to know that those caring for him are aware that he has anxieties which he cannot express.

Communication

Many times the unconscious or semiconscious person knows what is going on around them. People should avoid making any negative comments regarding the person's condition based on the false assumption that the unconscious or semiconscious person cannot hear or understand what is going on.

If the person is conscious but seems confused, talk to them and explain to them what is going on and what their condition is in a calm and simple way. Avoid using any phrases like, "you will be ok" or "don't worry". The person knows they are not ok and they have good reason to worry.

One of the greatest fears of a dying person is the fear of dying alone. Always have someone staying with a dying person. When you are with the dying person, share yourself, let your eyes meet their eyes, smile, touch them, hold their hand, make them feel that you care about them. Human contact means so much to the dying person.

If the dying person asks the direct question whether or not they are going to die, be aware that they really may not want the answer. What a person really wants is assurance that they are getting the best possible care and that they are not going to be left alone. You should not to try to fool a dying person who suspects the truth. On the other hand, in some cases it might be devastating to tell them the truth of their condition because it can take away any remaining hope. Once a person has been stripped of hope, they never do as well as a person who has some degree of hope. It might be better to say that you are all there fighting for him and you need to have him fight too. If the dying person still persists in asking whether or not they are dying, the best thing is to agree that this is a possibility but that everything possible is being done to help them.

Preparing Yourself for Death

When you are on the threshold of death your values seem to instantly change. Much of which was very important during life fades away and becomes very unimportant. A more profound set of values takes over.

To a certain extent, all of us relate to ourselves as our body. Most of the fear related to dying is due to this close identification with the physical body. In reality, the body is a vehicle which the real person, the spirit of the person, inhabits or wears. The body is what is discarded at death.

There is not much informational instruction available on the subject of going through the process of death because very few people go through the process and come back to tell about it. Some people view death as a veil which separates physical existence from spiritual existence. The problem we are faced with is that passing through this veil is usually a one-way trip. Most how-to books are based on someone's personal experience in a subject, but in this case most don't survive the experience. Thus there ins't much constructive advice left behind for those who will follow. If everyone who tried to climb a particular mountain never came back, then that mountain would develop a particularly ominous and mysterious reputation.

However, in recent years there have been a number of books on near-death experiences which have come onto the market. These books contain testimonies of individuals who survived near-death experiences. Typically these were people who were revived after drowning, cardiac arrest or death during surgery. Most of these recorded experiences contain common similarities which lend credence to their validity.

Some of the near-death survivors recalled a detachment from the body and a release from pain and confinement. This was described by one person to be "the most beautiful instant in the whole world." Many people recalled floating above their body with a conscious awareness of their own death. Some recalled viewing the body from a distance or even seeing it from many different directions at once. Most all experienced tranquility and delight, a timelessness, a limitless sense of

freedom and the feeling of being an unencumbered spirit. There seemed to be a strong sense of having reached reality. There was a general experience of detachment from earthly concerns and a release from the burdens of mortal existence. Many people also relate experiences of being met by beings of light who conveyed to them, without words, great comfort and love. There are also references to traveling down a dark tunnel to a spiritual realm. **In the final analysis it appears that the moment of death is more distressing for family, friends and relatives than for the dying person.**

By understanding death, we are better equipped to face and conquer fear. Another benefit to contemplating the death process is that it helps people reevaluate their present day priorities. What would you do different today if you knew you were going to die tomorrow? On the threshold of death many of the things which once were so important become trite and people lament things which they neglected during life.

Dannion Brinkley, the author of *Saved By The Light*, made a profound statement regarding death and the fear of dying, *"...once we're not afraid they cannot take our freedoms....all of the conspiracies in the world don't intimidate you."*

Three books on the market today which cover this subject are *Life After Life* by Raymond Moody, *Saved By The Light* by Dannion Brinkley, published by Villard Books, and *Embraced By The Light* by Betty J. Eadie, published by Gold Leaf Press.

Chapter 24
Management, Organization and Leadership

Many people decide to go it alone when it comes to shelter and survival projects. The main reason is the inherent difficulties involved in working with a group. This is unfortunate because ultimately the group can accomplish infinitely more through its collective effort than the individual himself can ever hope to accomplish alone. Also, the real key to survival in the future will most likely be community. An invaluable book for anyone attempting to organize a group effort is, *The Different Drum, Community Making and Peace* by M. Scott Peck. Scott Peck has some excellent instruction on group dynamics and understanding and appreciating differences.

The major problem with most of the failed or flawed group efforts always seems to go back to authoritarian management styles. The old saying that power corrupts and absolute power corrupts absolutely seems to be a real part of human nature. Checks and balances are a necessity whenever people vest and concentrate their power in any form of organization or leadership.

Thus, the first step in bringing a shelter project into the physical is creating an organization and management team. Every project should have a good management team or a Board of Directors which operates as a "Master Mind Alliance".

Napoleon Hill is the author of the "Master Mind Alliance" concept. Read Napoleon Hills' books, *Think and Grow Rich* and *Success Through A Positive Mental Attitude*. Don't be fooled by the title *Think and Grow Rich* — just scratch out the words on the cover which say Grow Rich, and replace them with the words, "Successfully Build a Shelter". The new title will be *Think and Successfully Build a Shelter*. The science involved in making money is the same science which applies to bringing a project successfully into physical reality.

The master mind alliance should be comprised of people with a complement of the skills necessary to fulfill the project, but the most important qualification for a member should be his ability to be harmonious and work as a team member. One characteristic of a team player is having the capacity to be objective and open to suggestion and feedback. This is characterized by not being attached to one's personal ideas and by being willing to submit one's ideas to the group's constructive evaluation.

All people by nature have unique talents and abilities but they also have their blind spots. By being able to work together in a group as a team player, the individual's blind spots can be mitigated through the group's collective scrutiny. This, coupled with the combination of diverse talents and skills, produces a superior management team which is capable of achieving a higher level of success and productivity.

Several other outstanding contributers to the field of management and interpersonal relations are Anthony Robbins, author of *Awaken The Giant Within* and Stephen R. Covey, author of *The Seven Habits of Highly Effective People*. Robins and Covey pick up where Napoleon Hill left off. Covey's book has some very valuable instruction on overcoming the interpersonal conflicts that we all inevitably face when undertaking any job or project. Personal conflict is one of the major causes of project failure.

A single individual cannot run a project as effectively as a management team. He can only spread his attention so far and due to his subjectivity, the project will, to a certain extent, become a victim of his blind spots.

The second requirement is formal agreements. This is a good idea especially whenever a project involves people outside of your immediate family. Most businesses fail at the point where they are trying to grow from a small mom and pop operation to a larger organization. The primary reason is their inability to formalize their business procedures. Formal organization and agreements are a major key to the success of any project.

Shoebox finances do not work for construction projects which involve the financial contribution of many people. A good bookkeeper and accurate accounting is a must.

Murphy's Law is a given truth, and based on my experience in the arena of human interactions, I have authored a few more statements of law after the Murphy tradition which is:

Murphy's 1st New Law

Where there is a nebulous agreement there will eventually be a disagreement.

Disagreement is always a product of miscommunication (a failure or breakdown in the communication process).

Most people will consent to a written formal agreement, with terms defining specific rights and responsibilities, as long as it is presented at the front end. But if a person makes a general unspecific agreement with another person leaving the interpretation of the parameters in terms of rights and responsibilities to the other individual's imagination, the potential for the development of differences in the imagination process are infinite and compound with the passage of time. In other words, one person initiated the agreement with a concept of what he felt the agreement was but didn't convey it clearly to the other person. That person was then left open to develop his own parameters and the difference eventually produces a conflict. Also, once someone pays money on a verbal agreement and gets their foot in the door, it is very difficult, if not impossible, to get them to sign a formal written agreement which encumbers them with responsibilities and limitations even if the contents of the agreement are reasonable and fair.

At the end of this chapter is a sample shelter space lease/membership agreement.

Murphy's 2nd New Law

The individuals who you make exceptions for and bend over backwards to help, will be the most ungrateful.

If the project is going to succeed, everyone involved in the project is going to have to carry their own share of the financial weight.

Experience has shown that whenever you cut a special deal with someone because they are a hardship case, they end up falling down at a later date and being totally ungrateful. You can't lower the hurdle, and you can't jump over it for people no matter how much you like them. It is their personal challenge. If you do intervene, you end up depriving them of the opportunity of developing their own strength and summoning their own resources. It is a challenge that life put on their doorstep for some sort of a motivational or developmental reason. I am not saying don't be compassionate, I am saying don't do it for them.

Another issue which will come up is people who want to trade services during construction in lieu of cash as payment for their share in a shelter project. This is fine as long as the person is charging a real rate that they would receive based on their real skill working out in the real world. Also these individuals should present invoices for their services which are not inflated, but represent real hour related services and last of all it has to be a service that is needed.

Murphy's 3rd New Law

Anyone who doesn't care enough to send a monthly statement obviously doesn't need the money.

Collecting money from people, especially friends and relatives, can be one of the most

unpleasant parts of a shelter project, but it has to be done!

Murphy's 4th New Law

The monthly bill which has the highest interest charge gets paid first.

Interest is an important issue. Interest is a motivator. It gives people a reason to make your monthly bill a priority over others they receive. Explained to your shelter members that the shelter project is in the business of being an emergency shelter, not a bank. Encourage people to shop around and borrow the money from someone else, possibly at a lesser interest rate, and then pay off the sum they owe the shelter project.

Murphy's 5th New Law

Any loan without repossessible collateral or enforceable collection will never be paid off.

Try to motivate your people to either get on or off. Interest charges are one way. Once they get back on line because of the fear of interest charges, and they show themselves to be trying hard to pay off the debt, you can always waive the interest charge. If interest doesn't work the next step is to threaten to throw them out of the project. The intent here is to motivate people. The squeaky wheel gets the grease.

All members of a shelter project who are not paid up in full should be required to sign a legally binding promissory note. If everyone knows the game rules up front, future disputes and conflicts will be reduced. At the end of this chapter is a sample promissory note.

Murphy's 6th New Law

Whoever takes the glory for a project also eventually ends up taking the financial credit for the failures.

In some instances the management of a shelter, subtly or overtly, relates to a shelter project as their project. Consequently, the members identify the project as belonging to the management. This is fine as long as things go well. If all of a sudden something goes wrong, the pattern of association continues and now from the members standpoint the problem is not seen as "our" problem, but "his" problem.

It is important to continually communicate and convey to the membership the concept of "our" shelter project. It is important to have the members identify with the project and have a strong sense of individual involvement and ownership. Ownership and credit for success needs to be passed around to all the members. If it is not, the members will not be inclined to take their share of the financial responsibility when things don't go so well.

Developing a sense of community is important and is not just an exercise in internal public relations. The management has to have right motivations, sincerity and fairness. If the management goes out of its way to communicate with the members, it will come back to them in the form of support from the membership. Without the solid support of the membership, a project can't ultimately succeed.

Leadership

Most people admire a competent leader and many of these admirers probably think that leadership is a quality someone is born with. To a certain extent this may be true, but any talent we have can be improved. Most leadership qualities are developed through hard work and experience. If we are to be responsible and competent leaders in crisis we need to understand and start developing the qualities which make a competent leader.

As we have previously discussed, during a crisis self control is important. If an individual loses control of their inner anxiety, they will be swept along with the group's emotional herd reaction. Such group reactions are usually without logic and more often than not, worsen an already bad situation. If the leader loses control and is swept away by an emotional wave, all the sheep will follow. For this reason, self control is of critical importance for the group leader.

It is important that a leader prepare himself mentally and emotionally in order to perform effectively under the pressure of a disaster or crisis. No matter how emotional an incident becomes the shelter manager must maintain control of his own emotions until the crisis is over.

A boat without a rudder will either move with the current or blow with the wind. If you let your mind wander, your tail will follow it. One way of maintaining control is to always be moving toward and concentrating on an objective. If you let your attention rest or dwell on an emotional situation, it will certainly ab-

294

sorb and overwhelm you. Try to dwell on the big picture during a crisis, always keeping one eye on the final goal.

A goal is a magnet that can draw you through the crisis. Goals are the another essential ingredient to human achievement. Without a plan or direction we go nowhere. Life's great achievements are a result of pursuing a goal.

A sense of mission is a goal on a higher level. It is a very personal thing. It is a goal which has meshed itself with emotion and spiritual values. A sense of mission is infinitely more powerful than a goal. A sense of mission can pull someone through unimaginable hardships.

Positive self-perception is a key to sound leadership. If an individual has a flawed self-perception, his functioning will be consistent with the flaw. Beliefs have immense power and people's lives are governed by their beliefs. If an individual does not really believe he can be a leader, or maintain emotional self-control in a crisis, then he can't. People fulfill their beliefs, be they positive or negative. An important key to leadership is consciously examining the beliefs you hold about yourself, the images you hold for your performance, and weeding out any self-defeating concepts you have unthinkingly assumed during your journey through life. Water seeks its own level and a person's achievement and self-control will always be consistent with his level of self-perception.

Humility and the ability to listen are important leadership qualities. Good leaders are individuals who have learned from experience that they are fallible. But a good leader balances this sense of fallibility with a sense of self esteem which enables him to learn and profit from mistakes. No one knows it all. A leader's success is dependent upon his ability to listen and use the valuable information he receives from the group.

Communication is all important to a good leader. If the leader does not communicate adequately with the group, he will lose the group's support. The leader can be honorably conducting himself, but the group members will fill in the void created by a lack of communication with vain imaginations about what is going on.

A successful leader is people-oriented as opposed to being task oriented. There is no substitute for personal contact on a one-to-one basis. Too many organizational structures end up with the leadership distancing itself from the members. When this happens the leader is deprived of an awareness of what is really going on. Middle managers tend to filter and suppress important information. A good leader does not cloister himself; he is constantly interacting with the members of the group.

Leadership is not license. Quality leaders realize that the power that they hold is conveyed to them by the individual members of the group. Leaders hold power by the group's consent. The authority which a leader holds is not his own and he is not law unto himself. Along with any vested authority comes an equal portion of accountability. The main reason the group vests the leader with power is to facilitate the resolution of problems and needs. A leader has a performance obligation. He needs to use power in the group's interest and to act in a timely manner.

Honesty, fairness and right motivation are the last keys to successful leadership. If the leader's primary motivation be the welfare of the group, he will succeed. If the leader's subconscious motivation is to exalt his ego, make himself look like someone of inordinate importance, or to make unfair gain, he will

SHELTER LEASE AGREEMENT

THIS PRIVATE LEASE AGREEMENT made and entered into this _____

day of _____, 199____, by and between _____shelter project,and_____

(hereinafter referred to as the "Lessee(s)" - list above, all individuals who are a part of this lease agreement including adults and children), residing at _____together with the Lessee's heirs, successors and assigns as determined according to this lease agreement and the Shelter Covenants for _____, in the following described shelter space(s) in an emergency shelter situated in_____ _____:

1. Description of Property. Lessee agrees to lease shelter space(s) in an EMERGENCY SHELTER located at_____.

2. Use. Said agreement entitles the above named Lessee the right to use this said shelter in time of war, disaster, or other national or local emergency. The lease includes space in the shelter for one person per share, approximately __ square feet per leased space in the personal rooms for personal storage, sleeping and family privacy, food for up to _____ months to be consumed during times of occupancy of the shelter, and use of one share's worth (as determined by the Board of Directors) of other community-leased supplies for the shelter as listed in Section 4; filtered air, unpolluted water, fuel, toilets, sinks, and other community equipment and shelter systems, but no other supplies, guarantees or encumbrances beyond those specifically designated inside the shelter. This agreement entitles the Lessee to no individual development rights whatsoever to the real property above, below, or surrounding the shelter space hereby leased. By execution of this document, Lessee holds harmless and releases from any liability the owners of the property on which or under which the shelter is located for any accidents should occur natural or manmade occur on the property. Lessee is not purchasing any piece of or rights to this property. The shelter group is leasing this land from the owners of the parcel.

3. Term. The term of this agreement is to be approximately _____ years from _____, 199____ and ending _____, _____.

4. Payments.   Lessee agrees to pay the shelter group the total lease amount based upon the following schedules. The lease amount will vary depending upon the time of entry into the shelter. All payments are due upon agreement to enter into the shelter.

Shelter Space Fee Schedule   _____, 199___

Adult: $_____   Youth: $_____ (4 to 18 years)   Child: $_____ (3 years and under - no additional floor or bunk space)

An individual who entered the shelter project as a youth and becomes 18 years of age before any emergency occupation of the shelter, will automatically move to a full adult status, and he or his parents will be liable to pay an additional fee of $_____ which is the difference between a youth space and a adult space.

Children three and under will be allowed in the shelter at the discounted prices quoted above, but there will be no additional floor space for the child in the family sleeping room. Only one discounted child rate is available per family.

All shelter members including all children will pay $130.00 to cover those items listed below and referred to as Appendix "A" items:

Plate, bowl, cup and eating utensils Toilet paper - 26 rolls per person Dosimeter - (radiation monitor) sufficient quantity for the shelter's needs Radiation suit - sufficient quantity for the shelter's needs One gas mask per person Kleenex - 7 boxes per person Foam mattress pad Chair Laundry soap - one load per person per week for one year

5. Long Term Storage Space Lessee has been offered the option of purchasing community storage space to be installed and maintained adjacent to the shelter. The purchase and use of this storage space is subject to the following:

The storage containers are to remain at the installation site.

The storage space is to be used for survival storage only including food, clothing and personal effects for seven years, and camping and survival equipment. This storage space is not to be considered as a local "U-Store-It" facility.

No explosives, fuel or other combustible substances may be stored in the storage space.

Storage space is to be accessible to the Board of Directors of the shelter group for emergency and inspection purposes. Lessee will provide the board of directors with a key to the storage container space.

6. Cost Overruns/Special Assessment Until the shelter is completed and ready for occupancy, the Board of Directors reserves the right to make assessments. If at a later date it is deemed advisable to upgrade the shelter, the Board will bring such plans and costs to the shelter shareholders for consideration. Upon a majority vote of the eligible voters who have voted, and collection of the funds, implementation of the plans will begin.

If the shelter is not used within a one-year period, the Board will inspect the quality of food and fuel stored at the shelter site. Thereafter annual inspections will take place. A rotation program will be implemented for food needing replacement. At that time, additional fees will be assessed to purchase new stock in food and, if necessary, fuel.

7. Maintenance Fees and Damages Lessee is to pay the shelter group regular monthly maintenance fees of $3.00 per share, subject to inflation or cost-of-living increase. Monthly fees begin _____, 199___ for the duration of the time when this shelter is being maintained in readiness for emergency uses. Monthly maintenance fees will not be levied during periods of emergency occupancy. The purpose of this levy is to keep the shelter and its systems properly maintained when it is not in use. Said fees will be deposited in an account and will accrue interest, if not immediately needed, for such time when said monies are needed.

Lessee is financially responsible for any damage to the shelter or its systems caused by himself or any member of his family. Lessee agrees to replace or otherwise return to good working order, as determined by the shelter group's Board of Directors, any system or other community-owned item, damaged by them or those for whom they are responsible.

8. Utilities. Utilities such as telephone and electricity will be paid from Shelter Community funds.

9. Lease Termination and Penalties The shelter group will allow Lessee(s) to terminate his lease for shelter space(s) only upon meeting the following conditions.

A. The shelter group locates a suitable replacement, i.e., an individual or individuals desiring to enter into a lease agreement with the shelter group for the said space(s). The Board of Directors shall return Lessee's shelter space fees within thirty (30) days of the execution of a Lease Agreement with a suitable replacement.

B. A fee of ten percent (10%) of the total membership cost is paid by Lessee to the shelter group, which will be withheld from the return payment.

C. Any sale of leased shelter space shall be from the Lessee to the shelter group and from the shelter group to an approved replacement. Lessee does not have the right to sell leased space directly to a third party.

D. Upon termination of this lease and the surrender of leased space(s), all personal improvements to said leased space shall be forfeited. The shelter group shall not be held responsible to reimburse Lessee for any such improvements.

13. Future Children Lessee(s) agrees to lease additional shares for any children born, adopted or coming under their legal care after the date of this agreement. The additional leased share will be according to the fee schedule in effect at the time of the family's entry into the shelter. If the family has one child three or under, the first child's space will be upgraded to a full share and a larger personal room will be provided as space becomes available.

14. Improvements, Alterations or Additions Lessee shall not make any improvements, alterations or additions to the leased premises without the prior, written consent of the shelter group's Board of Directors.

Individual rooms will have a door passage set which does not lock. Lessee may change the passage set on their personal room to a keyed locking type but must provide a key copy to the shelter office. All rooms must be made available for inspection as deemed necessary by the Board of Directors.

15. Tenantability/Liability In the event that all or any portion of the shelter is not tenantable due to any damage, natural or man-made, the Board of Directors of the shelter group shall not rebate to Lessee any portion of the lease payments, or be expected to replace any of the destroyed or lost personal effects of the Lessee or the community, stored or kept in the shelter.

16. Insurance As far as the Board of Directors of the shelter group is aware, it is not possible to insure this structure, equipment or its inhabitants in time of natural or man-made disaster of the type that would require its use. It is the responsibility of Lessee to insure the contents of their room and storage space.

17. Limitations This agreement, Lessee and the Board of Directors of the shelter group, are subject to the following encumbrances:

All building, use, zoning, sanitary and environmental regulations and restrictions of government agencies.

The additional terms and conditions as contained herein.

18. Additional Tenants of Lessee A. Lessee shall store within walking distance of the shelter a complete one-year supply of food (protein, beans, grains, fruits, vegetables, oil and medical supplies) for himself and all shares leased, along with a four-year supply of non-hybrid seeds for growing vegetables. Lessee shall continue to work toward purchasing a total of six (6) additional years of food for each member of this lease agreement.

B. Lessee will pay all assessments made for the construction, systems development and equipping of the shelter or otherwise make arrangements with the Board of Directors of the shelter group for such payments. Payments and assessments are described in Sections 4 and 6 of this document.

19. Waiver of Liability. The Board of Directors of the shelter group are not to be held legally responsible or liable for whatever accidents, natural or man-made disasters, war or malfunctioning of equipment or systems, which may occur to the shelter itself or to anyone living in, visiting, or trespassing upon the property above or in or around the shelter, before, during or after construction, or use. The shelter, plans and systems cannot be guaranteed to protect anyone in time of nuclear war or any other disaster and the Board of Directors of the shelter project, builders, and designers hereby waive all responsibility for said premises. By signing this lease agreement, Lessee(s) agrees not to sue the shelter project or its directors for any damages occasioned by being in or around said shelter premises. Lessee agrees not to sue the Corporation in case of any accident, natural or man-made disaster, malfunction, effect of war or other occurrence which may cause harm to him or family members or others for whom the space(s) in the shelter is/are purchased.

20. Causes of Default.   There are three ways in which the Lessee's or the Corporation's actions or lack thereof can bring about default:

A. Grave moral or legal offenses or Lessee's choice to depart therefrom, or extreme conflicts with the Board of Directors or other Lessees could bring about default.

B. Neglect or refusal to make any payment herein required to be made when due, including monthly installment payments, if any, maintenance fees, communal Shelter supply assessments or other financial payments, which may be contracted by the Board of Directors for the highest good of the Shelter Community as a whole.

C. Lessee's failure, neglect or refusal to keep or perform any covenant or agreement herein after notice from the Board of Directors is given of such failure, neglect or refusal.

D. Default is determined by the Board of Directors in consideration of these documents and the Shelter Community as a whole. If the question of default concerns a member of the Board, a replacement shall be appointed by the Board of Directors from the list of Lessees and shall be called in to sit on the Board in that person's place.

21. Default. In the case of a default of the lease, the Board of Directors shall serve written notice describing such default and options for its correction. If after one month Lessee fails to cure any default in performance of the terms and conditions herein contained, or otherwise work out an agreement to do so which is amenable to the Board of Directors, then and in that event, this lease agreement shall thereupon be deemed terminated, effective immediately.

The Board of Directors shall offer the position(s) to another party. Upon payment and the subtraction of any fees, damages, or other assessments deemed honorable by the Board of Directors, the defaulting party shall be able to recover the remainder of his financial interest in the shelter. However, should the defaulting party be deemed harmful or a menace to the shelter community as a whole, there shall be a total forfeiture of any funds heretofore paid.

If a person is not making payments on sums owed to the shelter project, they have two options. First option is that the person removes all of their belongings from the shelter, and signs an agreement stating that they understand that they have defaulted on their original agreement, and that they understand that they no longer have any right or use to their shelter space. In consideration for their cooperation the Board of Directors will agree to put their name on the waiting list for the space to be resold. If and when it can be resold the Board of Directors will give them back 80% of what they have paid against the space. If the defaulted member refuse to sign such an agreement the Board of Directors will move them into the default category. Such a defaulted member will be denied access to the shelter but will be allowed to make an appointment to pick up their personal things in the shelter. If they don't they pick up their personal belongings the Board of Directors will have their belongings packed up and moved them into a secured storage area away from the shelter. They will then have to make an appointment to pick them up.   The defaulted space will then be put on the waiting list to be resold.

If and when the space can be resold the Board of Directors will give the defaulted member back 50% of what they originally paid toward the space. Also in the event that the Board of Directors has to move their personal items out of the shelter to a secured storage area, the Board of Directors will deduct from the 50% refund a charge for having to pack up and move and put in storage their personal belongings.

Any disagreements which cannot be settled between the Lessee and the Board of Directors of the shelter group, shall be taken to binding arbitration as provided by a recognized arbitrator. All decisions rendered through this process are final.

22. Surrender of Premises by Lessee Upon termination of this lease agreement, Lessee shall promptly surrender the premises to the Board of Directors by vacating the premises, removing all personal property belonging to the Lessee and by leaving the premises in clean and good repair. Upon termination of the lease agreement, the Board shall have immediate claim to the shelter space(s).

23. Notice. Any notice to be given hereunder may be served upon the party personally or by registered or certified mail or commercial courier service, addressed to the party at the address first above given for each, or to the last known address that such party has provided in accordance with the terms of this paragraph. A notice served by mail shall be sent "return receipt requested" and shall be considered delivered when the return receipt is signed.

Notices of change of address are to be given in writing and served in the same manner as other notices, except that the mailings of such notices need not be registered or certified. Each of the parties agrees to notify the other of its current address within ten (10) days after any change.

24. Inspection. Lessee acknowledges and agrees that the shelter group has made no oral or other representations, express or implied, not contained in this Agreement, and the Agreement-Acknowledgment form, and that in signing this Agreement Lessee is not relying upon oral or other representations. Lessee has the responsibility and obligation to inform himself of all matters pertaining to the shelter including this Agreement and all other written documents governing the shelter project. Lessee has examined, inspected, and investigated the subject discussed, the plans, agreements and property on which the shelter is to be placed and makes this agreement in reliance upon his own independent investigations and judgment.

25. Costs and Expenses of Litigation. In the event that any legal action is filed to enforce or construe the provisions of this agreement, the losing party in the action shall pay to the prevailing party all costs and disbursements necessarily incurred therein, including a reasonable attorney's fee.

26. Binding Effect. All future transfers, conveyances, assignments and bequests with respect to this lease agreement shall also be subject to these provisions. This agreement shall apply to and bind the heirs, personal representatives, administrators, successors and assigns of the respective parties hereto.

Heirs for the purpose of this agreement shall mean members of Lessee's immediate family who originally had shelter space leased to them under this agreement. Their names, relationship and ages are listed at the end of this agreement.

27. Titles not Controlling. None of the titles, captions or headings to any paragraph or section within this agreement shall control, limit or expand the meaning as set forth. References to "he," "his," or "him" or to the male or female gender is also meant to include the opposite gender where appropriate.

This agreement supersedes any previous agreement between Lessee and the shelter group.

_____Date_____

President, Board of Directors (of the shelter group)

_____Date_____

Lessee

_____Date_____

Lessee

List all members of immediate family subject to this lease agreement for whom shelter space has been purchased.

Name	Relationship	Age

Property donated or transferred to the shelter group as part of purchase into shelter:

_____

_____

_____

_____

_____

PROMISSORY NOTE

I (WE)_____ do hereby agree to the following secured debt and obligation to _____Address_____. In consideration for _____ spaces in the _____ fallout survival shelter and all the provisions, supplies, and privileges received therewith as detailed in the Purchase Agreement of this said _____ and any and all subsequent revisions and amendments thereto.

The Total debt as of _____, 199_____ is $_____.

I (We) agree to make monthly payments starting _____, of $_____ per month until the balance of the principle of this debt is paid in full. I also acknowledge that I will be charged one and one-half percent (1 1/2%) interest per month on the unpaid balance as of _____, 199_.

Failure to honor this Promissory Note obligation could result in the forfeiture of my (our) membership in _____ and its accompanying rights, privileges, and use and access to facilities.

In the event of any legal action or litigation, by the Board of Directors, to enforce the terms of this Note, the prevailing party in any such action will be entitled to recover reasonable legal and court costs.

Dated this_____day of_____, 199_____.

_____   _____
Member/Witness   Member/Witness

For Shelter Project:

_____

Board Member

---

eventually fail. Motivation is like an initial compass reading at the inception of a voyage. If the reading is off, the ship will not meet its course and will end up a long way from its appointed destination.

Visualize and rehearse your reactions to crisis. Imagine yourself in control of your emotions for the betterment and welfare of the group. Mentally rehearse and prepare yourself for the emotional challenges you will most likely face in a crisis. Mental exercise strengthens the muscles of the mind which keep contained pressurized emotions. What man can conceive and believe man can achieve with grace.

Appendix

Appendix A

Heat Gain From People

Room Dry-Bulb Temperature

Activity	Average Metabolic Rate	78 F.		77 F.		76 F.		75F.	
		Sensible Btu	Latent Btu	Sensible Btu	Latent Btu	Sensible Btu	Latent Btu	Sensible Btu	Latent Btu
Rest	350	210	140	216	134	223	127	230	120
Seated	450	180	270	211	239	242	208	275	125
Walking slow	550	190	360	223	237	257	293	290	210
Light work	750	190	560	248	502	307	443	365	385
Mod. work	1000	270	730	333	667	397	603	460	540
Heavy work	1500	450	1000	475	1025	500	1000	525	925

APPENDIX B
Summer Outdoor Design Conditions

State and City	Dry Bulb	Wet Bulb	Moisture gr./lb.	State and City	Dry Bulb	Wet Bulb	Moisture gr./lb.
Alabama				**California**			
Anniston	95	78	117.5	San Bernadino	105	72	65
Birmingham	95	78	117.5	San Diego	85	68	75
Mobile	95	80	131	San Francisco	85	65	60
Montgomery	95	78	117.5	San Jose	91	70	76.5
Arizona				**Colorado**			
Flagstaff	90	65	81	Denver	95	64	60
Phoenix	105	76	94	Durango	95	65	70
Tucson	105	72	77	Grand Junction	95	65	62
Winslow	100	70	85	Pueblo	95	65	63
Yuma	110	78	93	**Connecticut**			
Arkansas				Bridgeport	95	75	99
Fort Smith	95	76	104.5	Hartford	93	75	102
Little Rock	95	78	117.5	New Haven	95	75	99
California				**Delaware**			
Bakersfield	105	70	54	Wilmington	95	78	117.5
El Centro	110	78	94	**Dist. of Columbia**			
Eureka	90	65	52	Washington	95	78	117.5
Fresno	105	74	76	**Florida**			
Long Beach	90	70	78	Apalachicola	95	80	131
Los Angeles	90	70	78	Jacksonville	95	78	117.5
Oakland	85	65	60	Key West	98	78	112.5
Pasadena	95	70	70	Miami	91	79	131
Red Bluff	100	70	62	Pensacola	95	78	117.5
Sacramento	100	72	73	Tampa	95	78	117.5

State and City	Dry Bulb	Wet Bulb	Moisture gr./lb.	State and City	Dry Bulb	Wet Bulb	Moisture gr./lb.
Georgia				**Kansas**			
Atlanta	95	76	109.5	Concordia	95	78	125
Augusta	98	76	100	Dodge City	95	78	132
Brunswick	95	78	117.5	Topeka	100	78	109.5
Columbus	98	76	100	Wichita	100	75	98
Macon	95	78	117.5	**Kentucky**			
Savannah	95	78	117.5	Louisville	95	78	117.5
Idaho				**Louisiana**			
Boise	95	65	54.5	New Orleans	95	80	131
Lewiston	95	65	44	Shreveport	100	78	109.5
Pocatello	95	65	61	**Maine**			
Illinois				Augusta	90	73	95
Cairo	98	78	112.5	Bangor	90	73	95
Chicago	95	75	99	Eastport	90	70	78
Moline	96	76	103	Portland	90	73	95
Peoria	96	76	103	**Maryland**			
Springfield	98	77	106	Baltimore	95	78	117.5
Indiana				Cumberland	95	75	99
Evansville	95	78	117.5	**Massachusetts**			
Fort Wayne	95	75	99	Boston	92	75	104
Indianapolis	95	76	104.5	Fitchburg	93	75	102
Terre Haute	95	78	124	Nantucket	95	75	99
Iowa				Springfield	93	75	102
Davenport	95	78	117.5	Worcester	93	75	102
Des Moines	95	78	123	**Michigan**			
Dubuque	95	78	117.5	Alpena	95	75	99
Keokuk	95	78	117.5	Detroit	95	75	99
Sioux City	95	78	124	Flint	95	75	99
				Grand Rapids	95	75	99

State and City	Dry Bulb	Wet Bulb	Moisture gr./lb.	State and City	Dry Bulb	Wet Bulb	Moisture gr./lb.
Lansing	95	75	104	**New Hampshire**			
Marquette	93	73	90	Concord	90	73	95
Saginaw	95	75	99	Manchester	90	73	95
Minnesota				Portsmouth	90	73	95
Duluth	93	73	96	**New Jersey**			
Minneapolis	95	75	103	Atlantic City	95	78	117.5
St. Paul	95	75	99	Bloomfield	95	75	99
Mississippi				East Orange	95	75	99
Meridian	95	79	124	Jersey City	95	75	99
Vicksburg	95	78	117.5	Newark	95	75	99
Missouri				Paterson	95	75	99
Columbia	100	78	109.5	Trenton	95	78	117.5
Kansas City	100	76	106.5	**New Mexico**			
St. Louis	95	78	117.5	Albuquerque	95	70	94.5
Montana				Roswell	95	70	87
Billings	90	66	70	Santa Fe	90	65	80
Havre	95	70	82	**New York**			
Helena	95	67	71	Albany	93	75	102
Kalispell	95	65	56	Binghamton	95	75	103.5
Missoula	95	66	49	Buffalo	93	73	90
Nebraska				Canton	90	73	95
Lincoln	95	78	124	New York City	95	75	99
North Platte	95	78	135	Oswego	93	73	90
Omaha	95	78	123	Rochester	95	75	102
Valentine	95	78	135	Schenectady	93	75	102
Nevada				Syracuse	93	75	102
Las Vegas	115	75	76	**North Carolina**			
Reno	95	65	62	Asheville	93	75	114.5
Winnemucca	95	65	62	Charlotte	95	78	117.5

State and City	Dry Bulb	Wet Bulb	Moisture gr./lb.	State and City	Dry Bulb	Wet Bulb	Moisture gr./lb.
Greensboro	95	78	123.5	**Pennsylvania**			
Raleigh	95	78	117.5	Harrisburg	95	75	99
Wilmington	95	78	117.5	Oil City	95	75	99
North Dakota				Philadelphia	95	78	117.5
Bismarck	95	73	95.5	Pittsburg	95	75	105
Devils Lake	95	70	77	Reading	95	75	99
Fargo	95	75	104.5	Scranton	95	75	99
Williston	95	73	96.5	**Rhode Island**			
Ohio				Block Island	95	75	99
Akron	95	75	99	Pawtucket	93	75	102
Cincinnati	95	78	117.5	Providence	93	75	102
Cleveland	95	75	99	**South Carolina**			
Columbus	95	76	104.5	Charleston	95	78	117.5
Dayton	95	78	123	Columbia	95	75	99
Sandusky	95	75	99	Greenville	95	76	104.5
Toledo	95	75	99	**South Dakota**			
Youngstown	95	75	99	Huron	95	75	106
Oklahoma				Rapid City	95	70	85
Oklahoma City	101	77	108	Sioux Falls	95	75	99
Tulsa	101	77	101.5	**Tennessee**			
Oregon				Chattanooga	95	76	104.5
Baker	90	66	71	Knoxville	95	75	103.5
Eugene	90	68	67	Memphis	95	78	117.5
Medford	95	70	76	Nashville	95	78	117.5
Portland	90	68	67	**Texas**			
Roseburg	90	66	57	Abilene	100	74	93
Pennsylvania				Amarillo	100	72	91.6
Altoona	95	75	99	Austin	100	78	109.5
Erie	93	75	102	Brownsville	95	80	131

State and City	Dry Bulb	Wet Bulb	Moisture gr./lb.	State and City	Dry Bulb	Wet Bulb	Moisture gr./lb.
Corpus Christi	95	80	131	Yakima	95	65	48
Dallas	100	78	109.5	**West Virginia**			
Del Rio	100	78	115	Bluefield	95	75	99
El Paso	100	69	73	Charleston	95	75	99
Fort Worth	100	78	109.5	Huntington	95	76	104.5
Galveston	95	80	131	Parkersburg	95	75	99
Houston	95	80	131	Wheeling	95	75	99
Palestine	100	78	109.5	**Wisconsin**			
Port Arthur	95	79	124	Green Bay	95	75	99
San Antonio	100	78	109.5	La Crosse	95	75	99
Utah				Madison	95	75	103.5
Modena	95	65	66	Milwaukee	95	75	99
Salt Lake City	95	65	61	**Wyoming**			
Vermont				Cheyenne	95	65	68.5
Burlington	90	73	95	Lander	95	65	66
Rutland	90	73	95	**Canada**			
Virginia				Calgary	90	66	71
Cape Henry	95	78	117.5	Edmonton	90	68	77
Lynchburg	95	75	99	Vancouver	80	67	78
Norfolk	95	78	117.5	Winnipeg	90	71	83.5
Richmond	95	78	117.5	Fredericton	90	75	107
Roanoke	95	76	111.5	Halifax	90	75	107
Washington				Ottawa	90	75	107
North Head	85	65	60	Toronto	93	75	102
Seattle	85	65	60	Montreal	90	75	107
Spokane	93	65	54.5	Quebec City	90	75	107
Tacoma	85	64	55.5	Regina	90	71	92.5
Walla Walla	95	65	47.5	Saskatoon	90	70	81
Wenatchee	90	65	52				

APPENDIX C

Heat Gain From Electric Motors

Brake Horsepower	Btu Per Hour
1/20	190
1/12	220
1/8	260
1/6	280
1/4	360
1/3	440
1/2	540
3/4	680
1	750
1 1/2	950
2	1,280
3	1,800
5	2,800
7 1/2	3,400
10	4,500

APPENDIX D

Heat Gain From Appliances

Appliance Heat	Sensible Heat	Latent Heat	Total
Coffee Brewer - 1/2 gal.	1,350	350	1,700
Coffee Warmer - 1/2 gal.	400	100	500
Open Flame Stove Burner	4,200	4,200	8,400
Water Distiller (5 gal./hr.)	1,700	2,700	4,400

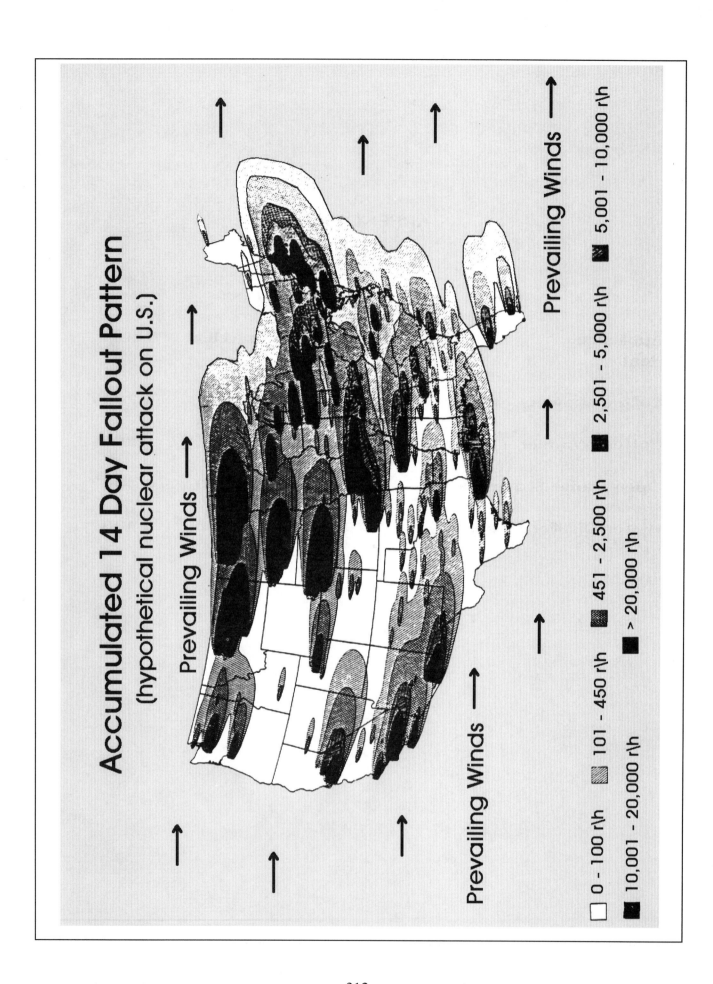

Accumulated 14 Day Fallout Pattern
(hypothetical nuclear attack on U.S.)

Prevailing Winds

Prevailing Winds

Prevailing Winds

Prevailing Winds

0 - 100 r\h
101 - 450 r\h
451 - 2,500 r\h
2,501 - 5,000 r\h
5,001 - 10,000 r\h
10,001 - 20,000 r\h
> 20,000 r\h

Appendix F

Steel Tank Size, Capacity and Specification Data

Capacity (Gallons)	Tank Diameter	Tank Length	Overall Length	Wall Thickness	Weight (Pounds)
4,000	5'4"	240"	24'10"	.313"	5.680
4,000	7'0"	14'0"	14'10"	.313"	4.910
4,000	8'0"	108"	126"	.384"	5.840
5,000	8'0"	13'4"	14'2"	.384"	6.920
6,000	6'0"	28'S"	29'3"	.384"	8.000
6,000	8'0"	16'0"	16'10"	.384"	9.600
8,000	8'0"	21'4"	22'2"	.384"	10.120
10,000	8'0"	27'0"	27'10"	.384"	12.220
10,000	10'6"	15'6"	16'4"	.429"	12.200
12,000	8'0"	32'0"	32'10"	.429"	16.000
12,000	10'6"	18'6"	194"	.429"	13.900
15,000	8'0"	40'0"	40'10"	.492"	22.400
15,000	10'6"	23'4"	24'2"	.492"	19.120
20,000	10'6"	31'0"	31'10"	.492"	24.200
25,000	10'6"	38'9"	39'7"	.554"	33.300
25,000	12'0"	30'0"	30'10"	.625"	34.700
30,000	10'6"	46'6"	47'4"	.554"	39.000
30,000	12'0"	36'0"	37'10"	.625"	40.800

Tanks listed above represent the industry standard.

Appendix G

Lighting Comparison

Name	Watts	Lumens	Lumens Per Watt	Heat Btu/Hr	Rated Life Hrs.
Fluorescent					
Philips SL Magnetic	18	750	41.66	61.4	9,000
Philips SL 18/27	18	1100	61.1	61.4	10,000
Philips SL 18/R40/27	18	800	44.44	61.4	10,000
Philips SL 18/G40/27	18	950	52.77	61.4	10,000
Philips PL5/27	5	250	50	17	10,000
Philips PL7/27	7	400	57.14	23.89	10,000
Philips PL9/27	9	600	66.66	30.72	10,000
Philips PL13/27	13	900	69.2	44.38	10,000
Philips PL5/27/SYS	9	250	27.77	30.72	10,000
Philips PL7/27/SYS	10	400	40	34.14	10,000
Philips PL9/27/SYS	12	600	50	40.96	10,000
Dayton 4 ft./F40CW	40	3150	78.75	136.56	20,000
Dayron 4 ft./F96/T12/CW	75	6300	84	256.05	12,000
Philips 4 ft./F40/CW/RS/EW11	34	2775	81.61	116.07	20,000
Philips 4 ft./F40/WW/RS/EW11	34	2850	83.82	116.07	20,000
Incandescent					
Sylvania 42MB/CAP/120	42	665	15.83	143.38	3,500
Sylvania 52MB/CAP/120	52	885	17.01	177.52	3,500
Sylvania 72MB/CAP/120	72	1300	18.05	245.80	3,500
G.E. Standard	60	870	14.5	204.84	1,000
G.E. Soft White	100	1710	17.1	341.40	750
Tungsten Halogen Quartz					
Philips 250Q/CL/DC	250	5,000	20	853.5	?
Philips 300T3Q/CL	300	6,000	20	1024.2	?
Philips 500T3Q/CL	500	11,100	22.2	1707	?
Philips 1500T3Q/CL	1500	35,800	23.86	5121	?
High Intensity Discharge (HID) Lamps					
Mercury Vapor	400	23,000	57.5	1365.6	24,000
High Pressure Sodium	400	50,000	125	1365.6	24,000
Metal Halide	400	34,000	85	1365.6	20,000

Appendix H

Wire Loss Table

Shown below is a wire loss table. This is a 5 percent table which means that at the given amperage ratings at the listed distances which are in feet. A 5 percent loss is normally acceptable in low voltage systems. For 48 volt systems double the 24 volt distances.

<u>Wire Size</u>

Amps in Wire	Wattage at 120 volts	#14	#12	#10	#8	#6	#4	#2	1/0	2/0	3/0	
1	120	--	--	--	--	--	--	--	--	--	--	1
2	240	563	875	--	--	--	--	--	--	--	--	2
4	480	250	438	688	--	--	--	--	--	--	--	0
6	720	188	300	438	750	--	--	--	--	--	--	
8	960	138	213	363	563	888	--	--	--	--	--	V
10	1,200	113	175	288	450	713	--	--	--	--	--	o
15	1,800	75.0	113	175	300	475	750	--	--	--	--	l
20	2,400	50.0	87.5	138	225	363	563	900	--	--	--	t
25	3,000	45.0	70.0	113.0	175	288	450	725	--	--	--	
30	3,600	37.5	60.0	87.5	150	238	375	600	963	--	--	s
40	4,800	--	--	70.0	113	175	288	450	725	900	--	
50	6,000	--	--	57.5	90.0	138	228	363	575	725	913	

Amps in Wire	Wattage at 24 volts	#14	#12	#10	#8	#6	#4	#2	1/0	2/0	3/0	
1	24	226	350	550	900	--	--	--	--	--	--	2
2	48	113	175	276	450	710	--	--	--	--	--	4
4	96	50	88	138	226	356	--	--	--	--	--	
6	144	38	60	88	150	238	376	--	--	--	--	V
8	192	28	43	72	113	178	288	--	--	--	--	o
10	240	23	35	58	90	143	226	360	--	--	--	l
15	360	15	23	35	60	95	150	240	386	--	--	t
20	480	10	18	28	50	73	113	180	290	360	458	
25	600	9	14	23	35	58	90	145	230	290	366	s
30	720	8	12	18	30	48	75	120	193	240	304	
40	960	--	--	14	23	35	58	90	145	180	228	
50	1,200	--	--	12	18	28	46	73	115	145	183	

Amps in Wire	Wattage at 12 volts	#14	#12	#10	#8	#6	#4	#2	1/0	2/0	3/0	
1	12	113	175	275	450	710	--	--	--	--	--	1
2	24	56.3	87.5	138	225	355	576	900	--	--	--	2
4	48	25.0	43.8	68.8	113	178	288	450	725	900	--	
6	72	18.8	30.0	43.8	75.0	119	188	300	481	600	760	V
8	96	13.8	21.3	36.3	56.3	88.8	144	225	363	450	570	o
10	120	11.3	17.5	28.8	45.0	71.3	113	180	290	360	457	l
15	180	7.7	11.3	17.5	30.0	47.5	75.0	120	193	240	304	t
20	240	05.0	08.8	13.8	22.5	36.3	56.3	90.0	145	180	229	
25	300	04.5	07.0	11.3	17.5	28.8	45.0	72.5	115	145	183	s
30	360	03.8	06.0	08.8	15.0	23.8	37.5	60.0	96.3	120	152	
40	480	--	--	07.0	11.3	17.5	28.8	45.0	72.5	90.0	114	

Appendix I

Weight & Measurement Conversions

Weight

1 Milligram (mg.)	= .001 Grams	= .0154 Grains	
1 Grain (gr.)	= .0648 Grams	= 64.8 Milligrams	
1 Gram (gm.)	= 1,000 Milligrams	= .035274 oz. Avo.	= 15.43 Grains
1 Ounce - (oz.) Avo.	= 437.5 Grains	= 31.1 Grams	
1 Ounce - Troy	= 480 Grains	= 28.35 Grams	
1 Kilogram (kg.)	= 1,000 Grams	= 35.27 Ounces Avo.	= 2.205 lb.
1 Pound (lb.)	= 16 Ounces	= 2.205 Kilograms	
1 Ton - U.S. (short)	= 2,000 Pounds	= 4,4410 Kilograms	
1 Ton - U.S. (long)	= 2,240 Pounds	= 4,939.2 Kilograms	
1 Ton - Metric	= 2,205 Pounds	= 1,000 Kilograms	

Length

1 Micron (micrometer)	= .001 Millimeter		
1 Millimeter (mm.)	= .1 Centimeters	= .03937 Inches	
1 Centimeter (cm.)	= 10 Millimeters	= .3937 Inches	= .03281 Feet
1 Inch (in.)	= 25.4 Millimeters	= 2.54 Centimeters	= .0833 Feet
1 Foot (ft.)	= 304.8 mm.	= 30.48 cm.	= 12 Inches
1 Yard (yd.)	= 914 mm.	= 91 cm. = 3 ft.	= 36 in.
1 Meter (m.)	= 1000 mm. =100 cm.=	39.37 in. = 3.2808 ft.=	1.094 yd.
1 Mile	= 5,280 ft.	= 1.609 Kilometers	= 1,609.36 m.
1 Kilometer (km.)	= 3,281 ft.	= .6215 miles	= 1,000 m.

Volume

1 Cubic Centimeter (cc.)	= .270 U.S. Fluid Dram		= .0610 Cubic Inches
1 U.S. Fluid Dram	= 3.7 cc.		= .226 Cubic Inches
1 Cubic Inch	= 16.38706 cc.		= 4.43 U.S. Fluid Dram
1 U.S. Fluid Ounce	= 29.574 cc.	= 1.805 cu. in.	= 8 U.S. Fluid Dram
1 Pint	= .473 L.	= 473 cc.	= 16 U.S. Fluid oz.
1 Quart (qt.) = 946.35 cc.	= 256 U.S. Fluid Dram	= 57.75 cu. in.	= 32 U.S. Fluid oz.
1 Liter (L.) = 1,000 cc.	= 270 U.S. Fluid Dram	= 61 cu. in.	= 33.81 U.S. Fluid oz.
1 Gallon = 8 Pints	= 128 U.S. Fluid oz.	= 4 qt.	= 3.785 Liters

Temperature

	Centigrade	Fahrenheit
Freezing Water Sea Level	0	32
	36.5	97.7
	37	98.6
	38	100.4
	39	102.2
	40	104
Boiling Water Sea Level	100	212

316

Publication Resource List

Books on Shelters and Nuclear War:

Nuclear War Survival Skills, by Cresson Kearny. Expedient Shelter Information: Oregon Institute of Science and Medicine, P.O. Box 1279, Cave Junction, OR 97523

Ten Feet Under - A Fighting Chance, by Arthur Robinson, Oregon Institute of Science and Medicine, P.O. Box 1279, Cave Junction, OR 97523

Nuclear Defense Issues, Sharon Packer, Civil Defense Volunteers of Utah, (801) 942-5638

Underground Bases and Tunnels, What is the government trying to hide?, by Richard Sauder, Ph.D., P.O. Box 81543, Albuquerque, NM 87198, ISBN# 0-9644979-0-5.

Books on Self Improvement & Master Mind Alliance:

Think and Grow Rich, by Napoleon Hill

Success Through Positive Mental Attitude, by Napoleon Hill and Bradford Stone.

Awaken the Giant Within, by Anthony Robbins

The Seven Habits of Highly Effective People, by Stephen R. Covey

The Different Drum: Community Making and Peace, M. Scott Peck, M.D.

Books With Information on Ionizing Radiation:

Diet for the Atomic Age, by Sara Shannon, Avery Publications, Wayne, NJ.

ReluctantSurvivors, by Wayne D. Le-Baron, Dream Garden Press, Salt Lake City, UT.

Fighting Radiation with Foods, Herbs and Vitamins, by Stephen Schaechter, MD.

Diet for a Small Planet, by Francis Moore Lappe, Ballentine Books, NY.

Nutrition Against Disease, by Roger J. Williams, Bantam Books, NY.

Remote Power Systems & Alternative Energy:

Alternative Energy Sourcebook, Real Goods Trading Corporation, 966 Mazzoni St., Ukiah, CA 95482,

Home Power Magazine, P0 Box 130, Hormbrook, CA 96044-0130, (916) 475-3179.

Backwoods Home Magazine, 1257 Siskiyou Blvd., #213, Ashland, OR 97520

General Survival Magazines:

American Survival Guide, P.O. Box 70015, Anaheim, CA 92825-0015.

Preparedness Journal, P.O. Box 12818 Ogden, UT 84412, (801) 265-8828.

E.M.P. Publications:

CPG 2-17, Volumes I, II, and III, F.E.M.A., E.M.P shielding, U.S. Government Printing Office (202) 646-3484.

Civil Defense:

Journal of Civil Defense, published quarterly by The American Civil Defense Association, P.O. Box 910, 118 Court Street, Starke, FL 32091, (904) 964-5397.

Publications on Military Science and Security:

Delta Press, P.O. Box 1625, 215S. Washington Street, EL Dorado, AR 71731, (501) 862-4984 or (800) 852-4445.

Paladin Press, P.O. Box 1307, Boulder, CO 80306, (303) 443-7250.

Lancer Military, Box 886, Mt. Ida, AZ 71957, (501) 867-2232.

Soldier of Fortune Magazine: A good source of information in regards to military science and security, P.O. Box 1307, Boulder, CO 80306 (303) 443-7250.

Books on Food Storage:

Making the Best of Basics, by James Stevens. Best book available on food storage, menu planning and food preparation. 15123 Little Wren Lane, San Antonio, TX 78255, (210) 659-4200.

Passport to Survival, by Ester Dickey. Great book on food storage and meal preparation.

Essentials of Home Production and Storage, published by the L.D.S., can be obtained through the Welfare Services, 7th Floor, 50 East North Temple St., Salt Lake City, UT 84150.

Magazines and Newsletters:

The New American, 800 Roosevelt Rd., Building B, Suite 407, Glen Ellyn, IL 60137, (708) 469-4611 or (800) 946-4611. Information on Russian activities and the global agenda.

Media Bypass, PO Box 5326, Evansville, IN 47716, (812) 477-8670. This is another source of information the global agenda and American politics.

The McAlvany Intelligence Advisor, by Donald McAlvany. P.O. Box 84904, Phoenix, AZ 85071.

Yellowstone River Review, Published monthly by Philip L. Hoag. Preparedness, civil defense, world events, and prophecy review, Yellowstone River Publishing, P.O. Box 206, Emigrant, Montana, 59027, (406) 333-4707, <http://www.nodoom.com>, e-mail <yrpub@alpinet.com>.

Earth Changes:

Nexus Magazine, P.O. Box 177, Kempton, IL 60946-0177, (815) 253-6464. This magazine focuses on earth changes, international black politics and suppressed advanced technology.

Atlantis Rising, P.O. Box 441, Livingston, MT 59047, (800) 228-8381. This magazine focuses on earth changes, suppressed archeology and suppressed advanced technology.

Fingerprints of the Gods, by Gram Hancock, The Bible of Earth Change books. ISBN # 0-517-88729-0.

Maps of the Ancient Sea Kings, by Charles Hapgood, More on Earth Changes. ISBN # 0-932813-42-9

Amateur Radio Books and Magazines:

Emergency Survival Communications, by Dave Ingram, Universal Electronics, 4555 Groves Rd., Suite #12, Columbus, OH 43232, (614) 866-4605.

Passport to World Band Radio, by Larry Magme, Shortwave Broadcast.

Shortwave Directory, by Bob Grove, Shortwave - Military Broadcast

Novice Antenna Notebook, by Doug DeMaw, simple Antennas.

Monitoring Times, Grove Enterprises, 140 Dog Branch Road, Brasstown, NC 28902, (800) 438-8155;

CQ, Monthly Amateur Magazine, 76 N. Broadway, Hicksville, NY, 11801, (800) 457-7373

Ham Trader Yellow Sheets, P.O. Box 15142, Seattle, WA 98115, Amateur Equipment Ads Nationwide, Subscription $15.00 a year, Bi-weekly. Also publishes an equipment buyer's guide.

Books on Medicine and Health:

Brady Emergency Care, by Grant, Murray, and Bergeron, Sixth edition, published by Brady-Prentice Hall, ISBN# 0-89303-155-0, (800) 638-0220.

Where There Is No Doctor, a Village Health Care Handbook, by David Werner, The Hesperian Foundation, P.O. Box 1692, Palo Alto, CA 94302. **Get this Book!**

Ditch Medicine, by Hugh L. Coffee, ISBN# 0-87364-717-3, Paladin Press, P.O. Box 1307, Boulder, CO 80306, (303) 443-7250.

Family Guide to Homeopathy, by Dr. Andrew Lockie, ISBN # 0-671-76771-2.

Healing Power of Herbs, ($17.95), **Encyclopedia of Natural Medicine** ($21.95) and **Natural Alternative to Over the Counter Drugs** ($30.00). These books can be ordered from Impact Communications (800) 477-2995.

Vaccines: Are They Really Safe and Effective? by Neil Z. Miller, ISBN # 1-881217-10-8, published by New Atlantean Press, PO Box 9638, Santa Fe, NM 87504.

Equipment Resource List

Chemical Warfare Agent Detectors:

BxICAD, individual chemical agent detector, made by Allied - Bendix Aerospace. This unit is being used by NATO forces. It has dual electrochemical sensor cells which simultaneously monitor for chemical agents. It runs about $3,000 per unit.

M256A1 Chemical Agent Detector Kit, Linde, National Specialty Gases Office, 100 Davison Ave., Sommerset, N.J. 08873, (201) 356-8000.

Survey Meters and Radiation Monitoring Equipment:

Rados RDS-110 & 120 Survey Meters, made in Finland by Alnor Oy. Manufactured to army specifications and built to withstand EMP effects. Yellowstone River Trading, P.O. Box 206, Emigrant, MT 59027, (406) 333-4707 or (800) 327-7656, E-mail <yrpub@alpinet.com>, Web site <http://www.nodoom.com>.

Dosimeter Corporation, 11286 Grooms Road, Cincinnati, OH 45242, (800) 322- 8258

Jordan Nuclear Company, 3244 Arroyo Seco Ave, Los Angeles, CA 90065, (213) 222-8141

Safe-Trek Outfitters, 17 West Main, Bozeman, MT 59715, (800) 424-7870.

E.M.P. Shielding Equipment:

Poly Phaser Corp., (800) 325-7170. Nevada residents call 782-2511.

Spikeguard Suppressors, 3121 W. 139 th St, Unit F, Hawthorne, CA 90250, (213) 644-0728.

General Semiconductor Industries, P.O. Box 3078, Tempe, AZ 85281 (602) 968-3101.

Semcon Components, Inc. 10 North Avenue, Burlington, MA 01803 (617) 272-9015.

Comprehensive Food Storage Programs:

Yellowstone River Trading, Source for AlpineAire Gourmet Reserves and SafeTrek Reserves, P.O. Box 206, Emigrant, MT 59027, (406) 333-4707 or (800) 327-7656, E-mail <yrpub@alpinet.com>, Web site <http://www.nodoom.com>.

Safe-Trek Outfitters, 17 West Main, Bozeman, MT 59715, (800) 424-7870.

B & A Products, Rte 1 Box 100, Bunch, OK 74931-9705, (918) 696-5998.

Ready Resources, 125 Country Rd. #585, Engelwood, TN 37329, (423) 746-9814.

American Freedom Network, P.O. Box 43, Johnstown, CO 80534, (800) 205-6245.

Nitro Pack, 151 N. Main St., Heber, UT 84032, (801) 654-0099, (800) 866-4876.

Survival Center, P.O. Box 234, MoKenna, WA 98558, (206) 458-6778 or (800) 321-2900.

Out N Back, P0 Box 1279, Provo, UT 84603, (800) 533-7415.

San Francisco Herb Co., 250 14th St., San Francisco, CA 94103 (800) 227-4530 or in CA (800) 622-0768

Hand Powered Grain Mills:

Yellowstone River Trading, source for Country Living Hand Mill, P.O. Box 206, Emigrant, MT 59027, (406) 333-4707 or (800) 327-7656, E-mail <yrpub@alpinet.com>, Web site <http://www.nodoom.com>.

R & R Mill Company, 45 West First North, Smithfield, UT 84335 (801) 563-3333

Lehman's Hardware, P.O. Box 41, 4779 Kidron Road, Kidron, Ohio 44636, (216) 957-5441.

Nitro Pack, 151 N. Main Street, Heber, UT 84032, (801) 654-0099 or (800) 866-4876.

Water Filtration Equipment:

Yellowstone River Trading, P.O. Box 206, Emigrant, MT 59027, (406) 333-4707 or (800) 327-7656, E-mail <yrpub@alpinet.com>, Web site <http://www.nodoom.com>.

Ultra Pure, ultraviolet and carbon filter systems, 1140 S. Aviation Dr., Provo, UT 84601, (801) 375-3900.

Nitro Pack, 151 N. Main Street, Heber, UT 84032, (801) 654-0099 or (800) 866-4876.

B & A Products, Rt. 1 Box 100, Bunch, OK 74931-9705, (918) 696-5998.

American Freedom Network, P.O.Box 43, Johnstown, CO 80534, (800) 205-6245.

Out N Back, P0 Box 1279, Provo, UT 84603, (800) 533-7415.

Major Surplus and Survival, 435 West Alonda Blvd., Gardena, CA 90248 (800) 441-8855.

Kerosene Stoves, Refrigerators and Freezers:

Lehman's Hardware, Perfection stoves, Siber refrigerators and freezers, and non-electrically powered equipment, P.O. Box 41, 4779 Kidron Road, Kidron, Ohio 44636, (216) 957-5441.

Diesel and Propane Generators and Battery Chargers:

Yellowstone River Trading, Balmar diesel and Generac propane generators and battery chargers, P.O. Box 206, Emigrant, MT 59027, (406) 333-4707 or (800) 327-7656, E-mail <yrpub@alpinet.com>, Web site <http://www.nodoom.com>.

China Diesel Imports, 15749 Lyons Valley Road, Jamul, CA 91935, (619) 669-1995.

Marine Heat Exchangers:

Bowman, Trans Cool Ltd, 7700 River Road, North Bergen, N.J. 07047.

Liquid to Liquid Heat Exchangers:

Young Company, 2825 Four Mile Road, Racine, WI 53404, (414) 639-1011.

Modine, 500 DeKoven Ave., Racine, WI 53401 (414) 636-1200.

Exhaust Heat Exchangers:

Vaporphase, 600 South Holmes, Saint Louis, MO 63122, (314) 821-7900

Exothermic Eclipse, (419) 729-9726

Homestead Equipment:

The Countryside Catalog, 312 Portland Road, Waterloo, WI 53594.

Lehman's Hardware, P.O. Box 41, 4779 Kidron Road, Kidron, OH 44636, (216) 957-5441.

Low Voltage and Alternative Energy Equipment:

Sunelco Inc., equipment and technical advice for power systems, P.O. Box 1499, Hamilton, MT 59840-1499, (406) 363-6924, order # (800) 338-6844

Real Goods, 555 Leslie St., Ukiah, CA 95482-5576, (800) 762-7325

 E-Motion, (Reconditioned Ni-Cad Batteries) P0 Box 556, McMinnville, OR 97128, (503) 434-4332.

Jade Mountain, P.O. Box 4616, Boulder, CO 80306, (800) 442-1972 or in CO (303) 449-6601

Photocomm, Inc., 7681 East Gary Road, P.O. Box 14670, Scottsdale, AZ 85267-4670, (602) 951-6329 or (800) 544-6466.

Shelter Kits and Plans:

Yellowstone River Publishing, (Newhouse System Plans). , P.O. Box 206, Emigrant, MT 59027. (406) 333-4707 or (800) 357-7656.

Terra-Dome Corporation, 8908 S. Shrout Rd., Grain Valley, MO 64029, (800) 481-3663 or (816) 299-6000

Monolithic Constructors, Inc. P.O., Box 479, Italy, TX 76651, (214) 483-7423

Rocky Mountain Dome Co., LLC, HRC 85 Box 170-M, Bonners Ferry, ID 83805, (208) 267-8596, Fax (208) 267-1037, e-mail <domes@dmi.net>.

Arthur Robinson, Oregon Institute of Science and Medicine, (Fuel Tank Shelter Plans) P.O. Box 1279, Cave Junction, Oregon 97523.

Arch Technology Corporation, (Steel Quonsets) P.O. Box 6, Plato Center IL 60170. (312) 464-5656

CO2 Monitors:

Gas Tech, 8445 Central Avenue, Newark, CA 94560-3431, (415) 745-6201.

Industrial Scientific Corporation, 355 Stubenville Pike, Oakdale, PA 15071-1093, (412) 788-4353 or (800) 338-3287.

MSA, P.O. Box 426, Pittsburgh, PA 15230, (800) MSA-2222.

Marine Supply:

West Marine, PO Box 50050, Watsonville, CA 95077-5050, (800)538-0775.

Doc Freemans Inc., 999 North Northlake Way, P.O. Box C-30314, Seattle WA 98103-8883.

Shure Marine Service Inc., 5320 28th Avenue N.W., Seattle, WA 98107, (206) 784-9903.

Marine Doors:

William Drury Company, Inc., 2220 East 11th Street, Tacoma, WA 98421, (206) 272-4109

R. J. Ballinger, P.O. Box 1057, Edmonds, WA 98020, (206) 670-1729.

Decontamination Equipment:

Safety Master, P.O.Box 30277, Billings, MT 59107-0277, (800) 825-7233.

Lab Safety Supply, (800) 356-0783.

International Safety Instruments, Inc., 922 Hurricane Shoal Rd., Lawrence, GA 30243-420 (800) 235-7677

Survival Supplies Equipment:

Major Surplus and Survival, 435 West Alonda Blvd, Gardena, CA 90248, (800) 441-8855 .

Safe-Trek Outfitters, 17 West Main, Bozeman, MT 59715, (800) 424-7870.

The Survival Center, 19223 Cook Road, Box 234, McKenna, WA 98558, (206) 458-6778 or (800) 321-2900.

B & A Products, Rt. 1 Box 100, Bunch, OK 74931-9705, (918) 696-5998.

Ready Resources, 125 Country Rd. #585, Engelwood, TN 37329, (423) 746-9814.

American Freedom Network, P.O. Box 43, Johnstown, CO 80534, (800) 205-6245.

Out N Back, PO Box 1279, Provo, UT 84603, (800) 533-7415.

Nitro Pack, 151 N. Main Street, Heber, UT 84032, (801) 654-0099 or (800) 866-4876.

Outdoor Clothing and Equipment:

Wear Guard, (800) 343-4406.

U.S. Cavalry, (800) 626-6171.

Eddie Bauer, (800) 426-8020.

R.E.I., (800) 426-4840.

LL Bean, (800) 221-4221.

Gander Mountain, (800) 558-9410.

Atlanta Army and Navy, (800) 241-9414.

Stan Sports, (818) 347-7353, (800) 421-6131.

Brigade Quartermasters, 1025 Cobb International Blvd., Kennesaw, GA 30144-4300, (800) 338-4327.

Ken Nolan Inc., P.O. Box C-19555, Irvine, CA 92713, (714) 863-1536.

Safe-Trek Outfitters, 17 West Main, Bozeman, MT 59715, (800) 424-7870.

B & A Products, Rt 1 Box 100, Bunch, OK 74931-9705, (918) 696-5998.

American Freedom Network, P.O. Box 43, Johnstown, CO 80534, (800) 205-6245.

Ready Resources, 125 Country Rd. #585, Engelwood, TN 37329, (423) 746-9814.

Out N Back, PO Box 1279, Provo, UT 84603, (800) 533-7415.

Knives:

Cutlery Shoppe, 5461 Kendall Street, Boise, ID 83706-1248, (208) 376-0430 or (800) 231-1272.

Special Projects, 2128-D Knoll Drive, Ventura, CA 93003, (800) 255-4922.

Security and Military Equipment:

Body Armor, P.A.C.E., 371 East Norris Road, Norris, TN 37828, (615) 494-9335.

Phoenix Systems Inc., P.O. Box 3339, Evergreen, CO 80439, (303) 277-0305.

Safe-Trek Outfitters, 17 West Main, Bozeman, MT 59715, (800) 424-7870.

B & W Trading Company, Box 692, Newark, OH 43055, (614) 344-2772.

Radio Equipment:

Falcon Direct, #36 20th Avenue N.W., Birmingham, AL 35215, (800) 489-2611.

Universal Shortwave Radio, 6830 American Parkway, Reynoldsburg, OH 43068 (800) 431-3439

Amateur Electronic Supply (AES), 5710 W. Good Hope Rd., Milwaukee, WI 53223, (800) 558-0411

Ham Radio Outlet, 933 N. Euclid St. Anaheim, CA 92801, (800) 444-9476

Grove Enterprises, 140 Dog Branch Road, Brasstown, NC 28902, (800) 438-8155

Antennas West, 1500 N 150 W, Provo, UT 84604, (800) 926-7373

Radio Works, Box 6159, Portsmouth, VA 23703, (804) 484-0140

Club: ARRL, (Amateur Radio Relay League) 225 Main St., Newington, CT 06111, (203) 666-1541

Carbon Air Filtration Equipment and Components:

RSC Inc., 34875 23 Mile Road, New Baltimore, MI 48047.

Barnebey - Cheney, P.O.Box 2526, Columbus, OH 43216, (614) 258-9501.

Heppa Filters:

American Air Filter, P.O. Box 35690, Louisville, KY 40232-5690.

Advanced Air Products,18868 72nd Ave. S. Bldg. B, Kent, WA 98032, (206) 251-5743.

Servadyne Corp., P.O. Box 6447, Santa Rosa, CA 95406, (707) 544-8720.

Donaldson Company, P.O. Box 1299, Minneapolis, MN 55440, (612) 887-3344.

ACS, 180786 El Camino Real, Millbrae, CA 94030, (800) 633-4007.

C. Goodman Co., 75 Spruce St., P.O. Box 2777, Patterson, NJ 07509.

Barnebey - Cheney, P.O. Box 2526 Columbus, OH 43216, (614) 258-9501.

Dear Friends,

    If you happen across any good suppliers, useful publications or information not found in this book, please contact us and we will see about including the information in our next addition and on our web site.
        Philip L. Hoag

Yellowstone River Publishing P.O. Box 206, Emigrant, Montana 59027, (406) 333-4707, Fax (406) 333-4344, E-mail <yrpub@alpinet.com>, Web site <http://www.nodoom.com>.